Ingrid Hoesl.

D0529772

**BASIC
INTEGRATED
CIRCUIT
ENGINEERING**

MOTOROLA SERIES IN SOLID-STATE ELECTRONICS

McGRAW-HILL
BOOK COMPANY
New York
St. Louis
San Francisco
Auckland
Düsseldorf
Johannesburg
Kuala Lumpur
London
Mexico
Montreal
New Delhi
Panama
Paris
São Paulo
Singapore
Sydney
Tokyo
Toronto

DOUGLAS J. HAMILTON
Professor and Director of the
Solid State Engineering Laboratory
Department of Electrical Engineering
University of Arizona

WILLIAM G. HOWARD
Group Operations Manager
Linear Integrated Circuits
Motorola Integrated Circuits Center

Basic
Integrated
Circuit
Engineering

This book was set in Times New Roman.
The editors were Kenneth J. Bowman and Madelaine Eichberg;
the cover was designed by J. Paul Kirouac, A Good Thing, Inc.;
the production supervisor was Sam Ratkewitch.
The drawings were done by Oxford Illustrators Limited.
Kingsport Press, Inc., was printer and binder.

Library of Congress Cataloging in Publication Data

Hamilton, Douglas J
    Basic integrated circuit engineering.

    (Motorola series in solid-state electronics)
    1. Integrated circuits.   I.   Howard, William G.,
joint author.   II.   Title.
TK7874.H345   621.381'73   74-23921
ISBN 0-07-025763-9

**BASIC
INTEGRATED
CIRCUIT
ENGINEERING**

Copyright © 1975 by McGraw-Hill, Inc.   All rights reserved.
Printed in the United States of America.   No part of this publication may be reproduced,
stored in a retrieval system, or transmitted, in any form or by any means,
electronic, mechanical, photocopying, recording, or otherwise,
without the prior written permission of the publisher.

    10   KPKP   832

To D. O. PEDERSON

for his own contributions to integrated circuits
and the contributions of his many students who
have become leaders in the field.

CONTENTS

PREFACE

The purpose of this book is to provide an introductory treatment, in context suitable for classroom use, of the four basic aspects of integrated-circuit engineering: fabrication, device behavior, small-signal (or linear) circuits, and digital circuits. The majority of engineering students who become involved during their professional careers with integrated circuits will do so as users rather than as process engineers. However, the interdependent technologies employed in fabrication so markedly influence the device, circuit, and even subsystem performance that the user must have some understanding of processing techniques in order to appreciate the limitations they impose on performance. We have therefore attempted to slant the material toward the user and at the same time maintain the proportion of emphasis we feel to be necessary on the fabrication and device aspects. In order to accomplish these objectives in a volume of reasonable size and to maintain an introductory level of treatment, we have emphasized first-order behavior and first-order analyses. Sufficient references are included to guide the student who wishes to go beyond first-order cases.

The level for which the book is intended is a senior elective or first-year graduate course; it is therefore assumed that the student has as prerequisites the usual undergraduate electrical engineering courses in electronics and circuit theory.

Understanding of device behavior at a level comparable to that of the well-known "SEEC Notes" (Wiley) is presumed.

More material is included in the book than can usually be covered in two semesters or three quarters of three lecture hours per week, and some selection by the instructor is necessary. It should be noted however, that the material dealing with linear circuits is basically independent of that dealing with digital circuits, so that one can be studied without the other. The chapters on fabrication and device behavior contain sufficient material for a one-quarter or one-semester course, should the instructor wish to emphasize these aspects.

Two types of problems are used in the book. Exercises are integral with the textual material; they are generally short and limited in scope and are designed to provide the student with a means for testing his understanding of the material just covered. Ordinary problems appear at the end of the chapters; these are generally broader in scope and require more effort than the exercises.

Material selected from the book has been used in a one-semester three-unit laboratory course and a one-semester three-unit lecture course at the University of Arizona; in a two-quarter three-unit lecture course at the University of California, Berkeley; and in a one-quarter three-unit lecture course at Montana State University.

ACKNOWLEDGMENT

The authors are grateful to the many people who have made helpful suggestions and criticisms during the development of this text. Special thanks are due J. A. Narud, D. O. Pederson, and G. A. Rigby, who helped initiate the project and who participated in many discussions of the organization of the material; P. R. Gray, who assisted in the organization of the material on linear circuits and temperature-stabilized systems; A. R. Fletcher, who supplied the material on electrothermal behavior; A. H. Marshak, who assisted with the material on diffusion-related processes; V. A. Wells, who provided helpful comments and references on processing; W. J. Kerwin, who made many helpful suggestions on linear circuits; J. G. Fossum, who provided much of the material on integrated JFETs; and C. S. Meyer, who assisted in the organization both of the material on MOS digital circuits and the material on applications of digital circuits.

The authors are particularly grateful to Motorola, Inc., Semiconductor Products Division, for its generous support of the project. Special thanks are due Mrs. S. A. McKeown for typing the manuscript.

DOUGLAS J. HAMILTON
WILLIAM G. HOWARD

**BASIC
INTEGRATED
CIRCUIT
ENGINEERING**

THE BASIC PROCESSING TECHNOLOGY AND LAYOUT FUNDAMENTALS

1-1 INTRODUCTION

Production of integrated circuits relies on close interaction among technologies from as diverse disciplines as solid-state physics, chemical engineering, and photography. In order to appreciate these interactions and their relevance to the integrated-circuit designer, we first develop perspective by listing some general facts about integrated circuits and drawing some conclusions about how these facts influence the role of the designer.

The Technologies

The technologies for producing integrated-circuit components are completely interdependent. All components and their interconnections are fabricated during a single sequence, and the technologies employed in this sequence must be compatible. This interdependence of technologies and the resulting limitation of available components impose some severe compromises on circuit design.

Economics and Production

A far greater capital investment is required for integrated-circuit production than, for example, the production of discrete-component etched-card circuits. Furthermore, many more man-hours of design time are required for a given integrated circuit than for a given etched-card circuit. Therefore large-volume high-yield production is essential to the economic success of integrated circuits.

Reliability and Performance

Reliability of integrated circuits is higher than that of corresponding discrete-component circuits because of the absence in the former of such low-reliability aspects as solder joints. Performance of integrated circuits is usually higher, because of the reduction of interconnection parasitic effects and because of matched-component capabilities. The increased reliability and performance make possible the realization of systems of much greater complexity and sophistication than was previously possible with discrete-component circuits. The designer can now begin to think in terms of subsystem "building blocks" instead of just in terms of circuits.

Lack of Component Adjustment

Because the integrated components are all fabricated as part of a monolithic structure, the designer loses the freedom of postfabrication adjustment of components that he enjoyed with discrete-component circuits. Although some trimming of components such as thin-film resistors is possible, it is very costly.

Matching of Components

The fact that all components are made during the same processing sequence ensures that those device parameters which depend on the technologies will be well matched in any given production run. Moreover, all devices on a chip will have nearly the same environmental operating conditions; this leads to temperature tracking of component values to a degree not achievable in discrete-component circuits.

The Role of the Designer

Because of the interdependence of the technologies in integrated circuits, the role of the integrated-circuit designer is quite different from and more complex than that of the discrete-component-circuit designer. The discrete-component-circuit designer requires only a thorough knowledge of the terminal behavior of the devices he uses, so that he can predict how a circuit will perform when these devices are interconnected. He has at his disposal a large variety of different devices, but he does not need to concern himself with the design of the devices themselves as long as he knows their terminal behavior. Not only is the variety of discrete components large, so in general is the range of their values. Resistors of approximately the same physical size range from fractions of an ohm to several megohms; capacitors range from several picofarads to many microfarads. Finally, economics dictates a hierarchy of devices: active devices such as transistors are usually most costly, while resistors are least costly. In a discrete-component-circuit design, the designer usually tries to minimize the number of transistors required.

The integrated-circuit designer, however, must be at once a circuit designer, device designer, and process designer. The actual interconnection of devices to form a circuit is a relatively small part of the fabrication sequence used in making an integrated circuit. The devices must also be designed as part of this sequence. Furthermore the fabrication processes must be specified to some degree. Thus the integrated-circuit designer must concern himself with, for example, the influence of geometry and fabrication processes in the terminal behavior of transistors.

In addition to designing devices, the integrated-circuit engineer must concern himself with a different sort of economics than does the discrete-component-circuit designer. Instead of dollar cost, the economic variable for an integrated circuit is chip surface area. As chip area for a given circuit is reduced, more chips can be obtained per wafer, and hence more circuits are produced per batch. As we shall later see, the entire fabrication sequence is optimized for the production of bipolar transistors. Transistors also occupy less chip area than capacitors or large-value resistors. Therefore the designer must recast his design philosophy in terms of using transistors to the exclusion of other devices, where possible.

Other aspects of the design philosophy for integrated circuits must differ markedly from those for discrete-component circuits. Only a few types of components can be conveniently fabricated: transistors, diodes, resistors, and capacitors. Furthermore, the range of values of these components is very limited. For example, one tries to avoid the use of resistors with resistance less than about 50 Ω or greater than about 100 kΩ. One also tries to avoid capacitors entirely if possible. When capacitors must be used, the total capacitance cannot exceed about 100 pF. These restrictions mean that considerable ingenuity must often be employed in the design of circuits which are to meet the requirements of (1) being acceptable as building blocks, and (2) being amenable to integrated-circuit fabrication techniques. One result of these restrictions is that the circuit diagram for an integrated circuit will generally appear very complex and even economically unfeasible by discrete-circuit standards.

Finally, in addition to being circuit, device, and process designer, and to designing circuits with a limited variety of components with restricted ranges of values, the integrated-circuit designer must always attempt to take advantage of those properties which the integrated-circuit technology offers that are not to be found in discrete-component circuits: availability of closely matched components, capability of temperature tracking of matched components, and useful interactions among components.

It is clear from the preceding discussion that the design of integrated circuits is more complicated than, and quite different from, the design of discrete-component circuits. (The experienced discrete-component-circuit designer will find that he has some well-developed prejudices which he must lose if he is to make the transition to integrated-circuit design.) To begin to be an integrated-circuit designer, one must develop a basic foundation of knowledge in the essential fabrication processes; one must understand how they influence the behavior of integrated devices, and how those devices which can be fabricated in integrated form can be used in the production of digital and linear circuits. It is the purpose of this book to provide such a basic foundation.

1-2 THE BASIC FABRICATION SEQUENCE

During the fabrication of an integrated circuit, the silicon wafers from which a large number of the circuit chips will be cut undergo over a hundred individual processing steps such as chemical cleaning procedures, rinses, and so on. All of these steps are, of course, essential; however, a detailed understanding of the sequence is not necessary for our purposes in this book. Rather, we are concerned only with the major milestones in the sequence, and at this stage we require only a qualitative and superficial understanding of them. This can be obtained by looking at the block diagram of Fig. 1-1, which shows the order in which the processing proceeds, and Fig. 1-2, which shows the cross section of a typical n-p-n transistor as the sequence unfolds.[1]†

The sequence begins with the formation of a single-crystal ingot of silicon. There are several ways to produce such ingots; however, the process usually used today consists of pulling the crystal from a "melt" of doped molten silicon. The crystal "pulling" process results in a cylindrical crystal varying in diameter roughly from $\frac{3}{4}$ to 4 in, depending upon the pulling rate, melt temperature, and other externally controlled factors. The crystal is doped p-type by placing a small amount of impurity such as boron in the melt.

Next the crystal is sawed into thin wafers. These wafers are then carefully polished to a mirror finish by first mechanically lapping the wafer, and then polishing using successively finer grits, ending with a combination chemical-mechanical polish which leaves the surface virtually free of scratches and imperfections.

A layer of silicon dioxide (SiO_2) is grown on the surface of the wafer by heating the wafer to a temperature of approximately 1100°C and exposing it to oxygen or steam. This oxide serves to protect the silicon surface from contamination and to prevent diffusion of impurities into the silicon during subsequent processing steps.

Special photolithographic techniques are employed to remove the oxide in certain selected regions where transistors are ultimately to be placed. The regions from which oxide has been removed now provide "windows" through which impurities can be deposited on the silicon surface. Now the wafer is placed in a diffusion furnace at a temperature of about 1000°C, and a gas containing an n-type impurity such as arsenic is passed over the surface of the wafer.‡ The gas de-

---

† Superscript numbers refer to references at the end of each chapter.
‡ The diffusion processes described here are called *gaseous-source* diffusions. Methods using solid sources and liquid sources can also be employed.

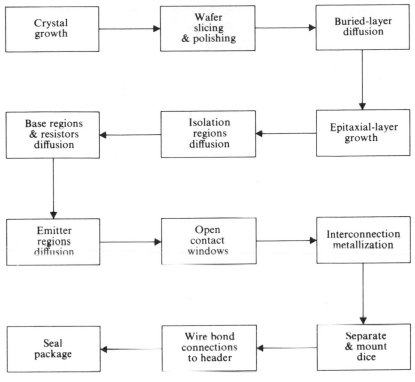

FIGURE 1-1
Basic silicon integrated-circuit fabrication sequence.

composes and the impurity atoms are deposited on the wafer. Where no windows exist in the oxide, the oxide prevents the impurities from reaching the silicon surface. Where windows exist, the impurities are deposited on the silicon surface and, because of the elevated temperature, they begin to diffuse into the silicon, thereby forming an n-type highly conductive layer, but only at the window locations, as shown in Fig. 1-2. This layer, called the *buried layer*, serves to reduce the series collector resistance of the transistor. Following the buried-layer diffusion, all oxide is removed from the wafer surface.

A new layer of n-type silicon is now grown on the surface of the wafer by means of epitaxial growth. In this step the wafer is exposed at high temperature to an atmosphere containing a silicon compound which decomposes at high temperatures, depositing silicon atoms on the wafer surface. Since these atoms are deposited in accordance with the crystal structure of the substrate, the epitaxial layer thus grown and the substrate are part of the same single-crystal structure. The epitaxial layer is doped by including small amounts of the desired impurity in the gas stream during growth. In this case, as Fig. 1-2 shows, the epitaxial layer is n-type.

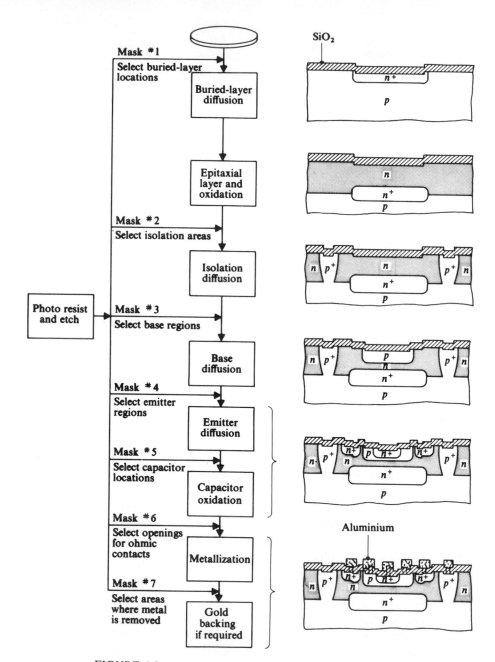

FIGURE 1-2
Flow chart of the sequence of processes used in the fabrication of a single-crystal monolithic circuit.

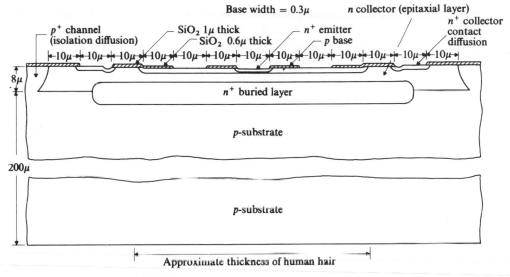

FIGURE 1-3
Cross section of an integrated-circuit transistor shown to correct scale.

Successive oxide growths, photolithography steps, and diffusions are now carried out to put isolation walls and base and emitter impurity distributions in place. The successive diffusion of isolation walls, base region, and then emitter, shown in Fig. 1-2, requires impurity compensation of the p-type base by the n-type emitter diffusion in order that the region containing the emitter be converted to a net n-type region.

Finally, contact windows are opened in the oxide layer for connections to be made to the device. A metal (usually aluminum) is evaporated over the entire wafer surface and the interconnecting contacts are formed by a photolithography step in which unwanted metal is etched away. Since the aluminum is evaporated into the contact windows, it also makes contact to the silicon itself.

It is important to note at this point that in the remainder of the book we shall make frequent use of drawings of device cross sections to explain various aspects of fabrication. These drawings are never scale drawings, but are distorted for the particular emphasis required. In order to give the reader some feeling for the aspect ratios involved between surface dimensions and junction depths, an exact scale drawing of a typical n-p-n transistor fabricated by the sequence just described is shown in Fig. 1-3. This drawing does not include the metallization.

Finally, the wafer is then broken up into small chips, or dice, each of which contains a single circuit, and the dice are then mounted on the case, or header. Wire leads are used to connect the aluminum pattern on the chip to the header leads. After sealing, the circuit is tested and is ready to be used.

Important conclusions about the design of integrated circuits can be drawn from the above summary of processing steps. First, since many components are made at the same time, utilizing the same set of diffusions, one can expect good matching of characteristics of similar devices on the same chip. Second, the same set of diffusions is used to make different types of components (for instance, the *p*-type base diffusion is also used to make resistors). Finally, since the entire circuit is fabricated from a single crystal of silicon, and since the component interconnections are accomplished by the metallization pattern, the only actual "wire" interconnections are those between chip and header. In some headers, even these are replaced by film stripes. Thus the integrated circuit should be more reliable than its discrete-component counterpart.

From the foregoing discussion, it can be seen that the cornerstones of the fabrication sequence are oxidation of silicon, photolithography, impurity diffusion, and epitaxial growth. All of these topics, with the exception of photolithography, are discussed in quantitative detail in Chap. 2. However, for purposes of layout and basic circuit design, one needs only the description given above, with additional qualitative details of photolithography.

1-3 PHOTOLITHOGRAPHY[2-5]

Integrated-circuit photolithography encompasses two major areas: photographic mask making and photoresist. In photographic mask making, two objectives are accomplished: the reduction of layout size and the production of multiple images of the layout. In the photoresist procedure, the multiple-image exact-size pattern required by the wafer is transferred from the photographic mask to the surface of the wafer. As we have seen in the preceding section, this is done several times in the fabrication sequence to open windows in the oxide, and finally to remove unwanted metal from the wafer surface.

Mask making begins with a large-scale layout called artwork. Once the designer has completed his circuit design, it is necessary to determine the locations of all the components of the circuit on the surface of the chip. Because the chip dimensions are of the order of 50 to 100 mils on a side,† the artwork must be made many times the actual chip size in order to avoid large tolerance errors and to be of size reasonable for human operators to deal with. The layout is usually carried out in the form of a drawing showing the position of the windows that are required for a particular major step of the fabrication sequence. Six or more layout drawings are required, as can be seen from Fig. 1-2. For very complex circuits, the layout process can be carried out by use of computer-aided graphics, and the computer can generate the drawing.

Artwork is now made from each drawing. This can be done on a plastic material called Rubylith, a clear mylar with a red plastic coating. Cuts are made in the red coating, and the coating is peeled off in appropriate regions. The Rubylith artwork is to be photographed, and the red areas will appear opaque to photographic film. Rubylith artwork for a simple circuit requiring no buried-layer diffusion is shown in Fig. 1-4.

† 1 mil = 0.001 in. Surface dimensions are usually given in mils, while cross-section depths are usually given in micrometers, abbreviated μm. 1 μm = 10^{-6} m.

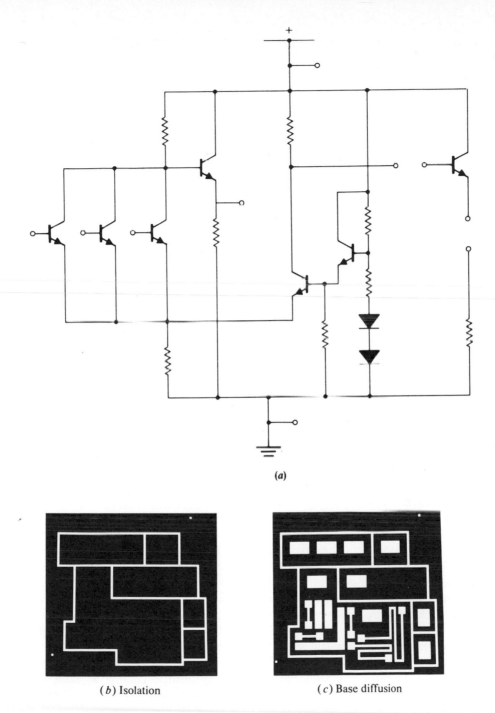

(a)

(b) Isolation   (c) Base diffusion

FIGURE 1-4
Rubylith artwork for a simple circuit. (a) Circuit diagram; (b) isolation
diffusion; (c) base diffusion.

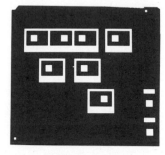
(d) Emitter diffusion

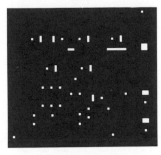
(e) Contact windows

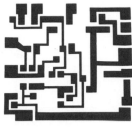
(f) Metallization

FIGURE 1-4 (continued)
(d) emitter diffusion; (e) contact windows; (f) metallization.

If the layout has been generated by computer, all layout information can be stored on punched or magnetic tape. The tape is then used to drive an automated light beam which exposes a large sheet of photographic film. The film is then used as artwork.

Next the artwork is photographed by a large camera. Typically the original artwork will be as much as 500 times the size of the final circuit chip, so for a chip 100 mils on a side, the artwork is 50 in on a side, and a large camera indeed is required. Successive photographs are taken to reduce the artwork size first to 100 times, then to 10 times, and finally to exact size on a master plate. The master plate is used in a precision step-and-repeat printer which produces multiple images of the layout on a high-resolution photographic plate. This plate is the mask which is used in the photoresist operation to transfer the layout pattern to the wafer surface.

The photoresist procedure enables small openings to be placed in the silicon dioxide layer covering a silicon wafer. Since the oxide acts as a barrier to impurity diffusion, impurities are deposited on the wafer surface only where there is an opening in the oxide. Photoresist is a light-sensitive coating which is placed on top of the oxide layer to be selectively removed, as shown in Fig. 1-5a. The coated wafer is then placed in contact with the glass mask containing the pattern of the oxide to be removed, as shown in Fig. 1-5b. During the development process the unexposed coating is dissolved, leaving an opening in the coating as

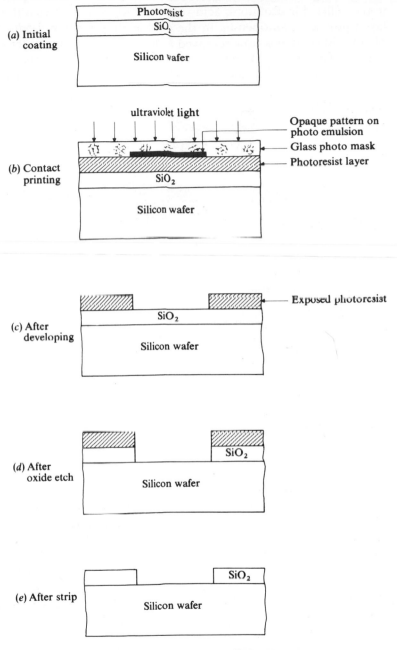

FIGURE 1-5
Steps in the photoresist process.

shown in Fig. 1-5c.   The photoresist coating which remains is chemically resistant to the buffered hydrofluoric acid solution which is used to etch through the oxide layer, producing an opening in the oxide as shown in Fig. 1-5d.   The remaining photoresist coating is now removed from the wafer, and the wafer is now ready to undergo a diffusion.   As has been previously mentioned, the photoresist procedure is used not only to etch silicon dioxide but also to remove metal after the metallization step.

1-4   LAYOUT FUNDAMENTALS

It is clear from the preceding discussion that circuit layout is the first step in integrated-circuit fabrication, since the layout determines where each of the diffusions is to take place on the wafer surface.   The discussion of circuit layout in this chapter is not complete, since layout considerations are often determined by parasitic effects allowable in the elements; however, the question of geometry and its effect on element values and tolerances can now be discussed.

Isolation by Reverse-Biased p-n Junctions

Since all components of the integrated circuit are to be fabricated simultaneously in a single-crystal silicon structure, the first problem which must be solved is that of providing some means for electrically isolating the components from each other. The method most compatible with the processing sequence is junction isolation; it is shown schematically in Fig. 1-6.   Basically the method involves producing islands of n-type material surrounded by p-type material.   Components which must be electrically isolated from each other are then fabricated in different n-type islands. As can be seen from Fig. 1-6c, each island is electrically isolated from the others by back-to-back diodes.

One begins with the wafer having an n-type epitaxial layer.   (If transistors are to be fabricated, the buried layers will already have been formed at the proper locations.)   A p-type diffusion is now performed from the surface of the wafer; sufficiently long drive-in time is used so that the acceptor impurity concentration is greater than the epitaxial-layer donor concentration throughout the entire thickness of the epitaxial layer.   Thus the region directly below the surface, at the location of the isolation diffusion, is changed to p-type from the surface to the substrate. As far as impurity *type* is concerned, the isolation region now becomes an extension of the substrate upward to the surface.   This causes the formation of a p-n junction everywhere around the n-type islands except at the surface.   All p-n junctions thus formed share electrically a common p region: the substrate.   As far as dc isolation is concerned, the n regions are as shown in Fig. 1-6c.   If the substrate is connected to a voltage which is always more negative than any of the n-region voltages, the junctions will all be reverse-biased and negligible dc current will flow.   It is to be understood, of course, that the diodes shown in Fig. 1-6c are actually distributed over all but the surfaces of the n regions.   There will also be a distributed capaci-

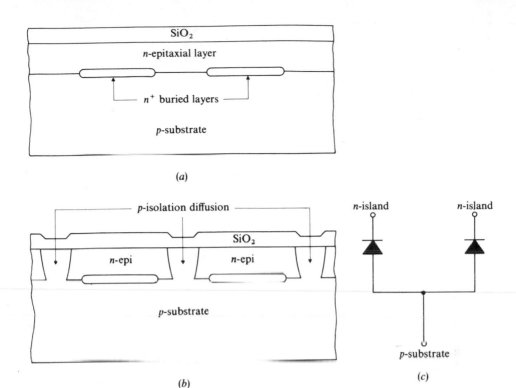

FIGURE 1-6

(a) Cross section of the wafer before isolation diffusion; (b) after diffusion of isolation walls; (c) dc electrical equivalent circuit.

tance associated with the isolation junctions; this distributed capacitance will produce parasitic effects at high frequencies.

MOS Capacitor Layout

The cross section of an MOS (metal-oxide-semiconductor) capacitor in an isolation region is shown in Fig. 1-7a; the effect of the substrate is shown in Fig. 1-7b. A simple parallel-plate structure forms the capacitor; the lower plate is made by an n^+ (heavily doped) diffusion performed at the same time as the emitter diffusion for all n-p-n transistors in the circuit. A thin layer of silicon dioxide forms the dielectric, and a metal pattern deposited at the same time as the metal interconnection pattern forms the upper plate.

If we assume that the isolation region has already been defined by an isolation diffusion and that a layer of oxide exists over the entire wafer surface, then the masks needed for an MOS capacitor are:

1 n^+ diffusion mask
2 Mask to define area for thin oxide

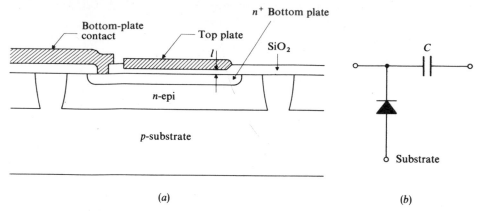

(a)   (b)

FIGURE 1-7
(a) Cross section of an isolated MOS capacitor; (b) equivalent circuit showing
the effect of the substrate.

3   Contact window mask
4   Metal plate and conductor mask

(As we shall later see, mask 1 also contains the emitter locations of *n-p-n* transistors,
mask 3 contains the locations of all contact windows for the entire circuit, and mask
4 contains the locations of all metallization interconnection stripes.)

The first mask is used to open a window in the existing oxide for the n^+ diffusion. During the n^+ diffusion cycle, a new layer of oxide is grown over the entire
surface; this layer will generally be thicker than that desired for the capacitor dielectric. The second mask is used to remove all the oxide at the capacitor location,
and a thin layer is regrown there. The third mask is used to select a window to be
opened in the oxide at the location where contact is to be made to the lower plate.
Metal, usually aluminum, is deposited over the entire surface, and the fourth mask
is used to remove *metal* from the proper locations, leaving the interconnection
pattern and the upper plate.

To determine what surface dimensions must be delineated by the masks, we
assume that fringing effects can be neglected, and we calculate the capacitance of
a parallel-plate capacitor by

$$C = \frac{K_0 \varepsilon_0 A}{l} \qquad (1\text{-}1)$$

where K_0 = relative dielectric constant of oxide
      l = its thickness
      A = effective area of one plate

Note that for the MOS capacitor, the bottom plate is larger than the top plate;
therefore if fringing effects are neglected, the area to be used is that of the *top*
plate.

EXAMPLE 1A *Layout of a 100-pF Capacitor.* It is clear from (1-1) that to consume the smallest surface area for a given capacitance, one should use the thinnest possible oxide. In practice, 500 Å is about the thinnest oxide that can be reproducibly obtained; therefore that value is used for l. For silicon dioxide,

$$K_0 \varepsilon_0 = 3.9\varepsilon_0 = 3.46 \times 10^{-11} \text{ F/m}$$

With $l = 5 \times 10^{-8}$ m, the area required on the chip is

$$A = 1.45 \times 10^{-7} \text{ m}^2 = 2.24 \times 10^{-4} \text{ in}^2 = 2.24 \times 10^2 \text{ mils}^2$$

Before we can draw the artwork for the masks, we must have additional information about the type of photographic system being used, the photoreduction ratio, and the type of photoresist. For this example we assume that

1 A two-step photoreduction system employing a total of two image reversals is used.
2 The total reduction ratio is 125×. (In practice, a larger ratio would be used.)
3 A *negative* photoresist is used. For negative photoresists, that part of the photoresist exposed to ultraviolet light through the mask is *not removed* by the developing process. Therefore areas of oxide which are to be opened as windows must lie under photoresist which is *not exposed.*

We also must know something about the minimum size of window openings which can be tolerated, and about the maximum registration errors that are likely to be encountered during the alignment of successive patterns atop the wafer. For this example, we assume that

1 Minimum window size is 1 × 1 mil.
2 Maximum registration errors are 1 mil.

These are somewhat larger than would actually be encountered in a practical situation. Carefully controlled production lines can achieve window sizes as small as 0.2 × 0.2 mil and registration errors less than 0.2 mil.

With two image reversals and a negative photoresist, the artwork must be cut and the Rubylith peeled so that areas representing *removal* of oxide or metal are *opaque.* The masks to be used are shown separately in the sketches of Fig. 1-8. Note that with 125× reduction, $\frac{1}{8}$ in on the artwork reduces to 1 mil on the chip. To realize 224 mils² on the chip, the artwork for the upper plate must be 3.5 in². The shape of the plate is arbitrary; however, long thin shapes are avoided since they produce excessive parasitic resistance in the n^+ diffusion of the lower plate.

Since registration errors as much as 1 mil must be allowed for, the size of the artwork must be increased so that misalignment of successive masks does not change the value of the capacitance. Accordingly, the n^+ region and the oxide regrowth areas are made larger than the upper plate. If the masks all register perfectly, their overlay is as shown in Fig. 1-9.   *////*

The errors in capacitor value from layout are due to errors in area. For a capacitance tolerance of ± 1 percent, the layout area must be held to within ± 1 per-

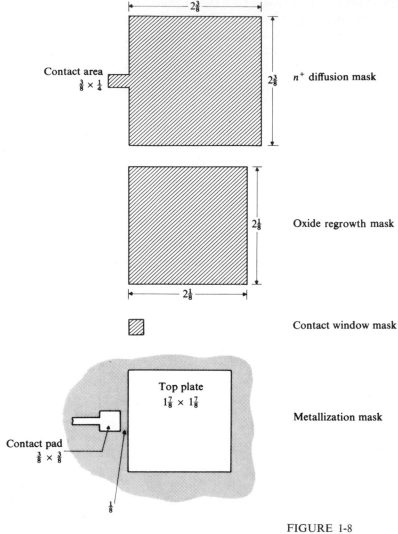

FIGURE 1-8
Capacitor mask layout.

cent. In order to achieve this, the individual capacitor plate dimensions must be held to within ±0.5 percent. In the above example, this means that the side dimensions of the artwork must be accurate to within approximately ±10 mils (0.01 in).

Diffused-Resistor Layout

Because a high value of resistance is generally desirable and because adequate control of resistance is essential, the *p*-type diffusion which is used to form the base of transistors is also used to form resistors. The cross section of the diffused

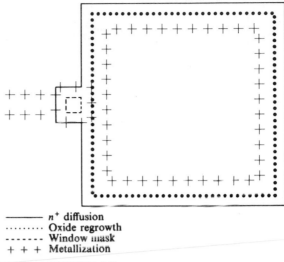

$\underline{\hspace{2cm}}$ n^+ diffusion
$\cdots\cdots\cdots$ Oxide regrowth
$------$ Window mask
$+ + +$ Metallization

FIGURE 1-9
Registration of the masks for the MOS capacitor.

resistor is shown in Fig. 1-10. The value of the resistance is controlled by the surface geometry (length and width) of the resistor as well as the characteristics of the diffused impurity profile. As we will later see, the diffusion characteristics are embodied in the surface resistance of one square (independent of the size of the square). This characteristic resistance, called the _sheet resistance_ of the diffusion, is easily measured and is fixed by the processing schedule. The resistance of a diffused resistor can be expressed in terms of this sheet resistance and the surface dimensions L and W:

$$R = \rho_s' \frac{L}{W} \qquad (1\text{-}2)$$

where $\rho_s' =$ characteristic sheet resistance
$\qquad L =$ resistor length
$\qquad W =$ resistor width

Note that L/W is the aspect ratio of the surface geometry, and is therefore the effective number of "squares" contained in the resistor.

If it is assumed that the isolation diffusion has already been performed and a layer of oxide grown over the entire wafer, then the masks required for a diffused resistor are

    _1_ p-type diffusion mask
    _2_ Contact window mask
    _3_ Metal conductor interconnection mask

The first mask is used to open a window in the oxide over the location where the resistor is to be formed. A p-type diffusion into this window forms the diffused

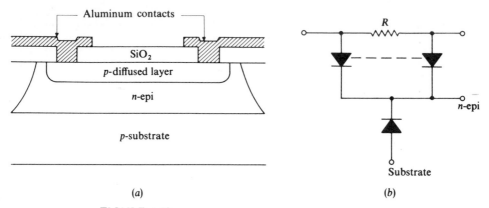

FIGURE 1-10
(a) Cross section (single diffused p-type resistor); (b) Equivalent circuit showing
effects of the substrate.

resistor and is accomplished at the same time that base regions are diffused for any
transistors in the circuit. The second mask is used to open contact windows in
the oxide over appropriate locations on the resistor. This step is carried out at
the same time that contact windows are opened for all other components in the
circuit. Metal is then deposited over the entire wafer and the third mask is used to
remove metal from the proper locations, leaving the interconnection pattern.

EXAMPLE 1B *Design of a 2-kΩ Diffused Resistor.* We again assume that the
photographic system uses two image reversals and that a negative photoresist is
employed. We also assume that the sheet resistance resulting from the p-type
diffusion is 200 Ω per square, a typical value for transistor base diffusions. Then
from (1-2) we find

$$\frac{L}{W} = \frac{R}{\rho'_s} = \frac{2 \times 10^3}{200} = 10 \text{ sq}$$

Only the aspect ratio is calculated; the value of either L or W must be chosen from
other considerations.

One such consideration is resistor layout tolerance. It should be noted that
the tolerance variations of ρ'_s may be as much as 20 percent, so that errors of, for
example, 1 percent in layout dimensions introduce insignificant errors in the *absolute
values* of resistances. However, in many circuits the performance can be made to
depend not upon absolute values of resistances, but rather on *ratios* of resistance.
If it is necessary to control resistor ratios to within 1 percent, artwork errors become
significant.

If both L and W are reproducible to within $\pm \Delta x$ in the layout drafting, the
resistance for a given ρ'_s is

$$R = \rho'_s \frac{L \pm \Delta x}{W \pm \Delta x} \approx \frac{L}{W} \rho'_s \left(1 \pm \frac{\Delta x}{L}\right)\left(1 \mp \frac{\Delta x}{W}\right)$$

For $\Delta x/L$ and $\Delta x/W$ sufficiently small (less than 0.05), the resistance becomes in the worst case

$$R \approx \rho_s' \frac{L}{W}\left[1 \pm \left(\frac{\Delta x}{L} - \frac{\Delta x}{W}\right)\right] \qquad (1\text{-}3)$$

In our example, $L/W = 10$ so $\Delta x/L$ is small compared with $\Delta x/W$. If resistance variations due to layout are to be less than 5 percent, then $\Delta x/W \approx 0.05$ or $W = 20\,\Delta x$. Suppose that the accuracy with which the artwork can be cut is 0.00625 in. Then $W = 0.125$ in. If the photoreduction is $125\times$, as was used for the capacitor example, the width of the resistor on the chip is 0.001 in or 1 mil, and its length would be 10 mils.

We cannot complete the design of the resistor without some knowledge of how the contacts are to be made, since the contact geometry also influences the total resistance of the resistor. A good aluminum-silicon ohmic contact has a conductance of approximately 0.08 $\mho/\text{mil}^2$; a contact made through a 1-mil square window will add about 12.5 Ω to the resistance of the resistor. Furthermore, the resistance added by the region where the contact is made must also be taken into account. The contact window must not overlap the edge of the resistor, or the contact will electrically connect the resistor to the n material in which it is embedded. Because of the registration tolerances encountered in mask alignment, the window must be placed well within the region to be contacted. In cases where wide resistors are used, this can be done as shown in Fig. 1-11a. When the region around the contact window size is comparable to the width of the resistor, which is more often the case, the contact region is enlarged as shown in Fig. 1-11b to allow for registration clearances. Here the contact region adds effectively 0.65 square to the number of squares of the resistor, as shown.

For our example, we can now calculate all the dimensions and draw the masks. If we assume that a 1-mil width is used for the resistor, and that 1-mil windows are used for contacts, the resistance contributed by the contacts themselves is 25 Ω, and the resistor must then have a resistance of 1975 Ω. For $\rho_s' = 200\ \Omega$ per square, the total number of squares of the resistor must be

$$\frac{L}{W} = \frac{1975}{200} = 9.88\ \text{sq}$$

Since each contact region, or end pad, contributes 0.65 square, the body of the resistor must contain $9.88 - 1.30 = 8.58$ squares. For $W = 1$ mil, the length of the resistor between end pads is 8.58 mils. The layout is then as shown in Fig. 1-12. /////

For this example, the layout was straightforward. However, if we require a much larger resistance, complications arise. Suppose that a 100-kΩ resistor is required on a 100 × 100 mil chip. If ρ_s' is 200 Ω per square, and if we must again use $W = 1$ mil, the length of the resistor must be approximately 500 mils. A layout of the type shown in Fig. 1-12 cannot fit on the chip. In such cases, maze geom-

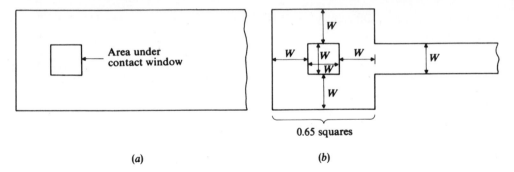

0.65 squares

(a)   (b)

FIGURE 1-11
(a) Contact area for a wide resistor; (b) enlargement of contact region for the
case in which contact window dimensions are equal to resistor width.

etry of the type shown in Fig. 1-13 is used.   However, at the bends the electric
field lines are not uniformly spaced across the width of the resistor, but are crowded
toward the inside corner.   A square at a bend of the resistor does not contribute
exactly 1 square, but rather 0.65 square.   If it becomes necessary to bend a resistor,
this fact must be taken into account in determining the total length of the resistor.

Exercise 1-1   For $\rho'_s = 200$ Ω per square, find the maximum resistance that
can be fabricated on a 100×100 mil chip for: (a) 1-mil lines and 1-mil spacing;
(b) 0.5-mil lines and 0.5-mil spacing.

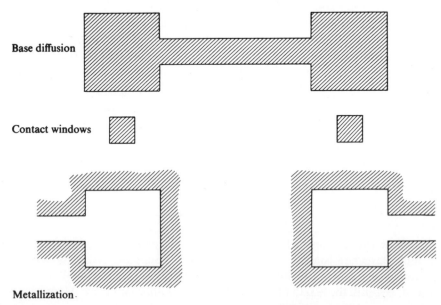

Base diffusion

Contact windows

Metallization

FIGURE 1-12
Layout for diffused resistor masks.

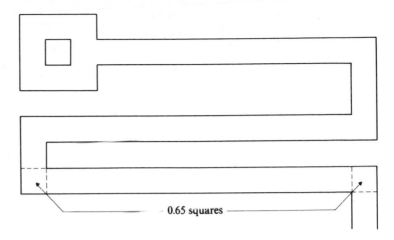

FIGURE 1-13
Maze geometry for long resistors.

n-p-n **Transistor Layout**

A number of different surface geometries can be used for the *n-p-n* transistor; only one will be treated here. The cross section of a single-base-stripe single-collector-stripe isolated transistor is shown in Fig. 1-14. The masks required for this device are:

1 n^+ buried-layer diffusion mask
2 p isolation diffusion mask
3 p base-region diffusion mask
4 n^+ emitter diffusion mask
5 Contact window mask
6 Metallization mask

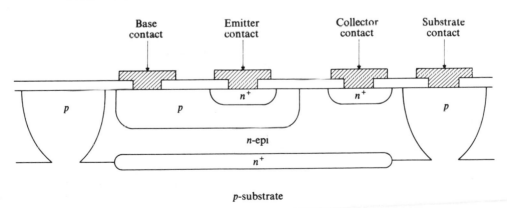

FIGURE 1-14
Cross section of a single-base-stripe single-collector-stripe isolated *n-p-n* transistor.

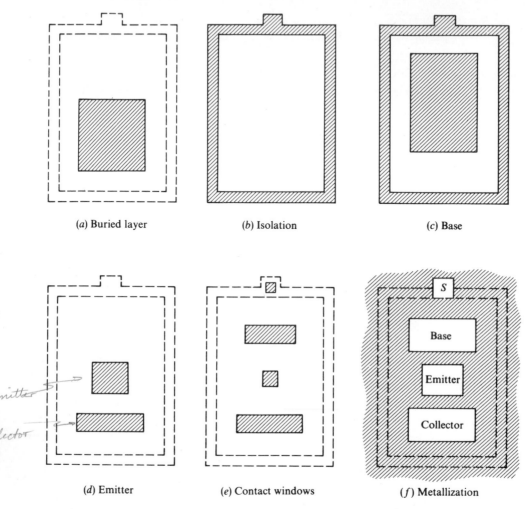

(a) Buried layer   (b) Isolation   (c) Base

emitter

collector

(d) Emitter   (e) Contact windows   (f) Metallization

FIGURE 1-15
A set of masks for an *n-p-n* transistor.

negative resist

In general, unless exceptionally close matching between transistors is required, the performance is not first-order-dependent upon the surface geometry. (Second-order effects related to surface geometry are discussed later.) Therefore the transistor most often used is one which requires the least amount of surface area. The amount of surface area required depends upon the minimum window size, the minimum line widths, and the registration tolerances for the processing sequence being used. For generality, let

I = minimum clearance from isolation window to any p diffusion window
IB = minimum clearance from isolation window to buried-layer window
W = minimum window size

R = maximum registration tolerance for successive masks
S = minimum spacing between lines on the same mask

A set of masks for an isolated *n-p-n* transistor is shown in Fig. 1-15.   To show the interrelationships among the quantities listed above, a superimposed set is given in Fig. 1-16.   There are some subtleties involved in the layout of the transistor, and

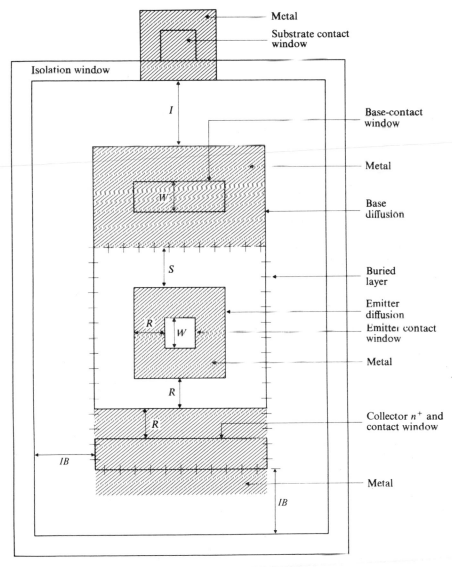

FIGURE 1-16
Layout of a single-base-stripe single-collector-stripe *n-p-n* transistor with buried layer.

some clearances are less critical than others.   These subtleties are best understood in terms of their effects on device terminal performance; they are discussed later in conjunction with device design.   A typical 1-mil-geometry device has all dimensions above equal to 1 mil except $I = 2$ mils, and $IB = 2$ mils.

Exercise 1-2   Find the area required for a 1-mil-geometry *n-p-n* transistor.

1-5   LAYOUT GUIDELINES

We have discussed thus far the layout of MOS capacitors, diffused resistors, and *n-p-n* transistors.   Other components, for example, *p-n-p* transistors and field-effect transistors, can be fabricated in monolithic form.   However, we have at this point sufficient knowledge to formulate some useful general guidelines for layout; the layout of other devices will be discussed in conjunction with their design.

1   Since isolation diffusions always occur at the periphery of device regions, and since isolation clearances are usually larger than other clearances, the isolation diffusion occupies a significant amount of area.   Therefore it is desirable to *minimize the number of isolated regions* that have to be used. This is achieved by such expedients as putting all resistors together in a single isolation region and putting all transistors together whose collectors can be electrically connected.

2   A contact should always be made to the substrate, so that it can be *connected to the most negative supply voltage* in the circuit.   Often this contact is most conveniently made on the surface of the chip by opening a window over the isolation diffusion.

3   A contact should always be made to the *n* material of a region containing resistors.   This *n* material should be *connected to the most positive supply voltage* in the circuit, thus ensuring that the *p-n* junctions surrounding resistors will never become forward-biased.

The above three guidelines have to do mainly with ensuring proper isolation, and can be deduced from what is already known about layout.   There are several other important factors which bear on layout and which are discussed in later chapters.

4   The base diffusion mask should also include the isolation windows, so that the *p*-type isolation material receives a base diffusion.   This is to prevent an *n*-type inversion layer from forming at the surface of the isolation diffusion.

5   Any parts of the epitaxial *n* material which are to be contacted should receive an n^+ emitter diffusion.   This is because aluminum is an acceptor impurity in silicon; hence the donor concentration of the region to be contacted must be larger than the acceptor concentration in order to prevent formation of a *p-n* junction.

6   Aluminum areas used as bonding pads for bonding wires which make connection with the header pins should be at least 3 × 3 mils.   If smaller pads are used, it will be difficult to position the bonding tool on the pad, and the bond will also extend outside the pad area.

7   It is often desirable to include some sort of alignment pattern on the artwork to make the task of registering successive masks easier.

EXAMPLE 1C *Layout of a Single-Transistor Inverter Stage with Speedup Capacitor.*   We can now illustrate the layout of a simple but complete circuit involving only resistors, capacitors, and transistors.   The circuit diagram is shown in Fig. 1-17.

First, some knowledge of the constraints of the processing system is necessary. We assume the following:

1   The photoreduction is 250 × .

2   A two-step photographic procedure with two image reversals is used.

3   A negative photoresist is used.

4   All clearances are 1 mil except the isolation clearance, which is 2 mils.

5   Base diffusion sheet resistance is $\rho'_s = 200\ \Omega$ per square.

6   MOS capacitor oxide thickness is $l = 900$ Å.

7   Resistor tolerance due to layout are ±2 percent.

8   Drafting accuracy is $\Delta x = \pm 5$ mils on the artwork.

Next we determine how many isolation regions are needed.   Both resistors can go in one region, and this region should have its n material connected to V_{CC}. The transistor requires its own region, as does the capacitor.   The number of regions required is three.   The substrate should be connected to ground.

The size of resistors and capacitors can next be determined.   For the capacitor, we have

$$A = \frac{Cl}{\varepsilon} = 403\ \text{mil}^2$$

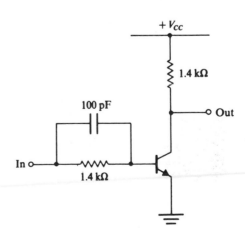

FIGURE 1-17
Single transistor inverter with speedup capacitor.

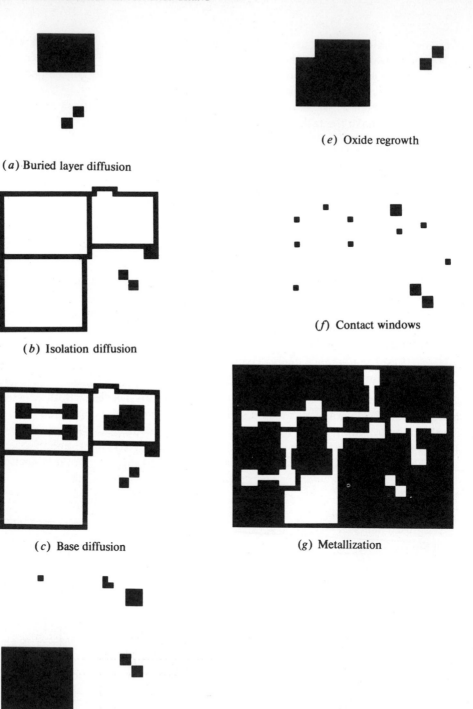

(a) Buried layer diffusion

(b) Isolation diffusion

(c) Base diffusion

(d) Emitter diffusion

(e) Oxide regrowth

(f) Contact windows

(g) Metallization

FIGURE 1-18
Rubylith artwork for the inverter circuit.

or an artwork area for the upper plate of 2.52 in². Both resistors have the same resistance, and the resistance is relatively low. To meet the tolerance requirement, it is necessary that the width of the resistor be

$$W = \frac{\Delta x}{0.02} = 0.250 \text{ in on the artwork}$$

On the chip the resistor will be 1 mil wide. Since the contacts have 1-mil windows, they contribute 25 Ω, so the number of squares must total 1375/200 or 6.88. Because the resistor width is 1 mil and the minimum contact window is also 1 × 1 mil, enlarged end pads must be used. These contribute a total of 1.3 squares, so the resistor body must be 5.58 squares, and the length between end pads is therefore 5.58 mils on the chip, or 1.39 inches on the artwork.

A standard 1-mil-geometry single-base-stripe single-collector-stripe buried-layer transistor is used. The artwork for the seven masks required in the fabrication of the entire circuit is shown in Fig. 1-18. ////

1-6 CROSSOVERS

In all our discussions of layout to this point, we have implicitly assumed that any circuit to be fabricated requires no crossovers of the metal interconnection pattern. For this to be the case requires that the circuit be able to be represented by a coplanar graph. Obviously not all circuits are coplanar; a case in point is the simple Eccles-Jordan multivibrator. Clearly we need some means for realizing crossovers.

One method which requires neither increase nor alteration of the fabrication steps is the so-called *buried crossover*. A cross section of a buried crossover, together with its surface geometry, is shown in Fig. 1-19. An n^+ diffusion is made and contact windows are opened at each end. Sufficient space is left between contacts so that a metal stripe can be deposited on the oxide above the n^+ diffusion. The diffused material serves as one conductor, and the metal above the oxide crosses it without making electrical connection. Since the n^+ diffusion can be performed

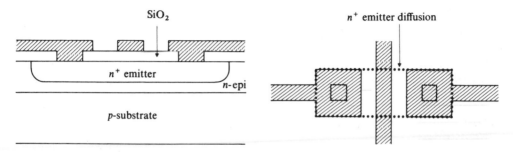

FIGURE 1-19
Cross section and surface geometry of a buried crossover,

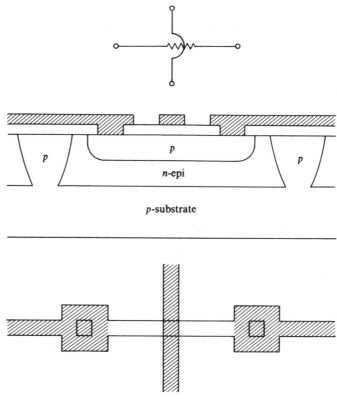

FIGURE 1-20
Use of a diffused resistor location to make a crossover.

at the same time as the emitter diffusion, no extra processing steps are needed. However, the n^+ material introduces some series resistance, so it should not be used in, for example, power supply lines or ground lines. Also it may be necessary to provide an isolation region for the n^+ diffusion; this will increase the area it consumes.

Before resorting to buried crossovers, one should determine that they are absolutely necessary. This involves more than a glance at the circuit diagram, however. Note, for example, that a diffused resistor provides a natural means for implementing a crossover. As long as the value of the resistance is large enough so that there is sufficient space between the end pads for a metal stripe, a crossover can be made there, as shown in Fig. 1-20. Furthermore, crossovers can sometimes be avoided by using double-base-stripe or double-collector-stripe transistors. Figure 1-21 shows a case in which effective crossing of emitter and base leads is accomplished by using a double-base-stripe transistor.

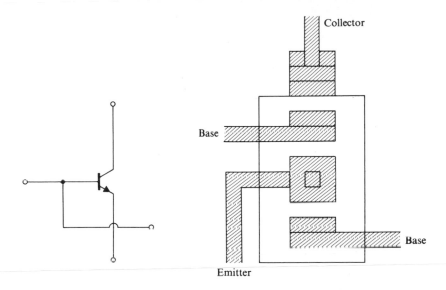

FIGURE 1-21
Use of a double-base-stripe transistor to implement crossing of base and emitter leads.

Exercise 1-3   For 1-mil geometry, find the resistance of a buried crossover if the sheet resistance of the n^+ diffusion is ρ'_s.

Exercise 1-4   For 1-mil geometry, what is the minimum resistance for a resistor to be used as a crossover?   Assume that the base diffusion sheet resistance is $\rho'_s = 200\ \Omega$ per square.

By making use of the above techniques it is usually possible to implement all necessary crossovers without unduly compromising circuit performance.   Should this not be the case, it is possible to use more than one layer of metal interconnect pattern; this is done as follows.   The circuit is fabricated as usual, but after the removal of unwanted metal, a layer of oxide is deposited at low temperature over the surface of the wafer.   This is typically done by pyrolytic decomposition of silane in a reactor at temperatures below 300°C.   The decomposition forms silicon dioxide on the wafer surface.   Because of the relatively low temperature involved, there is no damage to the existing metal pattern.   Windows are next etched in the oxide, a second layer of metal is deposited, and unwanted metal is removed in the usual way from the second layer.   A cross section of a crossover formed in this way is shown in Fig. 1-22.   Because several extra processing steps are required, this method of implementing crossovers should be avoided if at all possible.

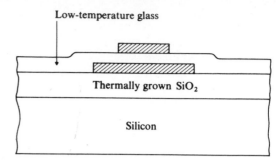

FIGURE 1-22
Multiple-layer metal crossover.

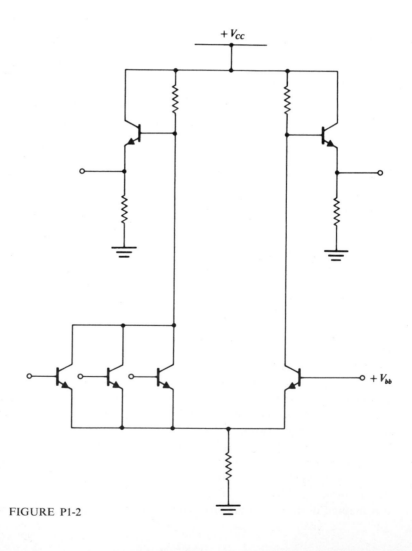

FIGURE P1-2

PROBLEMS

1-1 Suppose that all clearances and dimensions for the layout of an *n-p-n* transistor are 1 mil, except the isolation clearance, which is 2 mils. Find the area required for an isolated, non-buried-layer transistor for: (*a*) single-base-stripe single-collector-stripe geometry; (*b*) double-base-stripe double-collector-stripe geometry; (*c*) single-base-stripe collector-ring-contact geometry.

1-2 A three-input ECL gate with emitter-follower outputs is shown in Fig. P1-2. Determine how many isolation regions are required and what would be put in each region.

1-3 For 1-mil geometry, how much area is required for an isolated buried crossover?

REFERENCES

1 MEYER, C. S., D. K. LYNN, and D. J. HAMILTON: "Analysis and Design of Integrated Circuits," pp. 16–21   McGraw-Hill Book Company, New York, 1968.

2 BERRY, R. W., P. M. HALL, and M. T. HARRIS: Pattern Generation, "Thin Film Technology," chap. 10, D. Van Nostrand, Inc., Princeton, N.J., 1968.

3 STEVENS, G. W. W.: "Microphotography," John Wiley & Sons, Inc., New York, 1968.

4 "Techniques of Microphotography," publication P-52, Eastman Kodak Company, Rochester, N.Y.

5 "Kodak Photosensitive Resists for Industry," publication P-7, Eastman Kodak Company, Rochester, N.Y.

2

DIFFUSION-RELATED PROCESSES

Diffusion is a physical process which is important in several aspects of integrated-circuit fabrication, as well as in the operation of bipolar transistors. Because diffusion is such a basic process, we begin this chapter with a general treatment of diffusion theory, leading to Fick's laws for diffusion. These general results are then applied successively to impurity diffusion in silicon, oxidation of silicon, and epitaxial growth of silicon layers.

2-1 FIRST-ORDER DIFFUSION THEORY[1]

In obtaining a first-order model for diffusion, our purpose is not to provide a derivation, but rather to state the postulates which are to be assumed for the general case, and the results which follow from these postulates. We consider a collection of particles moving in some medium, and we assume the following:

> *1* The particles obey classical mechanics. This applies to the various types of diffusion we wish to consider in this book; it is possible, however, to remove this limitation and to account for particles obeying quantum statistics.[2]

2 The particle concentration is sufficiently low that collisions among particles are negligible in comparison with collisions between particles and the medium.

3 Collisions are adequately described by a linear relaxation model with relaxation time τ. Because of (2), τ is not a function of spatial coordinates.

4 A quasi-equilibrium description of the particle-distribution function is adequate; that is, the distribution function in six-dimensional (position and momentum) space differs only slightly from its equilibrium value.

By using these approximations, together with the Boltzmann transport equation,[3] one can show that the particle flux density f is given by

$$f(\mathbf{r}, t) = -\nabla DN + \mu NE \qquad (2\text{-}1)$$

where N = particle density
D = diffusivity, or diffusion coefficient, of particles
μ = mobility of particles
E = electric field
∇ = gradient operator in position space
$\mathbf{r}$ = position vector
t = time

Now let us also assume that:

5 The particles obey Maxwell-Boltzmann statistics.
6 Isothermal conditions exist.
7 The effects of the electric field can be neglected.

We can now obtain *Fick's first law*:

$$f = -D \nabla N \qquad (2\text{-}2)$$

One can also write a continuity equation relating the time rate of change of particle density to the divergence of flux density; this is known as *Fick's second law.*

$$\frac{\partial N}{\partial t} = -\nabla \cdot f \qquad (2\text{-}3)$$

or

$$\frac{\partial N}{\partial t} = -\nabla \cdot (D \nabla N) \qquad (2\text{-}4)$$

For our purposes, diffusion can usually be regarded on a first-order basis as one-dimensional, and the diffusion coefficient can be assumed to be constant. Fick's laws then become

$$f(x, t) = -D \frac{\partial N}{\partial x} \qquad (2\text{-}5)$$

$$\frac{\partial N}{\partial t} = -\frac{\partial f}{\partial x} = -\frac{\partial}{\partial x}\left(-D \frac{\partial N}{\partial x}\right) = D \frac{\partial^2 N}{\partial x^2} \qquad (2\text{-}6)$$

where $\mathbf{r} = x$.

The physical interpretation of (2-5) is that particles move by diffusion from regions of higher concentration to regions of lower concentration. One can also obtain a physical interpretation of (2-6) by considering a small volume $Z Y(x_2 - x_1)$. If one integrates (2-6) over this volume, the result is

$$\frac{dQ}{dt} = i(x_1, t) - i(x_2, t) \qquad (2\text{-}7)$$

where $Q(t) = $ total number of particles in the volume

$\qquad i(x, t) = particle$ current

The physical interpretation is embodied in (2-7): The rate of increase of the number of particles in the volume is equal to the *net* particle current flowing *into* the volume. [Note that if the particles have a charge q, qQ is the total charge in the volume, and $qi(x, t)$ is the electric current.]

2-2 DIFFUSION OF IMPURITY ATOMS IN SILICON

In planar processing technology, formation of *p*-type or *n*-type layers of silicon is usually accomplished by diffusing impurity atoms such as boron or phosporus into the silicon crystal from the surface of the wafer. If the crystal is originally of a given type, for example, *n*-type, it can be changed to the opposite type by diffusing in impurity atoms of opposite type, for example, acceptors, with density greater than that of the original impurity atoms. This technique is known as *compensation* and is used in forming base and emitter regions for bipolar transistors.

Impurity diffusion is generally carried out in two steps. In the first, or *predeposit* step, the wafers are placed on a quartz carrier in the quartz tube of a diffusion furnace. The region containing the wafers is heated to a high temperature (typically 950°C) by heating elements surrounding the tube. Impurity atoms are then deposited on the surface of the wafer. There are several ways of implementing the deposition; these are called *gaseous source, solid source,* and *liquid source,* according to the phase of the material originally supplying the impurity atoms. In these methods, the impurity atoms are carried to the wafer by the flow of a gas in the diffusion tube. A schematic drawing of a gaseous-source system is shown in Fig. 2-1. For boron diffusion, the impurity gas is diborane (B_2H_6), while for phosphorus, phosphine (PH_3) is used. The carrier gas is typically argon or nitrogen. Focusing attention on the boron case, we note that when the impurity gas enters the hot diffusion tube, it decomposes into several boron-containing compounds, which are carried to the wafer and react with the wafer surface. As a result, boron atoms are deposited on the surface, dissolve in the silicon, and begin to diffuse into the wafer.

It is important to note that there is a maximum density N_0 of impurity atoms which can dissolve in silicon; N_0 is known as the *solid solubility* limit. In general, N_0 differs for different impurities, but it is a weak function of temperature over a fairly wide range of temperatures. Graphs of N_0 versus T are given in Fig. 2-2,[4] and a table of values is given in Table 2-1. (Diffusivities of boron and phosphorus

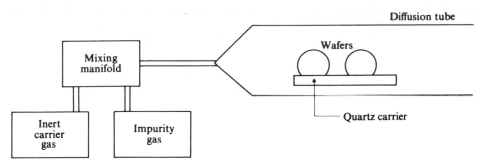

FIGURE 2-1
Schematic diagram of a gaseous-source diffusion system.

are also given in Table 2-1; note that D for boron is independent of surface concentration, while for phosporus it is not.) The existence of a solid solubility limit and its weak dependence on temperature are important to the success of the planar process, for they ensure that the number of impurity atoms deposited during the predeposition step is relatively independent of the partial pressure of the impurity gas in the gas stream. This solid solubility limit makes possible reasonable control of the predeposit results.

In the second, or *drive-in*, step the wafers are again placed in a diffusion furnace, but now the temperature is increased and the gas-stream constituents are changed. During this step, the impurities which were predeposited are diffused farther into the silicon wafer, usually for the purpose of producing a junction at a predetermined depth below the surface. Typical temperatures are of the order of 1100°C, and

Table 2-1   USEFUL VALUES FOR BORON AND PHOSPHORUS

	Boron		Phosphorus		
T, °C	N_0, cm^{-3}	D, μ^2/h	N_0, cm^{-3}	N_s, cm^{-3}	D, μ^2/h
950	4.5×10^{20}	1.6×10^{-3}	8.5×10^{20}	1×10^{19}	1.7×10^{-3}
				8.5×10^{20}	2.2×10^{-2}
1000		5.2×10^{-3}	1×10^{21}	1×10^{19}	6.4×10^{-3}
				1×10^{21}	6.0×10^{-2}
1050	5.0×10^{20}	1.7×10^{-2}	1.1×10^{21}	1×10^{19}	2.0×10^{-2}
				1×10^{21}	1.7×10^{-1}
1100		5.8×10^{-2}	1.2×10^{21}	1×10^{19}	7.3×10^{-2}
				1×10^{21}	3.7×10^{-1}
1150	5.2×10^{20}	1.6×10^{-1}	1.3×10^{21}	1×10^{19}	1.8×10^{-1}
				1×10^{21}	8.1×10^{-1}

Note: N_s = surface concentration, cm^{-3}
  N_0 = solid solubility, cm^{-3}
  $\mu = 10^{-6}$ m

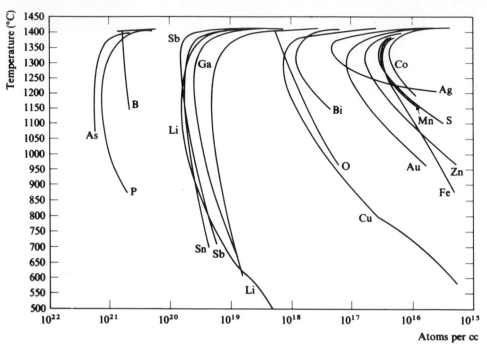

FIGURE 2-2
Temperature dependence of solid solubilities.

the gas stream is steam or oxygen for approximately the first one-half hour, followed by nitrogen for the remainder of the drive-in time.

Predeposit Diffusion

To analyze the predeposit step, we assume that Fick's laws adequately describe the diffusion of impurities inside the silicon wafer. It is also postulated that:

1 Quasi-equilibrium conditions exist at the wafer surface, so that whatever chemical reactions occur between the wafer surface and its surroundings cause the impurity concentration to be maintained at solid solubility N_0.
2 The wafer is semi-infinite in thickness.
3 The x direction is taken from the surface into the wafer.
4 No impurity atoms exist in the wafer for $t < 0$. This restriction is easily removed later.

We now solve (2-6) subject to the boundary and initial conditions which are embodied in the above postulates.[5] For concreteness we assume diffusion of acceptor atoms with concentration $N_A(x, t)$ cm^{-3}. The boundary conditions are

$$N_A(0, t) = N_0 u(t)$$

where $u(t)$ is the unit step function

$$N_A(\infty, t) < \infty$$

and the initial condition is

$$N_A(x, 0) = 0$$

To find the solution of (2-6) we first eliminate t by use of the Laplace transformation; the resulting ordinary differential equation is

$$sN_A(x, s) - N_A(x, 0) = D_p \frac{d^2 N_A(x, s)}{dx^2} \qquad (2\text{-}8)$$

where $N_A(x, s)$ is understood to mean the Laplace transform of $N_A(x, t)$. The subscript p is used to emphasize the fact that the diffusivity must be evaluated at the temperature of the predeposition.

The solution of (2-8) is

$$N_A(x, s) = A(s) \exp\left(-x\sqrt{\frac{s}{D_p}}\right) + B(s) \exp\left(x\sqrt{\frac{s}{D_p}}\right) \qquad (2\text{-}9)$$

Because $N_A(\infty, t) < \infty$, $B(s) = 0$. To evaluate $A(s)$ we note that $N_A(0, t) = N_0 u(t)$ which when transformed yields

$$N_A(0, s) = \frac{N_0}{s}$$

Thus we obtain

$$A(s) = \frac{N_0}{s}$$

and

$$N_A(x, s) = \frac{N_0}{s} \exp\left(-x\sqrt{\frac{s}{D_p}}\right) \qquad (2\text{-}10)$$

The inverse transform of (2-10) is

$$N_A(x, t) = N_0 \operatorname{erfc} \frac{x}{2\sqrt{D_p t}} \qquad (2\text{-}11)$$

where

$$\operatorname{erfc} y \triangleq 1 - \operatorname{erf} y = 1 - \frac{2}{\pi} \int_0^y e^{-\lambda^2} \, d\lambda$$

This impurity distribution is sketched in Fig. 2-3 for a predeposition time t_p.

It is also of interest to calculate the total number of impurity atoms in the wafer after a predeposition time t_p. If the surface area of the predeposit is A, then the total number of acceptor atoms $Q_A(t_p)$ is given by

$$Q_A(t_p) = A \int_0^\infty N_A(x, t_p) \, dx$$

$$= N_0 A \int_0^\infty \operatorname{erfc} \frac{x}{2\sqrt{D_p t_p}} \, dx \qquad (2\text{-}12)$$

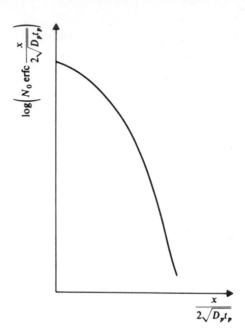

FIGURE 2-3
Sketch of impurity distribution resulting
from predeposition.

From a table of integrals, $\int_0^\infty \mathrm{erfc}\ y\ dy$ is found to be $1/\sqrt{\pi}$; therefore

$$Q_A(t_p) = \frac{2N_0 A}{\sqrt{\pi}} \sqrt{D_p t_p} \qquad (2\text{-}12a)$$

The quantity $\sqrt{D_p t_p}$ can be thought of as a characteristic length associated with a particular diffusion. If the thickness of the material is large in comparison with this length, the approximation of a semi-infinite wafer is a good one.

EXAMPLE 2A   *Calculation of the Characteristic Length.*   A boron predeposition is carried out at 950°C for 30 min. Using $D_p \approx 1.6 \times 10^{-3}\ \mu m^2/h$, we obtain a characteristic length

$$\sqrt{D_p t_p} = 2.8 \times 10^{-2}\ \mu m = 2.8 \times 10^{-8}\ m$$

If the wafer thickness is $L = 10\ \mathrm{mils} = 10 \times 25.4 \times 10^{-6}\ m$, we find $L/\sqrt{D_p t_p} \approx 10^4$, so the semi-infinite approximation is excellent for this case.   ////

Diffusion into Constant Background Concentration

The assumption that the wafer had no initial impurities is not a realistic one for most applications. For example, the base diffusion is performed into an epitaxial layer which, if *n*-type, has some distribution of donor atoms. It is convenient to regard this distribution of donor atoms as initially being uniform, although such an approximation is not valid near the substrate–epitaxial-layer junction. However, typical epitaxial-layer thicknesses are of the order of 8 to 12 μm, while device junctions of interest are fabricated within approximately 3 μm of the surface. Therefore the epitaxial layer near the surface is often assumed to have approximately a uniform distribution of impurities whose concentration is N_{BC}, the background concentration. Since a p-n junction will be formed where the diffusing acceptor density is equal to the donor density of the epitaxial material, it is relevant to ask what happens to the donor distribution during the diffusion of acceptors.

It must first be recalled that collisions among impurity atoms themselves during diffusion are assumed to be insignificant in comparison with collisions between impurity atoms and atoms of the silicon crystal. Therefore the diffusion of acceptors and the simultaneous diffusion of donors can be treated as independent problems.

It is assumed that during predeposition of acceptors, the flux of donor atoms is zero at the wafer surface; that is, donor atoms are neither added to nor subtracted from the wafer during diffusion of acceptor atoms. It is also assumed that the thickness of the epitaxial layer is semi-infinite, and that the initial donor distribution is $N_D(x, 0) = N_{BC}$. Again, the semi-infinite approximation is valid as long as the thickness of the epitaxial layer is large in comparison with the characteristic length for donors. The result of these approximations is that there is effectively no change of the donor distribution near the wafer surface, and

$$N_D(x, t) \approx N_{BC}$$

This approximation is not valid near the substrate junction since donors diffuse into the substrate.

During the acceptor predeposit, a p-n junction is formed; if the constant-background-concentration approximation is used, the junction depth x_j is given by

$$N_A(x_j, t_p) = N_{BC} \qquad (2\text{-}13)$$

Substituting (2-13) in (2-11), we obtain

$$N_{BC} = N_0 \, \mathrm{erfc} \, \frac{x_j}{2\sqrt{D_p t_p}} \qquad (2\text{-}14)$$

EXAMPLE 2B *Calculation of Predeposition Junction Depth.* A boron predeposition is performed in a wafer having $N_{BC} = 5 \times 10^{15}$ cm^{-3}, with predeposition duration of 30 min at 950°C. The resulting junction depth is given by (2-14), which yields

$$x_j \approx 0.2 \; \mu\text{m} = 0.2 \times 10^{-6} \; \text{m}$$

The total number of atoms predeposited is found from (2-12) to be

$$Q_A(t_p) = \frac{2N_0 A}{\sqrt{\pi}} \sqrt{D_p t_p} = 14.4 \times 10^{14} \times A$$

where A is the surface area in square centimeters.

/////

Drive-in Diffusion

After the predeposition has been carried out for a predetermined time, the wafer is placed in a different atmosphere, one not containing impurity-bearing gas, and at a higher temperature than was used during predeposit. Diffusion of the impurity atoms continues, but now with different boundary conditions. This procedure is used for two principal reasons:

1 It is necessary to grow a layer of SiO_2 on the surface for passivation and for use as oxide masking in subsequent photoresist steps. If this is done during preposition, the oxide growth will interfere with the deposition of impurity atoms.

2 As we shall see in later chapters, high-performance bipolar transistors require base-region surface concentrations lower than N_0. The two-step diffusion procedure lowers the surface concentration during drive-in.

Since the wafer is placed in an oxidizing atmosphere, it is reasonable to assume that no impurity atoms flow from the wafer to the atmosphere. For the present, we also assume that no impurities dissolve in the oxide layer. This means that the total number of impurity atoms remains constant, and that the flux is zero at the surface. To find the resulting impurity distribution, we first define a new time variable

$$t' \triangleq t - t_p$$

Next we must solve (2-6) subject to the following conditions:

1 $N_A(\infty, t') \to 0$

(The wafer is assumed to be semi-infinite in extent.)

2 $-D_d \dfrac{\partial N_A(0, t')}{\partial x} = 0$

(Note that the subscript d is used to indicate that the diffusivity must be evaluated at the drive-in temperature.)

3 $N_A(x, 0) = N_A(x, t = t_p)$

Condition (3) is simply a statement of the fact that the acceptor distribution at the beginning of the drive-in is that distribution which exists at the end of the predeposit diffusion.

Equation (2-6) can be solved subject to conditions (1) through (3); the result is[5]

$$N_A(x, t') = (4\pi D_d t')^{-1/2} \int_0^\infty N_A(\alpha, t_p)(e^{-(x-\alpha)^2/4D_d t} + e^{-(x+\alpha)^2/4D_d t}) \, d\alpha \quad (2\text{-}15)$$

Unfortunately, when (2-11) is substituted for $N_A(\alpha, t_p)$ in (2-15) the integral must be evaluated by numerical methods. We therefore wish to investigate the use of approximations for $N_A(x, t_p)$ which would

1 Enable the integration of (2-15) to be carried out
2 Yield tractable results which can subsequently be used in evaluating the profile and calculating junction depths

The Step Approximation

In the interest of finding an approximation which will make possible the integration in (2-15), we first consider

$$N_A(x, t_p) = N_0 \qquad 0 \le x \le h$$

$$N_A(x, t_p) = 0 \qquad x > h$$

$$N_0 h A = Q_A(t_p)$$

This approximation is shown pictorially in Fig. 2-4; note that it conserves the total number of predeposited atoms. Use of this approximation in (2-15) yields

$$N_A(x, t') = \frac{N_0}{2} \left(\mathrm{erf}\, \frac{x - h}{2\sqrt{D_d t'}} - \mathrm{erf}\, \frac{x + h}{2\sqrt{D_d t'}} \right) \qquad (2\text{-}16)$$

While the step approximation satisfies the requirement that the integration be carried out, the results can hardly be considered tractable.

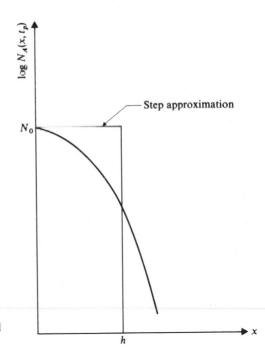

FIGURE 2-4
The step approximation for the initial distribution in drive-in diffusion.

The Delta-Function Approximation

A mathematically simpler approximation is to assume that all the atoms pre-deposited are concentrated at the wafer surface; that is,

$$N_A(x, t_p) \approx \frac{Q_A}{A} t_p \, \delta(x) \qquad (2\text{-}17)$$

where $\delta(x)$ is the unit delta function.   Now approximation (2-17) is used in (2-15); the delta function in the integrand makes the integration simple, with the result that

$$N_A(x, t') = \frac{Q_A(t_p)}{A\sqrt{\pi D_a t'}} \, e^{-x^2/4 D_a t'} \qquad (2\text{-}18)$$

Furthermore, (2-18) is certainly more tractable than (2-16).

To evaluate the profile after a predeposit followed by a drive-in, one need only

1   Use (2-12a) to calculate $Q_A(t_p)$.
2   Use the delta-function approximation, substituting $Q_A(t_p)$ in (2-18).

It is intuitively clear that the delta-function approximation is worst for $t' \approx 0$.   For characteristic lengths $\sqrt{D_a t'}$ large compared with the predeposit characteristic length $\sqrt{D_p t_p}$, the approximation is reasonably good, as is shown by Fig. 2-5.[6]

EXAMPLE 2C   *Drive-in Diffusion.*   Suppose that the wafer having the pre-deposit diffusion of Example 2B is now subjected to a drive-in diffusion in a nitrogen atmosphere for 2 h at 1150°C. $Q_A(t_p)$ has already been calculated.   For a drive-in time $t' = t_d$, the delta-function approximation yields a profile from (2-18) of

$$N_A(x, t_d) = \frac{Q_A(t_p)}{A\sqrt{\pi D_a t_d}} \, e^{-x^2/4 D_a t_d}$$

A *p-n* junction is formed at a depth x_j, given by

$$N_A(x_j, t_d) = N_{BC}$$

For $N_{BC} = 5 \times 10^{15}$ cm^{-3}, the resulting junction depth is

$$x_j = 3.1 \times 10^{-6} \text{ m}$$

The profile after drive-in, and the junction location, are sketched in Fig. 2-6.   ////

Note that the calculations show that the junction depth is more than an order of magnitude greater after drive-in than it was after predeposit.   This provides an affirmative consistency check on the validity of the delta-function approxima-tion.

Convenient normalized curves for erfc y and e^{-y^2} are given in Fig. 2-7.

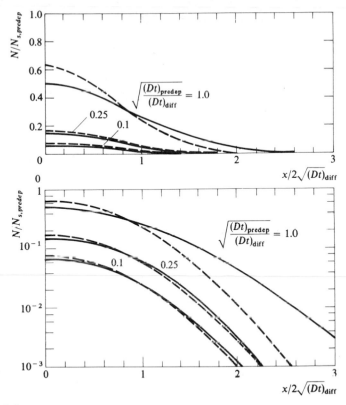

FIGURE 2-5
Accuracy of the delta-function approximation (after Grove[15]). Two-step diffusion: predeposition followed by drive-in diffusion. Solid lines are exact final distributions. Dashed lines correspond to case when predeposited distribution is approximated by delta function at $\alpha = 0$.

Variation of the Diffusivity

In order for the crystalline structure of the semiconductor to be maintained, the impurity atoms must enter substitutionally in the lattice; that is, a silicon atom must be displaced by an impurity atom. This means that impurity atoms must have some minimum energy ΔE_a in order to enter substitutionally into the lattice, and we therefore expect the diffusivity to be a function of temperature. It is assumed that the diffusivity is of the form

$$D(T) = D_0\, e^{-\Delta E_a/kT}$$

The proportionality constant D_0 and the activation energy ΔE_a can be determined experimentally. Measurements indicate that the functional form proposed for

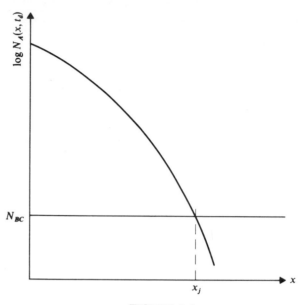

FIGURE 2-6
Drive-in profile and junction depth.

$D(T)$ is accurate. Graphs and values of $D(T)$ for several impurities in silicon are given in Fig. 2-8[7, 8] and Table 2-1.

During the several diffusion cycles performed on a wafer during a given process sequence the temperature varies quite drastically, causing an attendant variation of the diffusivities of all impurity species in the wafer. There are two principal variations of temperature which are of concern:

1 The variation of temperature as the wafer is inserted in and removed from the diffusion furnace. The rate of variation of temperature in this case is a function of the thermal characteristics of the diffusion boat holding the wafer and the transient response of the furnace.

2 Long-term variations of temperature resulting from successive steps being carried out at different temperatures, and from variations of furnace temperature during a given step.

If the diffusivity varies with time because of temperature variations, the continuity equation (2-6) is a partial differential equation with time-varying coefficient. We now demonstrate that its solution can be obtained from the solution which we have already found for the continuity equation with constant coefficients.[9] If we assume negligible electric field, and diffusivity independent of concentration, the equation which must be solved is

$$\frac{\partial N}{\partial t} = D(t) \frac{\partial^2 N}{\partial x^2}$$

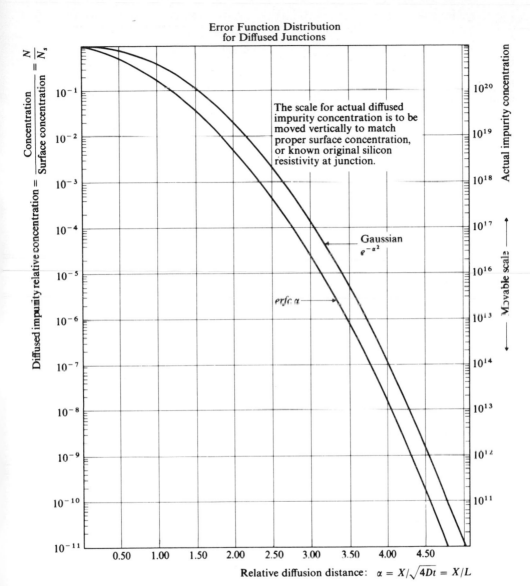

Error Function Distribution
for Diffused Junctions

The scale for actual diffused impurity concentration is to be moved vertically to match proper surface concentration, or known original silicon resistivity at junction.

Gaussian $e^{-\alpha^2}$

$erfc\ \alpha$

Relative diffusion distance: $\alpha = X/\sqrt{4Dt} = X/L$

FIGURE 2-7
Convenient normalized curves for erfc α and $e^{-\alpha^2}$.

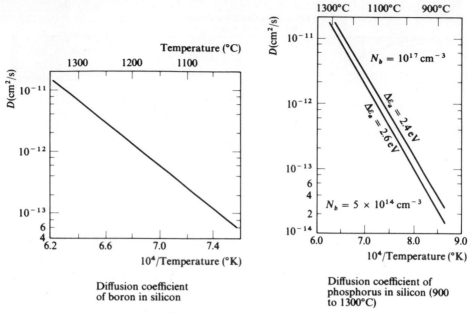

FIGURE 2-8
Diffusivity versus temperature for boron and phosphorus in silicon.

where N is the concentration of the impurity under consideration. We make a change of variable

$$\alpha \triangleq \alpha(t)$$

Now we have

$$\frac{\partial N}{\partial t} = \frac{\partial N}{\partial \alpha} \frac{\partial \alpha}{\partial t}$$

Since α was not restricted, we can let it be defined by the relation

$$\frac{d\alpha}{dt} = D(t)$$

from which we obtain

$$\alpha = \int_0^t D(t)\, dt = \bar{D}t$$

where $\bar{D}$ is the time average of $D(t)$. In terms of x and α, the partial differential equation to be solved is now

$$\frac{\partial N}{\partial \alpha} = \frac{\partial^2 N}{\partial x^2}$$

But this equation is of the same *form* as the continuity equation with *constant* diffusivity. Thus we see that for any given boundary conditions and any given initial profile $N(x, 0)$, we can find $N(x, t)$ for the case when temperature varies by:

1 Finding $N(x, t)$ for D assumed independent of time
2 Replacing Dt by $\bar{D}t$

For most diffusion systems, furnace temperatures are controlled to $\pm 1°C$, and transient temperature variations subside within a few minutes. Therefore for diffusion times greater than several minutes, the most important temperature variations are those which occur because successive steps are carried out at different temperatures. Since the temperature is essentially constant during each step, it is easy to obtain $\bar{D}$.

EXAMPLE 2D   *Use of the Average Diffusivity for Calculation of Boron Diffusion.* In the fabrication of integrated circuits the phosphorus emitter drive-in is usually carried out at a lower temperature than the boron base drive-in. Since the boron atoms are also subjected to the temperature change during the emitter drive-in, further diffusion of boron takes place, causing a deeper collector-base junction than would occur without the emitter drive-in. A sketch of $D(t)$ for the diffusion cycle is shown in Fig. 2-9.

Let the drive-in times be t_B and t_E for base and emitter, respectively, and let the diffusivity of boron be D_{B1} and D_{B2} at base drive-in and emitter drive-in temperatures. The transient effects will be neglected. Note also that when the wafer is not in the furnace, $D \approx 0$ for all impurities. Diffusion of boron during phosphorus predeposition is also neglected.

The total diffusion time for boron atoms is $t_1 \triangle t_B + t_E$. The average diffusivity is

$$\bar{D}_B = \frac{D_{B1} t_B + D_{B2} t_E}{t_B + t_E}$$

and the resulting boron profile at the end of the emitter drive-in is

$$N(x, t_1) = \frac{Q_A(t_p)}{A} \frac{1}{\sqrt{\pi \bar{D}_B t_1}} e^{-(x^2/4\bar{D}_B t_1)}$$

The collector junction occurs at x_{jc}, where

$$N(x_{jc}, t_1) = N_{BC} \qquad ////$$

Exercise 2-1   The wafer of Example 2C is further subjected to a phosphorus predeposit of 30 min at 950°C and a drive-in of 2 h at 1100°C. Find the new location of the *collector* junction after this cycle. Neglect the phosphorus concentration near the collector junction.

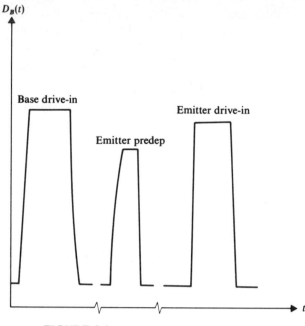

FIGURE 2-9
Boron diffusivity $D_B(t)$ for emitter and base diffusions.

Location of the Emitter and Collector Junctions

We have seen that prior to the emitter diffusion the collector junction occurred where $N_A(x_j, t_d) = N_{BC}$. When the phosphorus drive-in for the emitter diffusion is completed, the epitaxial layer contains the original donor atoms of that layer, the boron acceptor atoms of the base diffusion, and the phosphorus donor atoms of the emitter diffusion. These profiles are all sketched in Fig. 2-10. Junctions will occur at any values of x where the net impurity concentration $N_D(x) - N_A(x)$ is zero.

For most practical cases, the variation of boron and phosphorus concentrations with depth is so large that in the vicinity of the collector junction the phosphorus concentration is much less than the background concentration. Therefore the approximation

$$N_D(x) \approx N_{BC}$$

can be used in calculating the collector junction depth, resulting in

$$N_A(x_{jC}, t_1) \approx N_{BC} \qquad (2\text{-}20)$$

Furthermore, near the emitter junction the phosporus concentration is often much larger than the background concentration, and in this case the contribution of the background concentration to the total donor concentration is negligible. If this is true the emitter junction depth satisfies the relation

$$N_D(x_{jE}, t_E) \approx N_A(x_{jE}, t_1) \qquad (2\text{-}21)$$

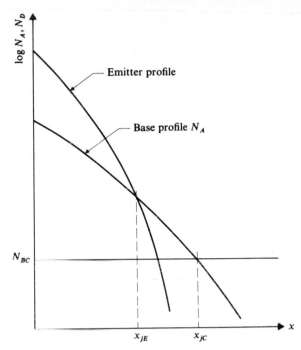

FIGURE 2-10
Impurity profiles after base and emitter diffusions.

where $N_D(x_{jE}, t_E)$ is the phosphorus profile after a drive-in time t_E. Since both $N_D(x_{jE}, t_E)$ and $N_A(x_{jE}, t_1)$ are of the same functional form, (2-21) can be solved for x_{jE}. It should be noted, however, that if N_{BC} is not negligible in comparison with $N_D(x_{jE}, t_E)$, a transcendental equation results:

$$N_D(x_{jE}, t_E) + N_{BC} - N_A(x_{jE}, t_1) \qquad (2\text{-}22)$$

Exercise 2-2 A double diffused *n-p-n* transistor has been made with the following diffusion schedule:

    1 Boron: Predeposition 30 min at 950°C; drive-in 2 h at 1150°C
    2 Phosphorus: Predeposition 30 min at 950°C; drive-in 2 h at 1100°C

The background concentration is $N_{BC} = 5 \times 10^{15}$ cm^{-3}. Find the emitter junction depth.

Lateral Diffusion

All diffusion calculations so far have been based on the one-dimensional form of Fick's second law. Consider the diffusion of an impurity at the edge of an oxide window, as shown in Fig. 2-11. It is unrealistic to assume that the diffusion will

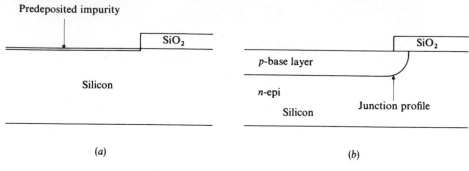

Predeposited impurity

SiO₂

p-base layer

Silicon

n-epi

Silicon

Junction profile

(a)   (b)

FIGURE 2-11
Diffusion at an oxide edge: (a) after predeposit; (b) after drive-in.

proceed only straight down from the oxide window. There will also be a diffusion tendency parallel to the surface, causing the impurity to spread out under the oxide as shown in Fig. 2-11b. To a first-order approximation the lateral diffusion will be approximately equal to the vertical diffusion, resulting in a junction profile which is approximately a quarter circle in the region under the oxide, if the background impurity concentration is uniform. Because of this lateral diffusion, the actual junction reaches the silicon surface at a point under the protective oxide layer. Since the oxide layer remains in place, it shields the junction against surface damage and contamination, and the oxide layer is often called a passivating layer.

Mathematical analysis of the three-dimensional diffusion problem suggests the approximation that diffusion occurs laterally in approximately the same way that it does vertically for most practical cases of interest. Detailed analysis confirms the validity of the approximation.[10]

When the emitter is diffused into the previously diffused base layer, the fact that the acceptor profile is nonuniform causes the locus of the emitter junction under the oxide to be no longer a section of a circle. Because the acceptor concentration is highest near the surface, the distance from the corner of the window to the junction is less at the surface than it is below the surface. This gives rise to a junction locus similar to that sketched in Fig. 2-12. It can be shown that if N_{BC} is negligible in comparison with the diffused donor concentration at a depth x_{jE}, the lateral distance to the junction is given by

$$y_j(\theta) = x_{jE} \cos \theta \left(\frac{D_{dA} t_{dA}}{D_{dD} t_{dD}} - 1 \right)^{1/2} \left(\frac{D_{dA} t_{dA}}{D_{dD} t_{dD}} - \sin^2 \theta \right)^{-1/2} \qquad (2\text{-}22)$$

where D_{dA} = acceptor diffusivity during base drive-in
  t_{dA} = base drive-in time
  D_{dD} = donor diffusivity during emitter drive-in
  t_{dD} = emitter drive-in time

The effects of lateral diffusion are particularly important in the layout of isolation regions. Typical epitaxial-layer thicknesses are $\frac{1}{4}$ to $\frac{1}{2}$ mil. To ensure isolation the concentration of the isolation diffusion acceptors must be greater than

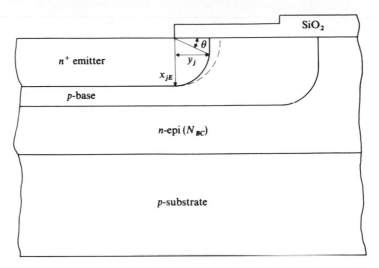

FIGURE 2-12
Effects of lateral diffusion on location of the emitter junction after diffusion into the base layer.

N_{BC} at a depth of $\frac{1}{2}$ mil. This means that the lateral location of the isolation junction will be at least $\frac{1}{2}$ mil from the isolation window. As a consequence, *the spacing between the isolation window and a neighboring window for a p diffusion must be at least $\frac{1}{2}$ mil greater than the minimum spacing* allowed for registration or clearance tolerances. For this reason, a 1-mil-geometry device usually requires 1.5 to 2 mil isolation clearances in the layout.

Out Diffusion

The emphasis of the discussion of diffusion has been primarily on base and emitter diffusions. It must be recognized, of course, that at elevated temperatures, all impurity atoms in the wafer are diffusing. This means that epitaxial-layer donors diffuse into the substrate, substrate acceptors diffuse into the epitaxial layer, and buried-layer donors diffuse into both epitaxial layer and substrate. In cases where undesirable diffusion occurs, such as for buried layers, its effects can sometimes be minimized by choosing impurities such as antimony or arsenic, which have small diffusivities.

Other conditions for diffusion occur when it is not possible to ensure that $\partial N(0, t)/\partial x \approx 0$. This can be caused by impurities dissolving in the oxide layer during drive-in, or by impurities reacting with the gas stream and leaving the wafer during predeposition. For example, if donors from the epitaxial layer leave the wafer during predeposition of base acceptors, the profile of the epitaxial-layer donors will no longer be constant near the surface. The result of this out diffusion is sketched in Fig. 2-13.

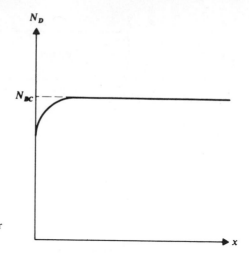

FIGURE 2-13
Effects of out diffusion of epitaxial-layer
donors.

Sensitivity to Process Variables

The kinetics of the gas-phase reactions in the diffusion tube, and of the gas-solid-phase reactions at the wafer surface, are not well understood.   Fortunately, the weak dependence of the solid solubilities of impurities on temperature mitigates this circumstance.   For solid-phase diffusion in the wafer, the temperature is the most important variable.   One can define a sensitivity of the diffusivity to temperature as

$$S_T{}^D \triangleq \frac{\partial D}{\partial T} \frac{T}{D} \qquad (2\text{-}23)$$

Since we have postulated that D is of the form

$$D = D_0 \, e^{-\Delta E_a/kT}$$

we find

$$S_T{}^D = \frac{\Delta E_a}{kT}$$

Measurements indicate that for boron at 1000°C,

$$\Delta E_a \approx 2.69 \text{ eV}$$

Using this value, we find

$$S_T{}^D = 24.5 \qquad \text{at } 1000°\text{C}$$

For a 2.5 percent tolerance on total number of atoms predeposited at a given time, D must be held to ± 5 percent.   This means that the maximum temperature variations are ± 2.0 percent or $\pm 2.0°$C.   Electronic controllers are generally used on diffusion furnaces to maintain these temperature tolerances.

It is, of course, possible to calculate sensitivities other than that of D to temperature; several possibilities are considered in Probs. 2-4 to 2-6.

2-3 EVALUATION OF DIFFUSION RESULTS

Since the results of a fabrication sequence are quite sensitive to the process variables, it is desirable to have some method of evaluating results during the fabrication sequence so that adjustments in subsequent steps can be made if necessary. Since it is very difficult to measure the actual impurity profiles, other parameters must be measured. The two parameters most often used are junction depth and sheet resistance. Typically, a monitor wafer is processed with the wafer containing circuits. This permits the use of destructive measurements without damage to the integrated circuits.

Measurement of Junction Depth

Junction depth is measured by lapping and staining a piece of the monitor wafer. One commonly used method is to mount the wafer on a post which is beveled at an angle of about 1°. The surface of the wafer is then lapped with a suitable abrasive until several micrometers of material have been removed. The resulting lapped surface is now at a 1° angle relative to the original surface, and the location of the junction is exposed by the lapping. A stain such as concentrated HF in the presence of intense light is applied, causing p regions to be darker in color than n regions. This color difference can be observed visually through a microscope. To measure the junction depth, one uses a partially reflecting mirror, as shown in Fig. 2-14, and monochromatic illumination.[11] The light reflected from the bottom surface of the mirror produces an interference pattern with the light reflected from the beveled surface of the wafer. The junction depth is measured by counting the number of interference fringes that appear between the actual wafer surface and the junction location, and multiplying by half the wavelength of the illumination used. The result is independent of the beveling angle, thus making accurate determination of the angle unimportant.

The stain is somewhat sensitive to the impurity concentration as well as to the space charge occurring in a junction depletion region. Calculations indicate

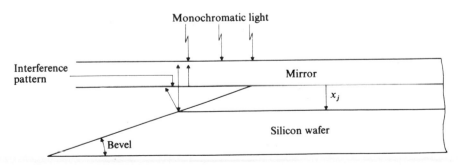

FIGURE 2-14
Angle-lap method of measuring junction depth.

that for typical integrated-circuit junctions, the stain described above causes a color change to occur at the p side of a depletion region, rather than at the metallurgical junction.   This should be taken into account when measuring junction depths.

While lap and stain methods are relatively quick and convenient to use, they are not extremely accurate.   For typical monochromatic illumination used for this measurement, the distance between fringes is approximately 0.3 μm.   It is therefore difficult to make measurements that are accurate to better than 0.15 μm.

Nondestructive electrical measurements can be made to determine the collector junction depth.   This is done by diffusing a V-shaped pattern during the base diffusion.   An angle of only a few degrees is used for the V, and the resistance of the pattern is measured.   This resistance will be a function of the amount of lateral diffusion that takes place inside the V.   Since the vertical diffusion is approximately the same as the lateral diffusion, if the lateral diffusion is found from the resistance measurement the junction depth is known.   Junction depth measurements made with this method correlate well with angle-lap measurements.[12]

Measurement of Sheet Resistance

The sheet resistance ρ_s' can be measured on a monitor wafer by the four-point probe method.   Four equally spaced probes are brought into contact with the wafer. A known dc current is applied to two probes, as shown in Fig. 2-15.   The voltage between the remaining probes is measured and the sheet resistance is computed from

$$\rho_s' = \frac{\pi}{\ln 2} \frac{V}{I} = 4.53 \frac{V}{I} \qquad (2\text{-}24)$$

While (2-24) applies only for an infinite wafer surface, it is a good approximation if the wafer dimensions are large in comparison with the probe spacing.

An alternate and superior method is to include a pattern for sheet-resistance measurement in the layout of the integrated circuit.   This method has the advantage

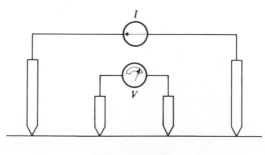

Silicon test wafer

FIGURE 2-15
Four-point probe method for measuring
sheet resistance.

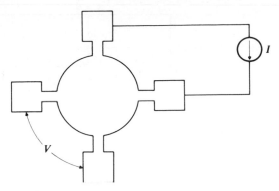

FIGURE 2-16
Van der Pauw method for measuring sheet resistance.

that it provides a measurement on the actual chip rather than on a monitor wafer. Furthermore, if the Van der Pauw method[13] shown in Fig. 2-16 is used, the sheet resistance measured is independent of the chip size and the pattern dimensions. The sheet resistance in this case is also given by (2-24).

The Van der Pauw method has the further advantage that the sheet resistance of, for example, the base layer *after* the emitter diffusion can easily be measured by performing an emitter diffusion over all but the contact areas.

Irvin's Curves[14]

The sheet resistance of a diffused layer between the depths x_1 and x_2 is related to the average conductivity of the layer by

$$\rho'_s = \frac{1}{\bar{\sigma}(x_2 - x_1)} \qquad (2\text{-}25)$$

where σ is the average conductivity and is given by

$$\bar{\sigma} = \frac{1}{x_2 - x_1} \int_{x_1}^{x_2} \sigma(x)\, dx = \frac{1}{x_2 - x_1} \int_{x_1}^{x_2} q\mu(N)N(x)\, dx \qquad (2\text{-}26)$$

where $\mu(N)$ = majority carrier mobility
$N(x)$ = majority carrier concentration

If a junction is formed at x_2, all lateral current flow in the layer is confined to the region above the junction; if depletion regions are assumed to be negligible we can let $x_2 = x_j$, and for $0 \le x < x_j$ the sheet resistance of any layer can be determined from (2-26). Curves have been plotted by J. C. Irvin to facilitate the determination of sheet resistance for commonly encountered impurity profiles diffused into a uniform background concentration. These curves are designed for calculation of the sheet resistance of the general layer depicted in Fig. 2-17. They

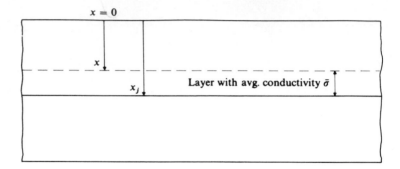

FIGURE 2-17
General layer for which Irvin's curves yield sheet resistance.

take into account the dependence of mobility on concentration, and are available for a wide range of background concentrations. Correspondingly, if the sheet resistance and the junction depth can be measured, Irvin's curves can be used to find the surface concentration. Typical curves are given in Fig. 2-18.

EXAMPLE 2E   *Calculation of Surface Concentration.*   A base diffusion has been performed into a uniform background concentration $N_{BC} = 10^{15}$. From an angle-lap measurement the junction depth is found to be $x_j = 3$ μm. From a four-point probe measurement, the sheet resistance of the base layer is found to be $\rho'_s = 200$ Ω per square. The surface concentration is found by using Irvin's curves as follows.

The base diffusion is known to produce approximately a gaussian impurity distribution after drive-in. Since the four-point probe measures sheet resistance at the surface, we wish to use the particular curve for $x/x_j = 0$. Furthermore, the average conductivity is

$$\bar{\sigma} = \frac{1}{\rho'_s(x_j - x)} = \frac{1}{\rho'_s x_j} = 16.7 \, \mho/cm$$

We enter the family of curves at $\bar{\sigma} = 16.7$ and move vertically until an intersection is obtained with the $x/x_j = 0$ member. Moving horizontally from that intersection we obtain

$$N_s = 3 \times 10^{18} \text{ cm}^{-3} \qquad ////$$

EXAMPLE 2F   *Sheet Resistance of a Partial Layer.*   Suppose that the surface is now etched to a depth of 2.7 μm. If depletion regions are neglected, we can calculate the sheet resistance which would be measured at the new surface as follows.

We wish the sheet resistance of the layer *originally* located between $x = 2.7$ μm and $x = x_j = 3.0$ μm. Entering the family at $N_s = 4 \times 10^{18}$ cm^{-3}, we move

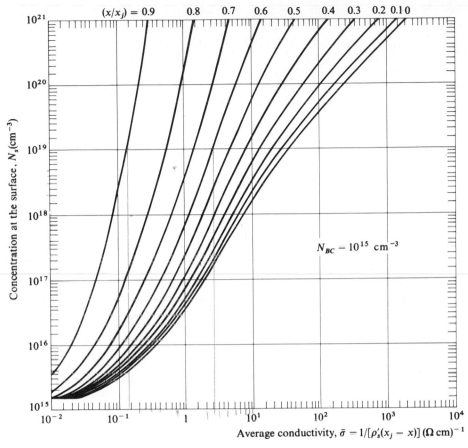

FIGURE 2-18
Irvin's curves for p-type guassian layers in a uniform n-type background of 10^{15} cm^{-3}.

horizontally until an intersection occurs with the $x/x_j = 2.7/3.0 = 0.90$ curve. Moving vertically to the axis, we find $\bar{\sigma} = 1.1 \times 10^{-1}\ \mho/cm$. The sheet resistance can now be calculated from

$$\rho'_s = \frac{1}{\bar{\sigma}(x_j - x)} = \frac{1}{1.1 \times 10^{-1} \times 0.3 \times 10^{-4}} = 3.0 \times 10^5\ \Omega/sq \qquad ////$$

Exercise 2-3 If the depletion region extended 0.1 μm into the p material in Example 2F, how would you calculate the sheet resistance of the layer after 2.7 μm are etched from the surface?

EXAMPLE 2G *Concentration Below the Surface.* Suppose we must determine what will be the new surface concentration after 2.7 μm is etched from the surface, that is, what the original concentration is at a depth of 2.7 μm. The average conductivity of the layer between $x = 2.7$ μm and $x_j = 3.0$ μm has been found to be $\bar{\sigma} = 1.1 \times 10^{-1}$. When the 2.7 μm is etched away, this will be the average conductivity measured at the new surface. Therefore the new surface concentration can be found by entering the family of curves at $\bar{\sigma} = 1.1 \times 10^{-1}$, moving vertically until an intersection with the $x/x_j = 0$ curve is obtained, and then moving horizontally to read $N_s = 3.5 \times 10^{15}$ cm^{-3}. ////

It is to be emphasized that Irvin's curves are for diffusions into constant background concentration. They cannot be used for emitter diffusions into diffused base regions for two reasons:

1 The background for the emitter diffusion is the gaussian base diffusion profile. One might be tempted to use Irvin's curves to obtain an estimate for emitter layers, except that

2 The curves do not take into account the degradation of mobility which occurs as a result of successive compensation of the material to form the *n*-type emitter.

2-4 OXIDATION†

Oxidation is the process of creating the protective silicon dioxide passivating layer on the wafer surface. The silicon dioxide layer serves to mask silicon against impurity diffusion in predeposition steps and to protect the surface against contaminants (particularly the junctions), and it serves as an insulating layer over which interconnections, bonding pads, capacitor plates, and thin-film elements may be placed. Silicon dioxide is, in many ways, a principal ingredient in the success of the planar process. It is indeed fortuitous that silicon has an easily formed protective oxide, for otherwise we would have to depend upon deposited insulators for surface protection—a necessity which has held back germanium integrated-circuit technology until just recently. We consider here only thermally grown oxide.

Silicon undergoes thermal oxidation in the presence of either oxygen or water vapor. In order to obtain a sufficiently rapid rate of oxide formation, the process is carried out at elevated temperatures of the order of the drive-in temperatures used for diffusion steps. For this reason, and because an oxidizing atmosphere can be inert as far as the impurity distribution is concerned, thermal oxidation is often done at the same time as the drive-in diffusion.

Figure 2-19 shows a silicon wafer covered by a layer of SiO_2. The oxide forms according to the chemical reaction

$$Si + O_2 \longrightarrow SiO_2$$

† Our treatment of oxidation follows closely that of Grove.[15]

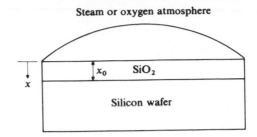

FIGURE 2-19
Growth of silicon dioxide on a silicon
wafer.

In order that the reaction take place, however, either silicon or oxygen, or both, must traverse the SiO_2 barrier. This means that either silicon or oxygen must diffuse through the dioxide. Experiments conducted using radioactive tracers have shown that thermal oxidation takes place as oxygen diffuses through the oxide layer. We therefore treat the problem of oxide growth as a diffusion problem governed by Fick's laws.

It is important to recognize that there are three distinct regions which must be considered: the gas stream in the diffusion tube, the oxide layer on the surface of the wafer, and the silicon wafer. Note that as oxide is grown, silicon atoms from the surface of the wafer are consumed, with the result that the actual surface of the silicon is changing position. We are thus confronted with diffusion in the presence of a time-varying boundary. To expedite the solution, we make the important *quasi-static* approximation that all reactions reach their steady-state conditions essentially instantaneously. The effects of the moving boundary on the diffusion process are neglected.

To simplify the analysis, it is assumed that the region outside the wafer consists of a thin stagnant film of gas contiguous with the wafer surface, and a gas stream contiguous with the stagnant layer. The concentration of oxygen molecules in the gas stream is assumed to be constant and of value N_G.

The oxygen-molecule concentration profiles in the gas stream, stagnant layer, and oxide are shown in Fig. 2-20. Note that the molecules diffuse across the stagnant layer and through the oxide to the silicon surface, where they react with silicon to form silicon dioxide. Since we have assumed static conditions to apply, Fick's second law becomes, in one dimension,

$$\frac{d^2 N}{dx^2} = 0$$

indicating that the concentration profile is linear. Furthermore, the quasi-static approximation also implies that the fluxes of oxygen molecules in the stagnant film, in the oxide, and at the silicon surface must all be equal. Since the fluxes in the stagnant film and the oxide are diffusion fluxes, we can write

$$F_1 = h_G(N_G - N_S) \qquad (2\text{-}27)$$

$$F_2 = \frac{D}{x_0}(N_S' - N_i) \qquad (2\text{-}28)$$

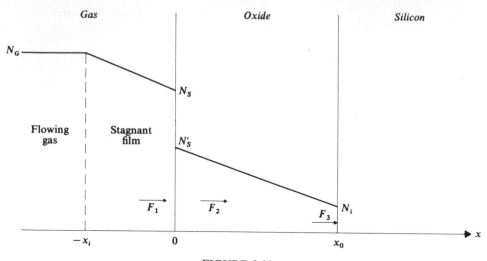

FIGURE 2-20
Concentrations of oxidizing molecules in the three regions.

where $N_G = O_2$ concentration in gas stream
$N_S = O_2$ concentration at gas-oxide interface
$N_S' = O_2$ concentration inside oxide surface
$N_i = O_2$ concentration at oxide-silicon interface
$h_G =$ gas-phase mass transfer coefficient
$D =$ diffusivity of O_2 in oxide
$x_0 =$ oxide thickness

Now we assume that the reaction rate at which oxygen molecules are used in the formation of oxide is proportional to the O_2 concentration at the oxide-silicon interface; that is

$$F_3 = k_S N_i \qquad (2\text{-}29)$$

where $k_S =$ chemical surface reaction constant.

We must also postulate a relationship between N_S and N_S'. To do this we make use of *Henry's law*, which states that *in equilibrium* the concentration of a species just inside the solid surface is proportional to the partial pressure of that species in the gas at the surface. Since Henry's law applies only in equilibrium, we assume that *quasi-equilibrium* quasi-static conditions apply for the oxidation process. The validity of the quasi-equilibrium quasi-static approximation is verified by the fact that good agreement is obtained between theoretical and experimental data.[16]

If we assume the carrier gas to be an ideal gas, we can write the concentration N_S as

$$N_S = \frac{p_S}{kT} \qquad (2\text{-}30)$$

where p_S = partial pressure of O_2 at the surface. By Henry's law we also have

$$N'_S = Hp_S \qquad (2\text{-}31)$$

where H = Henry's law constant. Combining (2-31) and (2-32), we obtain

$$N'_S = H'N_S \qquad (2\text{-}32)$$

where $H' = HkT$.

It is instructive to analyze an electrical analog model for the oxidation process. Let flux be analogous to current, and concentration analogous to voltage. From (2-27) and (2-28), we note that conductances can be used to represent flux in the stagnant layer and in the oxide. A conductance can also be used to represent flux at the silicon surface. Henry's law can be represented by two dependent sources. The analog circuit is shown in Fig. 2-21.

Noting first that $F_1 = F_2 = F_3 \triangleq F$, we see by inspection that

$$F = \frac{N'_S G_2 G_3}{G_2 + G_3} \qquad (2\text{-}33)$$

and that

$$F = G_1(N_G - N_S) = G_1\left(N_G - \frac{N'_S}{H'}\right) \qquad (2\text{-}34)$$

Combining (2-33) and (2-34), we obtain

$$F = \frac{G_1 N_G}{1 + G_1(G_2 + G_3)/H'G_2 G_3} = \frac{k_S H'N_G}{1 + (H'k_S/h_G) + (k_S/D)x_0} \qquad (2\text{-}35)$$

Now, at the silicon surface, let the number of oxidant molecules required to form a unit volume of oxide be N_1. The flux of molecules at the silicon surface is then

$$F_3 = N_1 \frac{dx_0}{dt} = F \qquad (2\text{-}36)$$

Combining (2-36) and (2-35), we obtain the differential equation for oxide growth:

$$N_1 \frac{dx_0}{dt} = \frac{k_S H'N_G}{1 + (H'k_S/h_G) + (k_S/D)x_0} \qquad (2\text{-}37)$$

Let the oxide thickness at $t = 0$ be L_0. Then the solution of (2-37) with this initial condition is

$$x_0 = \sqrt{\frac{2DH'}{N_1}N_G t + \left[L_0 + \frac{D}{k_S}\left(1 + \frac{H'k_S}{h_G}\right)\right]^2} - \frac{D}{k_S}\left(1 + \frac{H'k_S}{h_G}\right) \qquad (2\text{-}38)$$

It must be emphasized that (2-38) will be considerably in error when both L_0 and t are small, since for this condition one cannot neglect the effects of the moving boundary.

A limiting case of (2-38) is of particular interest. Note that for large t,

$$x_0 \rightarrow \sqrt{\frac{2DH'N_G}{N_1}t}$$

Experimental data confirm the square-root dependence of x_0 on t.[16]

Thus far we have considered oxidation only by oxygen molecules. It is also

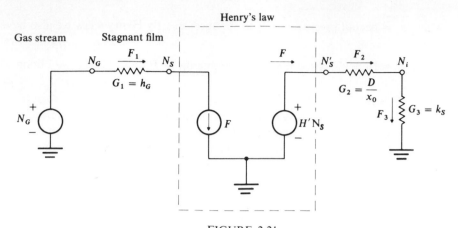

FIGURE 2-21
Electrical analog model for the oxidation process.

possible to produce silicon dioxide by exposing the wafer to steam instead of dry oxygen. For this case the chemical reaction at the silicon surface is

$$Si + 2H_2O \longrightarrow SiO_2 + 2H_2$$

Our simplified model for diffusion through the oxide is not quite correct in this case, since the hydrogen molecules must diffuse out through the oxide to the gas stream. However, this diffusion of hydrogen can safely be neglected because the diffusivity for hydrogen in oxide is so large. Typical diffusivities in silicon dioxide are

$$D_{O_2} \approx 2.82 \times 10^{-18} \text{ cm}^2/\text{s}$$
$$D_{H_2O} \approx 9.5 \times 10^{-10} \text{ cm}^2/\text{s}$$
$$D_{H_2} \approx 2.2 \times 10^{-6} \text{ cm/s}$$

Inspection of these diffusivities shows that:

1 Compared with O_2 or H_2O, the diffusion of hydrogen through the oxide is almost instantaneous, and it can therefore be neglected.
2 The H_2O diffuses much faster than O_2; therefore the growth rate of silicon dioxide in steam is much faster than in dry O_2.

Charts for determining oxide at various temperatures are shown in Fig. 2-22 for dry O_2 and in Fig. 2-23 for steam.[17]†

Exercise 2-4 A wafer has an oxide of thickness 2000 Å. A window is opened for a deep n^+ diffusion. After the predeposit, a 3-h drive-in is performed at 1200°C in an atmosphere of dry O_2. Find the oxide thickness over the window and over the remainder of the wafer.

---

† An excellent table of properties of Si, SiO_2, Ge, and GaAs is given by Grove.[18]

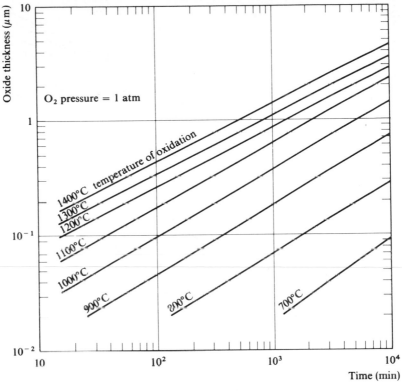

FIGURE 2-22
Oxide growth rate in dry O_2 (after Burger and Donovan[4]).

Effect of Oxide Growth on the Silicon Surface

When silicon dioxide is grown on the wafer, the silicon atoms which are used in the formation of oxide are taken from the surface of the wafer. Consider a wafer having no initial oxide. If an oxide layer of thickness x_0 is grown, the surface of the wafer will no longer be at its original location, because of this removal of silicon atoms. As is shown in Fig. 2-24, the new surface is at a depth x_2 below the original surface.

The number of silicon dioxide molecules per square centimeter is $N_{SiO_2} x_0$ where N_{SiO_2} is the density of oxide molecules per cubic centimeter. Since the number of silicon atoms in a given layer of oxide is equal to the number of oxide molecules,

$$N_{Si} x_2 = N_{SiO_2} x_0$$

where N_{Si} is the density of silicon atoms per cubic centimeter in the silicon crystal. Thus the change of the new surface from its original location is

$$x_2 = \frac{x_0 N_{SiO_2}}{N_{Si}}$$

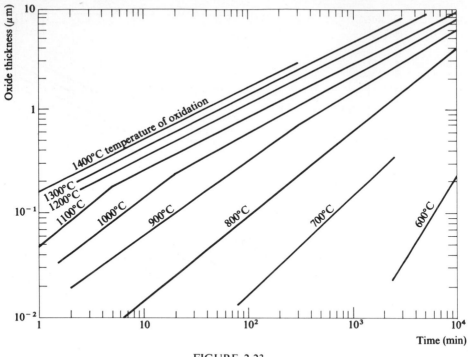

FIGURE 2-23
Oxide growth rate in steam (after Burger and Donovan[4]).

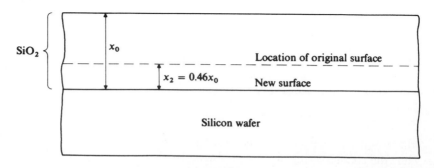

FIGURE 2-24
Effect of oxide growth on the silicon surface.

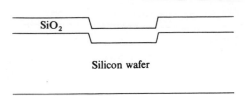

FIGURE 2-25
Formation of steps on the silicon surface
as a result of oxide growth over a
window.

Silicon wafer

Values for N_{Si} and N_{SiO_2} are

$$N_{Si} = 5.00 \times 10^{22}$$

$$N_{SiO_2} = 2.3 \times 10^{22}$$

We therefore obtain

$$x_2 = 0.46x_0$$

As windows are opened and oxides of various thicknesses regrown on different parts of the wafer, steps are formed on the silicon surface as is shown in Fig. 2-25.

Exercise 2-5   A window is opened for an emitter diffusion. After emitter predeposit, a drive-in is carried out for 40 min at 900°C in a steam atmosphere. Find the depth below its original position of the silicon surface under the emitter window.

Masking of Impurity Diffusion by Silicon Dioxide

One of the principal functions of the oxide layer is to serve as a mask against pre-deposited impurities. The oxide must present an effective impurity diffusion barrier at the temperatures of both predeposit and drive-in; such a masking effect is made possible by the fact that both boron and phosphorus have diffusivities of the order of 10^{-15} cm^2/s in silicon dioxide, compared with 10^{-12} cm^2/s in silicon.

In order to ensure effective masking it is necessary to provide an oxide layer thick enough to prevent appreciable amounts of impurity from reaching the silicon. The diffusion of boron and phosphorus in silicon dioxide takes place by a mechanism different from impurity diffusion in silicon. Since thermally grown silicon dioxide is not crystalline, but instead is amorphous, there can be no solid solubility in the same sense as there is in silicon. Instead, silicon dioxide, when exposed to heavy concentrations of boron or phosphorus, forms a surface layer of borosilicate glass or phosporus glass. These doped glasses result when impurity atoms become a part of the amorphous glass structure. As time progresses, the layer of contaminated glass becomes progressively thicker, and, if this is allowed to continue, the doped glass layer will eventually reach the silicon surface, introducing impurities into the silicon.

Unlike the case of diffusion in silicon, there is a well-defined division between the borosilicate or phosphorus glass and the underlying silicon dioxide. This

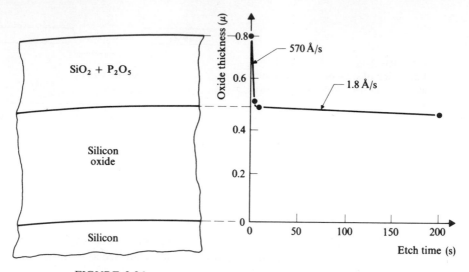

FIGURE 2-26
Etch rate of an oxide layer after exposure to phosphorus (after Burger and Donovan[4]).

sharp demarcation is shown by the etching-rate graph of Fig. 2-26. In this experiment, an oxide layer covered by a phosphorus glass layer was etched in a dilute solution of hydrofluoric acid, and the oxide thickness periodically determined. Since phosphorus glass etches much more rapidly than pure silicon dioxide, the sharp divison between the two glasses is clearly shown. Diffusion of impurities beyond this dividing line is generally negligible (this holds well for phosphorus and is adequate to explain the boron case). The graphs of Figs. 2-27 and 2-28 give the minimum masking thicknesses for boron and phosphorus at various times and temperatures.[19, 20]

While diffusion of boron and phosphorus through silicon dioxide is slow, other impurities diffuse very rapidly through both silicon dioxide and silicon. Among these are acceptors such as gallium, indium, and aluminum, which diffuse 400 to 700 times faster through silicon dioxide than boron or phosphorus. The implication of this is that furnace tubes used for predeposit and drive-in must be kept clean and free from cross-contamination from other processes, regardless of whether or not oxide windows are open while the wafer is in the furnace.

Effect of Oxide on Impurity Redistribution[21]

The thermal growth of undoped silicon dioxide layers by the process discussed above often affects the impurity distribution in the underlying silicon substrate. Thermal oxidation of silicon takes place at the silicon–silicon dioxide interface; as

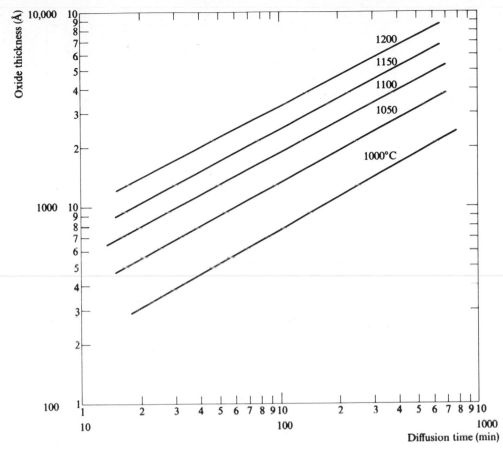

FIGURE 2-27
Minimum oxide thickness required to mask boron at various temperatures
(after Burger and Donovan[4]).

we have seen, this requires that some of the substrate silicon be incorporated into
the silicon dioxide film as the process continues.   The impurity atoms which were
present in the layer of silicon consumed must therefore be redistributed between the
silicon dioxide and the remaining silicon.   How this redistribution takes place
depends upon the relative solubility of the impurity in silicon and silicon dioxide.
In the case of boron, for example, the impurity is more soluble in the oxide than
the silicon.   As a result, the concentration of boron on the silicon side of the
silicon–silicon dioxide interface is reduced.   This process, known as *boron gettering*, is often employed to reduce the total number of impurities during the drive-in
step in order to increase the final sheet resistivity.   A danger implicit in boron
gettering is the possibility of inversion of the impurity type of a lightly boron-
doped surface to *n*-type because of the surface depletion of boron.   To avoid

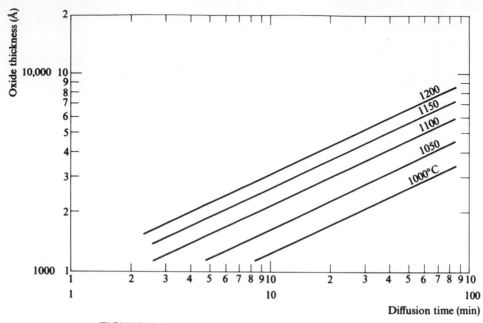

FIGURE 2-28
Minimum oxide thickness required for phosphorus diffusion masking as a function of time (after Burger and Donovan[4]).

inversion, the surface concentration of boron profiles must remain relatively high. For this reason, base and resistor p-type diffusion sheet resistances are limited to approximately 200 Ω per square.

Huang and Welliver[22] have investigated the problem of boron redistribution during oxide growth, and have found that in extreme cases as much as 80 percent of the predeposited boron can dissolve in the oxide. The boron profile which results from oxide growth during drive-in diffusion is shown in Fig. 2-29.

The opposite effect, an impurity buildup in the silicon, occurs when the impurity is less soluble in silicon dioxide than in silicon. This situation, which pertains to the case of phosphorus, causes the so-called "snowplow" effect and leads to a steadily increasing surface concentration as oxide growth proceeds. One practical use of snowplow effect is to produce a reverse bend in diffusion profiles of varactor diodes where "hyperabrupt" junctions are necessary to achieve a large capacitance variation with voltage.

In addition to their use for masking against impurities, oxides may also be employed for doping. Instead of relying on a gaseous predeposit, this process involves the low-temperature deposition on the silicon surface of an oxide layer containing the desired dopant. During the subsequent drive-in, the dopant present in the oxide layer diffuses across the oxide-silicon interface and into the silicon.

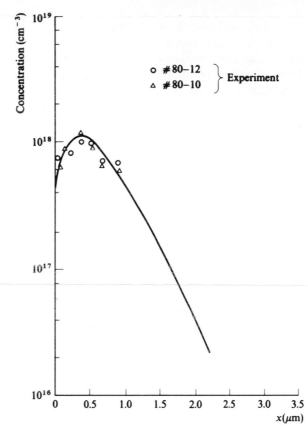

FIGURE 2-29
Boron profile after oxide growth (after Huang and Welliver[22]).

2-5 EPITAXIAL GROWTH OF SILICON†

As was discussed in Chap. 1, epitaxial growth of n-type silicon follows buried-layer diffusion.   The buried layer under an epitaxial-layer diffusion must be part of the integrated transistor structure in order to simultaneously meet requirements of high breakdown voltage and low series collector resistance.   In a transistor without a buried layer, the breakdown voltage is proportional to N_D^{-1}, the collector donor density, while the collector resistance is approximately proportional to N_D^{-1}. Thus, there is no way to simultaneously achieve high breakdown and low series resistance without modifying the collector region.   The buried layer is a region of high donor concentration, providing a low-resistance path for collector current. The thin layer between base-collector junction and buried layer has low enough donor concentration to provide reasonably high breakdown voltage, yet it is sufficiently thin that it contributes very little series collector resistance.[23]

† Our treatment of epitaxial growth closely follows that of Grove.[15]

Epitaxial growth of silicon takes place in a reactor at approximately 1200°C. A mixture of gases containing silicon tetrachloride or silane is passed over the wafer, and a chemical reaction occurs at the wafer surface which results in silicon atoms being deposited on the silicon wafer surface.   Because of the high temperature, the deposited atoms are quite mobile, and move across the surface until a favorable binding energy situation occurs, whereupon the silicon atom is deposited according to the crystal structure of the underlying substrate.   Nucleation sites are thus formed, other atoms attach themselves in preferred orientations, and the crystal grows.

Crystal Defects

Since the epitaxial layer is to grow along the crystal structure of the silicon substrate, the condition of the wafer is extremely important, and the growth must proceed without inducing defects in the epitaxial-layer crystal structure.[24]   Crystal defects are undesirable, since they cause diffusion anomalies and device failures in the general region of the defect.

There are two principal types of crystal defects common in epitaxial growth: stacking faults and edge dislocations.   Stacking faults occur as a result of a plane of atoms missing from the normal structure as the epitaxial layer grows vertically. Such faults usually originate at the epitaxial-layer–substrate interface, and propagate through the epitaxial layer to the surface.   In silicon having a (111) surface orientation, stacking faults appear as triangular structures on the surface.   The edge length of the triangle is approximately 1.22 times the epitaxial-layer thickness.

As the epitaxial layer grows horizontally between nucleation sites, atoms may occasionally be missing from the crystal structure at the edge of the layer.   Absence of these atoms disrupts the periodicity of the layer and causes edge dislocations. Edge dislocations are detrimental to the electrical behavior of junctions made in the region of the dislocation.   Viewed from above, an edge dislocation appears in the crystal structure as shown in Fig. 2-30.   If the substrate contains edge dislocations, these will propagate into the epitaxial layer.

Since defects in the crystal structure are detrimental to electric performance of devices made in the epitaxial layer, and since crystal defects propagate through a growing epitaxial layer, it is essential that the substrate have as nearly perfect a surface as possible before the epitaxial layer is grown.   Wafers are normally polished both mechanically and chemically to the highest possible degree of perfection before the oxide growth and buried-layer diffusion are carried out.   However, these procedures are not adequate to provide the surface perfection required in epitaxial growth.

The oxide growth prior to the buried-layer masking is thought to result in precipitation of impurities at the wafer surface, so that the removal of the oxide layer prior to epitaxial growth leaves numerous sites which cause lattice defects. To remedy this problem, the back of the wafer is left unpolished to provide alternate sites for precipitation where the integrated-circuit devices are not damaged. In addition, the initial step in growing an epitaxial layer is to etch the substrate

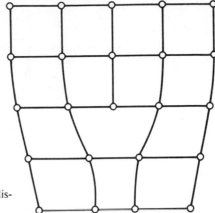

FIGURE 2-30
Schematic representation of an edge dislocation.

wafer surface in the reactor just before growth begins. Etching is accomplished
by introducing anhydrous hydrochloride (HCl) gas into the reaction chamber with
the wafer inductively heated to 1200°C. As the graph of Fig. 2-31 shows,[25]
removal of approximately 2 μm of the substrate material produces a reduction of
three orders of magnitude in the stacking fault density.

Growth of an Epitaxial Layer from Silicon Tetrachloride

We treat the growth of an epitaxial layer in much the same way as the growth of a
silicon dioxide layer, making a number of simplifying approximations in order to
obtain a first-order model for the growth process. As before, we adopt the artifice
of assuming a stagnant layer to exist near the surface of the wafer as shown in
Fig. 2-32. The incoming gas stream maintains a concentration C_G of silicon
(SiCl$_4$) molecules in the gas stream outside the stagnant layer, and these molecules
diffuse across the stagnant layer to the wafer surface. At the surface of the growing
layer, the concentration of molecules is C_S. A chemical reaction is occurring at
this surface in which the molecules are converted to silicon atoms and HCl molecules.

Again we use a quasi-static approximation: It is assumed that the film growth
rate is sufficiently slow in comparison with the chemical reaction that the system
reaches steady state essentially instantaneously and remains essentially in steady
state as the film grows. Then the flux F_1 of molecules through the stagnant layer
must be equal to the flux F_2 of molecules into the surface, i.e., being consumed
in the chemical reaction. Thus we have

$$F_1 = F_2$$

Since steady-state conditions are assumed, F_1 is found from Fick's first law
to be

$$F_1 = h_G(C_G - C_S)$$

where h_G is the gas-phase mass transfer coefficient.

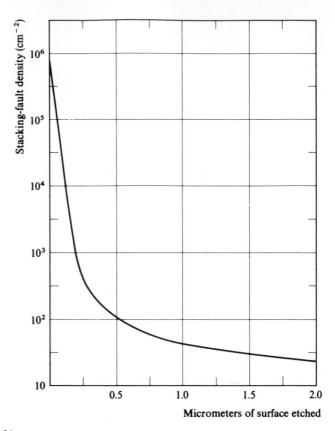

FIGURE 2-31
Effect of surface etching in reactor on epitaxial-layer stacking-fault density.

If we assume that the steady-state chemical reaction rate is proportional to the surface concentration of reacting molecules, we can write

$$F_2 = k_S C_S$$

where k_S is the chemical surface reaction-rate constant.

Again we note that an electrical analog circuit can be used to represent this process. Letting flux be analogous to current and concentration be analogous to voltage, we obtain the circuit of Fig. 2-33. From this circuit we see by inspection that

$$C_S = \frac{C_G(1/G_2)}{1/G_2 + 1/G_1} = \frac{C_G}{1 + k_S/h_G} \qquad (2\text{-}39)$$

If the density of silicon atoms required in the silicon crystal is $N_1 (N_1 = 5.00 \times 10^{22} \text{ cm}^{-3})$, the rate at which the film grows is

$$\frac{dx_0}{dt} = \frac{F_2}{N_1} \qquad (2\text{-}40)$$

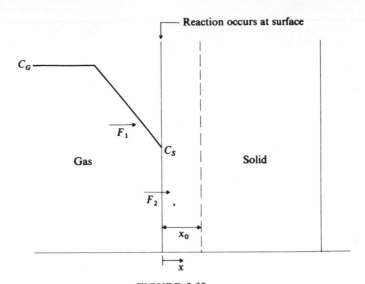

FIGURE 2-32
$SiCl_4$ concentrations during epitaxial growth.

Substituting for F_2 in (2-40) and combining the result with (2-39), we obtain

$$\frac{dx_0}{dt} = \frac{k_S C_S}{N_1} = \frac{k_S h_G C_G}{N_1(h_G + k_S)} \qquad (2\text{-}41)$$

Next we incorporate information about the gas mixture. Let C_T be the total concentration of molecules in the gas stream, and let Y be the mole fraction of $SiCl_4$ molecules. Then we obtain

$$\frac{dx_0}{dt} = \frac{k_S h_G}{k_S + h_G} \frac{Y C_T}{N_1} \qquad (2\text{-}42)$$

Two limiting cases are of interest; these can be deduced easily by in-spection

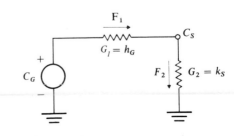

FIGURE 2-33
Electrical circuit analog for the epitaxial
growth process.

of the circuit analog. First, suppose $h_G \gg k_S$. Then $C_S \to C_G$, and the flux will be determined almost entirely by k_S. Then

$$\frac{dx_0}{dt} \approx k_S \frac{YC_T}{N_1} \qquad \text{for } h_G \gg k_S$$

This case is called the *surface-reaction-limited* case.

Second, suppose $k_S \gg h_G$. Now $C_S \to 0$, and the flux is determined almost entirely by h_G. Then

$$\frac{dx_0}{dt} \approx \frac{h_G YC_T}{N_1}$$

This case is called the *mass-transfer-limited* case.

As (2-42) shows, the temperature dependence of the growth rate can be found if the temperature dependences of k_S and h_G are known. Note that k_S is the rate of chemical reaction. We assume that a simple Boltzmann relation governs the reaction and that an activation energy E_a is required for the reaction to take place. We therefore make the approximation

$$k_S \approx K e^{-E_a/kT} \qquad (2\text{-}43)$$

where K is a constant.

Experimental results show that this is a reasonable approximation, and that

$$K \approx 10^7 \text{ cm/s}$$
$$E_a \approx 1.9 \text{ eV}$$

Experiment also shows that h_G is a weak function of temperature.

By combining (2-42) with (2-43), one can obtain growth rate as a function of temperature, with h_G as a parameter. Growth-rate curves are plotted in Fig. 2-34.[15]

Effect of the Gas Mixture on Growth Rate

Our simple model for growth rate predicts a linear dependence of growth rate on Y, the mole fraction of $SiCl_4$. This is because we have assumed only the single chemical reaction

$$SiCl_4 + 2H_2 \rightleftharpoons Si + 4HCl$$

to occur. There is, however, a second reaction which can also occur:

$$SiCl_4 + Si \rightleftharpoons 2SiCl_2$$

For low concentrations of $SiCl_4$ the first, or growth, reaction dominates; but if the $SiCl_4$ concentration is very large, the second, or etching, reaction takes place even without HCl in the gas stream. Observed growth rates as a function of Y are shown in Fig. 2-35.[26] It is common practice to operate with small values of Y and

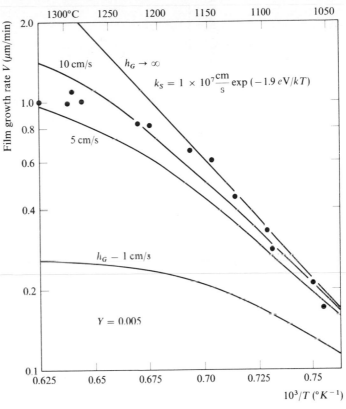

FIGURE 2-34
Growth rate versus temperature.

growth rates of about 1 μm/min. While it would be advantageous to operate with larger growth rates in order to reduce diffusion during growth, it is found that more surface defects occur at higher rates.

Impurity Doping of Epitaxial Layers

One of the advantages of epitaxial growth is that the impurity concentration in the epitaxial layer can be controlled by the composition of the gas stream in the reactor. The gases most commonly used for impurities are diborane for acceptor doping and phosphine for donor doping. The curves of Fig. 2-36 indicate the effects of impurity gas concentration on impurity concentration of the silicon layer.[27]

Diffusion During Epitaxial Growth

Since the epitaxial layer is grown at a high temperature, diffusion of impurities will occur. The buried layer will diffuse into the epitaxial layer, impurities in the

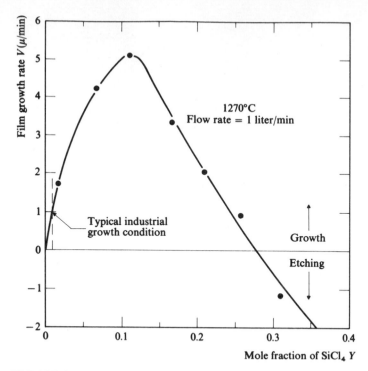

FIGURE 2-35
Observed growth rates of epitaxial layers (after Burger and Donovan[26]).

epitaxial layer will diffuse into the substrate, and substrate impurities will diffuse into the epitaxial layer. The most important of these is the diffusion of the buried layer into the epitaxial layer, since the impurity concentration is much greater in the buried layer than in any of the other regions. We thus confine our discussion to this case.

Consider first the case shown in Fig. 2-37, where the concentration in one region is assumed to be uniform and in the other region it is assumed to be zero. For simplicity assume each region to be semi-infinite. After a given diffusion time t, the impurities will have redistributed and the profile will be as shown by the solid line. In region 2 the profile is given by

$$N(x, t) = \frac{N_S}{2} \operatorname{erfc} \frac{x}{2\sqrt{Dt}}$$

In the case of an epitaxial layer growing on a substrate, the substrate can be considered to be semi-infinite, but the growing film represents a finite region with a time-varying boundary, as shown in Fig. 2-38. This moving boundary modifies the diffusion process. One can see qualitatively that for a given diffusion time, $N(x, t)$ will be smaller in the layer than it would be if the layer were semi-infinite. One can also see intuitively that once the thickness of the layer becomes large com-

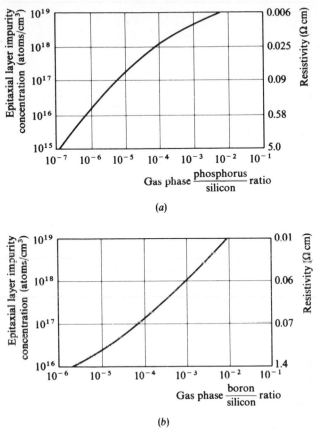

FIGURE 2-36
Effects of impurity gas concentration on impurity concentration of the silicon layer: (a) phosphine used as dopant, (b) diborane used as dopant. (after Warner and Fordemwalt[27]).

pared with $2\sqrt{Dt}$, the diffusion process proceeds as though the layer were semi-infinite. It can be shown that for $x_e \gg 2\sqrt{Dt}$, the profile in the epitaxial layer resulting from buried-layer diffusion is[28]

$$N(x, t) = \frac{N_S}{2}\, \varepsilon_1\, \text{erfc}\, \frac{x}{2\sqrt{Dt}}\,.$$

where ε_1 is a correction factor with value $0 < \varepsilon_1 < 1$.

Near the surface, the impurity concentration is essentially N_e, the concentration produced by the gas stream in the reactor. A reasonable approximation is to add N_e to $N(x, t)$ to obtain the profile for $x \gg 2\sqrt{Dt}$.

In order to minimize diffusion of the buried layer, impurities having low diffusivities, such as arsenic or antimony, are generally used for buried-layer impurities.

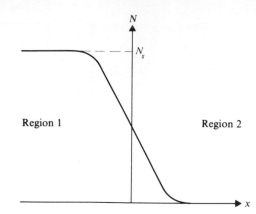

FIGURE 2-37
Diffusion from region of uniform concentration into a region of zero concentration.

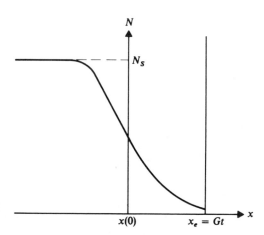

FIGURE 2-38
Time-varying boundary from epitaxial growth.

PROBLEMS

2-1 For a predeposition diffusion, find the profile $N_A(x, t_p)$ if it is assumed that the wafer is not semi-infinite but has thickness L. Show that for $L \gg \sqrt{D_p t_p}$ the semi-infinite approximation is valid.

2-2 Suppose that the donor concentration in the wafer is as shown in Fig. P2-2 at the beginning of an acceptor diffusion. Let the wafer thickness be semi-infinite, and the thickness of the epitaxial layer be L. Calculate the *donor* impurity profile $N_D(x, t_p)$ after the acceptor diffusion. How far below the surface is the approximation

$$N_D(x, t_p) \approx N_{BC}$$

valid?

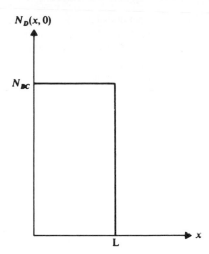

FIGURE P2-2

2-3 Consider the predeposited atoms to have been deposited by an initial delta function of flux at the beginning of the drive-in. Solve Eq. (2-6) subject to

$$N_A(x, 0) = 0, \qquad x \geq 0$$

$$\frac{\partial N_A}{\partial x}(0, t') = -\frac{Qt(_p)}{AD_d}\delta(t)$$

$$N_A(\infty, t) < \infty$$

to obtain Eq. (2-18).

2-4 Calculate the sensitivity $S_{D_p}^{Q(t_p)}$ of the number of predeposited atoms $Q(t_p)$ to the predeposit diffusivity D_p.

2-5 Calculate the sensitivity $S_{D_d}^{x_j}$ of the junction depth x_j to drive-in diffusivity D_d. Consider both the case of diffusion into constant background concentration and the case of diffusion into a gaussian profile. Show that for diffusion into constant background concentration, $S_{D_d}^{x_j}$ can be made zero by proper choice of x_j and $D_d t_d$.

2-6 If during a boron predeposit and drive-in we can hold temperature to $\pm 100\Delta$ percent, and we can measure time to within $\pm 100\,\delta$ percent, find the variation of junction depth which can occur. Let $\Delta = \delta = 0.01$, use typical values for base diffusion times and temperatures, and calculate the variation of x_j.

2-7 Select a reasonable processing schedule to fabricate a transistor with collector junction depth $x_{JC} = 3.0\ \mu m$, emitter junction depth $x_{JE} = 2.3\ \mu m$, in 1-$\Omega \cdot$ cm epitaxial material. Design for final base surface concentration of $6 \times 10^{18}\ cm^{-3}$ after drive-in.

2-8 A window is opened for a base diffusion. During the base drive-in a steam oxide is grown for 33 min at 1100°C. Next a window is opened for the emitter diffusion. During the emitter drive-in a steam oxide is grown for 17 min at 1000°C.

If the collector regions had 4000 Å of oxide before the base diffusion, find the oxide thicknesses over collector, emitter, and base regions, and the step sizes in the wafer surface.

2-9 An epitaxial layer is grown on a substrate having a uniform boron concentration of 10^{18} cm^{-3}. The epitaxial layer is grown with a SiCl$_4$ mole fraction $Y=0.03$ at 1250°C for 5 min. During growth the layer is doped by using a gas-phase phosphorus/silicon ratio of 10^{-7}.

After epitaxial growth, an isolation diffusion is to be performed. A predeposit of boron is carried out at 1050°C for 1 h. Estimate the drive-in time required for the isolation diffusion. Ignore the effects of the moving boundary during epitaxial growth.

2-10 A wafer has epitaxial-layer concentration $N_{BC} = 10^{15}$ cm^{-3}. A base drive-in diffusion is performed at 1100°C for 1 h, and a piece of the wafer is removed. Angle-lap measurements show the metallurgical junction depth to be 1 μm.

The wafer is returned to the furnace for 3 additional hours. What is the final base sheet resistance? No oxide is grown during the diffusion.

2-11 The process engineer on the day shift started an isolation diffusion, but left before it was completed. You arrive to take over the night shift, and find that he left no information on what he has done. You do not know what predeposit or what time and temperature he used for his drive-in diffusion.

On a monitor wafer, you make measurements and find that the original wafer had a 12-μm epitaxial-layer thickness with $N_{BC} = 10^{15}$ cm^{-3}, and the substrate has $N_A = 10^{14}$ cm^{-3}. His diffusion, which you assume to be boron, shows a junction depth $x_J = 6$ μm and surface concentration $N_{sA} = 10^{19}$ cm^{-3}. You assume his profile to be gaussian.

You decide to put the wafers in for further drive-in at 1150°C for 6.5 h. What junction depth do you expect to get at the end of that time?

2-12 A boron predeposit is made into a wafer having $N_{BC} = 10^{15}$ cm^{-3}. A drive-in diffusion is performed at 1150°C for 3.16 h. The junction depth is $x_J = 3.16$ μm.

Suppose that instead of the above drive-in, the same predeposit had been subjected to a drive-in of 17.6 h at 1050°C. What would the junction depth have been?

REFERENCES

1 MARSHAK, A. H.: Synthesis of General Impurity Distributions by Solid-State Diffusion, appendix A, Ph.D. dissertation, University of Arizona, Tucson, Arizona, 1969.

2 BLATT, F. J.: Theory of Mobility of Electrons in Solids, "Solid State Physics," Vol. 4, pp. 199–366, Academic Press, Inc., New York, 1957.

3 UMAN, M. A.: "Introduction to Plasma Physics," chap. 14 and appendix A, McGraw-Hill Book Company, New York, 1964.

4 BURGER, R. M., and R. P. DONOVAN: "Fundamentals of Silicon Integrated Device Technology," vol. 1, p. 254, Prentice-Hall, Inc., Englewood Cliffs, N.J., 1967.

5 MARSHAK: op. cit., appendix B.

6 KENNEDY, D. P., and P. C. MURLEY: Impurity Atom Distribution from a Two-Step Diffusion Process, *Proc. IEEE*, vol. 54, p. 620, 1964.

7 MACKINTOSH, I. M.: The Diffusion of Phosphorus and Boron in Silicon, *J. Appl. Phys.*, vol. 25, pp. 1439–1440, 1956.

8 KURTZ, A. D., and R. YEE: Diffusion of Boron in Silicon, *J. Appl. Phys.*, vol. 31, pp. 303–305, 1960.

9 MARSHAK: op. cit., p. 22.

10 KENNEDY, D. P., and R. R. O'BRIEN: Analysis of the Impurity Atom Distribution Near the Diffusion Mask for a Planar *p-n* Junction, *IBM J. Res. Dev.*, vol. 9, pp. 179–186, 1965.

11 BURGER and DONOVAN: op. cit., pp. 310–319.

12 MAR, J.: V-shaped Test Patterns for Measuring Lateral Diffusion, *IEEE J. Solid-State Circuits*, vol. SC-6, pp. 419–421, 1971.

13 VAN DER PAUW, L. J.: A Method of Measuring Specific Resistivity and Hall Effect of Discs of Arbitrary Shape, *Phillips Res. Repts.*, vol. 13, pp. 1–9, 1958.

14 IRVIN, J. C.: Resistivity of Bulk Silicon and of Diffused Layers in Silicon, *Bell System Tech. J.*, vol. 41, pp. 387–410, 1962.

15 GROVE, A. S.: "Physics and Technology of Semiconductor Devices," chaps. 1 and 2, John Wiley & Sons, Inc., New York, 1967.

16 DEAL, B. E., and A. S. GROVE: General Relationship for the Thermal Oxidation of Silicon, *J. Appl. Phys.*, vol. 36, p. 3770, 1965.

17 BURGER and DONOVAN: op. cit., pp. 41, 49.

18 GROVE: op. cit., pp. 102–103.

19 SAH, C. T., H. SELLO, and D. A. TREMERE: Diffusion of Phosphorus in Silicon Oxide Film, *J. Phys. Chem. Solids*, vol. 11, pp. 288–298, 1959.

20 HARDING, M. L.: Diffusion of Boron in Silicon Oxide, *J. Electrochem. Soc.*, vol. 110, p. 265C, 1963.

21 GROVE, A. S., O. LEISTIKO, and C. T. SAH: Redistribution of Acceptor and Donor Impurities during Thermal Oxidation of Silicon, *J. Appl. Phys.*, vol. 35, p. 2695, 1964.

22 HUANG, J. S., and W. C. WELLIVER: On the Redistribution of Boron in the Diffused Layer during Thermal Oxidation, *J. Electrochem. Soc.*, vol. 117, pp. 1577–1580, 1970.

23 JACKSON, D. M., JR.: Advanced Epitaxial Processes for Monolithic Integrated Circuit Applications, *Trans. Met. Soc. AIME*, vol. 233, pp. 596–601, 1965.

24 WARNER, R. M., and J. N. FORDEMWALT: "Integrated Circuits: Design Principles and Fabrication," pp. 282–284, McGraw-Hill Book Company, New York, 1965.

25 BURGER and DONOVAN: op. cit., p. 441.

26 Ibid., p. 380.

27 WARNER and FORDEMWALT: op. cit., pp. 284–285.

28 BURGER and DONOVAN: op. cit., pp. 412–420.

3

OTHER PROCESSING METHODS

In the preceding chapters we have described and analyzed the basic planar integrated-circuit fabrication sequence. This basic sequence has the advantage of minimizing the number of processing steps and therefore minimizing the cost and maximizing the yield. As one would expect, however, adherence to the requirements dictated by a minimum-complexity process often necessitates rather severe compromises in device and circuit performance. If additional processing methods compatible with the basic process are made available, they can introduce additional degrees of freedom which permit the designer to improve circuit performance. In this chapter we describe briefly and qualitatively some of the more important processing options which are compatible with the basic process. In nearly all cases, use of these options increases the cost of fabrication, so the designer must always weigh the performance improvement against the cost increase.

We begin by discussing alternatives to the diffused junction isolation method. Here the benefits are not only improved performance but also decreased area and thus increased component density. Next we consider alternatives to the diffusion of single-crystal monolithic components. Modifications of the diffusion process are described, and ion-implantation techniques are discussed. Alternatives to single-crystal monolithic devices are described in terms of compatible thin-film devices. Finally, interconnection methods, including beam-lead techniques, are discussed.

3-1 ISOLATION METHODS

Conventional isolation by diffused *p-n* junctions has the following disadvantages:

1 The time required for the isolation diffusion is considerably longer than any of the other diffusions.

2 Because lateral diffusion is significant during the isolation diffusion, considerable clearance must be used for isolation regions. Since isolation diffusions occur at the periphery of isolated regions, the area used for isolation purposes is a significant portion of the chip area. This isolation area must be considered wasted as far as component density is concerned.

3 The relatively deep sidewalls and large area of isolation regions contribute significant parasitic capacitance which degrades circuit performance. Note that not only the isolation sidewalls but also the bottom epitaxial-substrate junction of isolated regions contribute parasitic capacitance. The epitaxial-substrate junction must also be considered to be part of the isolation method.

Several isolation methods have been developed which avoid the use of an isolation diffusion; these are the Fairchild Isoplanar II method, the Raytheon V-ATE process, and the Motorola Multiphase Memory process. All of these circumvent the difficulties of large area and sidewall capacitance, but still suffer from epitaxial-substrate capacitance. A process known as dielectric isolation avoids this latter problem.

The Isoplanar II Process[1,2]

Fabrication of the double-diffused epitaxial *n-p-n* transistor by the Isoplanar II process begins, as in the basic process, with a buried-layer diffusion and the growth of an epitaxial layer as shown in Fig. 3-1*a*. Next a layer of silicon nitride is deposited over the surface of the wafer, and photoresist is used to remove the nitride in areas where isolation is to occur. An etch is now used to remove the *n*-epitaxial silicon as shown in Fig. 3-1*b*. Silicon dioxide is thermally grown to fill in the regions that have been etched; the nitride prevents oxide growth elsewhere on the wafer. A "sink" diffusion is performed for the collector contact, and silicon nitride is removed, leaving the wafer as shown in Fig. 3-1*c*. Note that oxide isolation is used not only to isolate the transistor but also to separate the collector contact region from the remainder of the device. This is done to eliminate the clearances that would ordinarily have to be observed when the emitter is diffused into the base region. From this point, the processing is similar to the basic process; the completed device is shown in Fig. 3-1*d*.

The reduction of surface geometry accompanying the Isoplanar II process

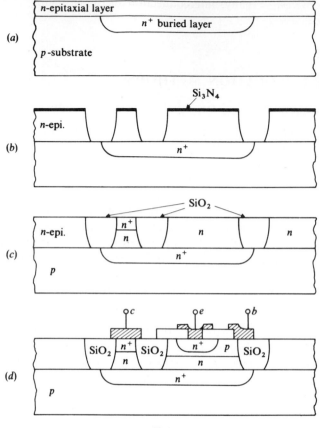

FIGURE 3-1
The Fairchild Isoplanar II process.

can be seen from Fig. 3-2*a*.   Since the entire base region is surrounded by oxide
rather than epitaxial collector material, there is no danger of misalignment causing
a short at the surface of the base-collector junction or an overlap of the emitter
diffusion and collector material.   The only clearances that must be observed are
those between emitter and base contact regions.   The double-base-stripe device
of Fig. 3-2*a* consumes less than half the area of the conventional device of Fig. 3-2*b*
having the same photolithography constraints.

     While the Isoplanar II process is compatible with the basic planar process, it
has the disadvantages of requiring thin epitaxial material in order that the etched
regions which have to be filled with oxide not be too deep, and of requiring long
oxidation times.   The thin epitaxial layer means that all diffusions have to be
shallow.   The process has shown much promise for large-scale memories, where
high component density is very important.

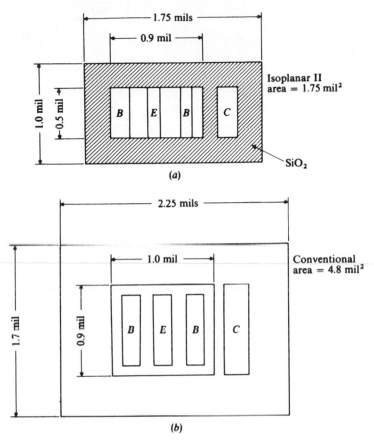

FIGURE 3-2
Reduction of size achieved with the Isoplanar II process: (a) Isoplanar II device;
(b) conventional device.

The V-ATE Process[3]

The V-ATE (Vertical Anisotropic Etch) process makes use of a preferential etch technique which etches 30 times faster along the $\langle 100 \rangle$ crystal plane than along the $\langle 111 \rangle$ plane.[4] If the wafer used has a $\langle 100 \rangle$ surface, etching of the silicon after a window has been opened in the oxide forms a V groove, as shown in Fig. 3-3. The preferential nature of the etch causes the groove to form a $54°$ angle with the surface. Since the angle is fixed, the depth of the groove is determined by the width of the opening at the surface. There is no undercutting of the oxide.

In the ordinary uniform etch, if channels several micrometers deep are etched and metal is deposited over them, voids in the metal will occur because of the steep walls of the channels. It is for this reason that the etched regions must be refilled with oxide in the Isoplanar II process. However, in the V-groove case,

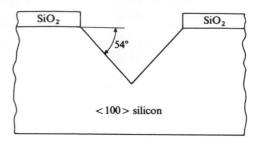

FIGURE 3-3
V groove produced by anisotropic etching
of ⟨100⟩ silicon.

the angle of the groove provides a sufficient taper to the walls to permit metal deposition without voids.

The V-ATE process begins with the buried-layer diffusion and growth of the epitaxial layer as in the basic process.   Next the base region is diffused as shown in Fig. 3-4a.   Windows for the isolation grooves are opened in the oxide and the grooves are etched.   Since there is nothing in this method analogous to the lateral diffusion of the isolation region in the basic process, the isolation grooves consume very little space.   The oxide is now removed, and a sandwich of oxide-nitride-oxide is deposited, as shown in Fig. 3-4b.   Windows for all contacts are opened down to the nitride, and those for emitter and collector are then etched through the remaining nitride and oxide to the silicon.   An emitter diffusion is performed, and the base contact is then etched through the nitride and oxide to the silicon, as shown in Fig. 3-4c.   All contact windows are now open.

A beam-lead type of metal interconnect system (to be discussed in a later section) is used.   This scheme consists of three layers of metal—titanium, platinum, and gold—and is used because it permits close spacing of well-defined metal patterns, leading to small geometries.   In the V-ATE process, two layers of the metallization process are used; the first is employed to make the thin, narrow lines required for contacting the small-geometry devices as shown in Fig. 3-4d.   A layer of oxide is then deposited and the second layer of metallization is deposited above the oxide.   The second layer of metal is electroplated to twice the thickness of the first, and is used for those interconnections which must carry larger current.

As can be seen from the foregoing discussion, the V-ATE process requires many steps and has a complicated metallization scheme.   One advantage of the metallization process is that it is compatible with the requirements for fabrication of Schottky-barrier diodes.†

The Motorola Multiphase Process

The Motorola Multiphase process, also known as VIP, makes use of a V groove formed by anisotropic etching, as in the V-ATE case.   However, the Motorola process is simpler.   After grooves have been etched for isolation, a standard oxide layer is grown.   The grooves are then filled to the surface, not with oxide, but rather

† Schottky-barrier diodes are discussed in Chap. 7.

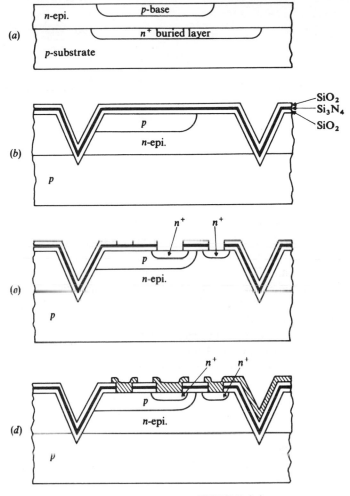

FIGURE 3-4
The Raytheon V-ATE process.

with polycrystalline silicon grown by standard epitaxial methods. The metallization is aluminum; a cross section of the completed wafer is shown in Fig. 3-5. This
process has the advantages of being simpler than V-ATE and of retaining the
planar-surface characteristics of Isoplanar II. Although metal can be deposited
over V grooves, as was previously pointed out, the reliability of the metal interconnections is improved if the surface is planar.

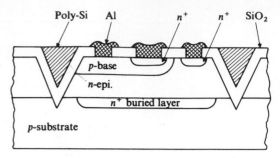

FIGURE 3-5
The Motorola Multiphase Memory process (VIP).

Dielectric Isolation[5]

In all of the isolation methods thus far discussed, the isolation diffusion has been eliminated, thereby reducing the device surface area and eliminating the isolation sidewall capacitance.   Note that the bottom capacitance of the isolation junction is still present.   This capacitance can be reduced, and the isolation diffusion also eliminated, by dielectric isolation.   The basic dielectric isolation process begins with an n-type wafer having the resistivity that would normally be selected for the n-type epitaxial layer in the standard process.   Buried-layer diffusions are performed at the desired locations, and channels are now etched to a depth of about 8 μm, as shown in Fig. 3-6a.   Oxide is grown over the entire surface of the wafer.

Next a layer of silicon several mils thick is grown on the surface by standard epitaxial methods as shown in Fig. 3-6b.   Since this silicon grows over the oxide layer rather than on the silicon wafer, it will be polycrystalline.   However, as will be seen, its only function is to provide mechanical support for the devices; that is, it will play the role of substrate.   The entire assembly is now turned over and the back side of the n-type wafer is lapped away until what was originally the bottom of the etched channels is reached.   What remains are islands of the original n wafer, surrounded by oxide and polycrystalline silicon.   Standard diffusion processing is now used to diffuse base and emitters into these islands, and contact windows and metallization are accomplished in the usual manner, as is shown in Fig. 3-6c.   The polycrystalline material forms a substrate, and the islands of n material are surrounded, on the bottom as well as the sides, by oxide.   Since the oxide is usually thicker than the depletion region of the substrate junction, and since the dielectric permittivity of oxide is one-third that of silicon, the parasitic capacitance is greatly reduced.   Moreover, no junction is involved in the isolation, so the biasing constraints of conventional isolation do not apply.   Also the isolation system is less susceptible to degradation from radiation, since no junction is involved.

This method of isolation uses extra processing steps, and it is difficult to perform the lapping operation with the precision that is required.   Dielectric isolation is therefore used only in cases where the low parasitic capacitances or radiation hardening it offers are essential to circuit performance.

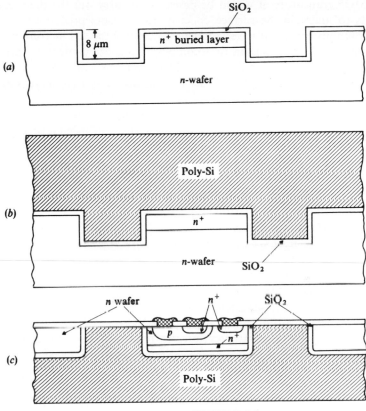

FIGURE 3-6
The basic dielectric isolation process.

3-2 FORMATION OF IMPURITY DISTRIBUTIONS

Diffusion from Doped Oxides[6]

In the standard two-step diffusion process, the predeposit is usually carried out with sufficient boron compounds present to ensure that the solid solubility limit of boron will be reached at the silicon surface. Typically the boron-containing gas in the diffusion tube decomposes, forming B_2O_3 on the silicon surface. This reacts with the silicon, producing silicon dioxide (SiO_2) and boron, the latter diffusing into the wafer. If too much B_2O_3 is present, not all the boron can be dissolved in the silicon, and a silicon-boron phase is produced at the surface. This phase is difficult to remove, and interferes with proper contacting during metallization. If the amount of B_2O_3 is still further increased, pitting of the silicon surface results, causing a considerable degradation of surface properties and hence of device behavior.

In some cases where devices require low surface concentrations, for example,

MOS transistors, it would be desirable to carry out the predeposit in such a way as to maintain the surface concentration during predeposit well below the solid solubility limit. It is very difficult to control the gas mixture in the diffusion tube with sufficient precision to accomplish adequate control of surface concentration below the solid solubility limit. Moreover, the walls of the tube become enriched with boron and themselves act as a diffusion source, increasing the difficulty of control.

One alternative to the standard diffusion method is to deposit a boron-containing glass on the wafer surface instead of using a predeposit. This can be done by introducing silane (SiH_4), O_2, and diborane (B_2H_6) into a chemical reactor, in which the silicon wafers are heated to a temperature of 300°C. The resulting chemical reaction produces on the wafer surface a glass (SiO_2) containing B_2O_3. By controlling the gas-flow rates during the reaction, the molar percent of B_2O_3 in the glass can be controlled.

Next the wafer is subjected to the standard drive-in diffusion; during the process the B_2O_3 in the glass reacts with the silicon as previously described and produces boron, which diffuses into the wafer. Figure 3-7 shows the surface concentration of boron, following diffusion, as a function of the molar percent of B_2O_3 in the glass. Table 3-1 gives junction depths, sheet resistances, and surface concentrations for different molar percents of B_2O_3.

Selective diffusions can easily be performed by depositing the boron glass and using photoresist procedures to remove it from areas where no diffusion is desired, or by depositing boron glass over oxide in which windows have been opened. Phosphorus-doped glass can also be deposited in a chemical reactor and used as the diffusion source for n-type diffusions.

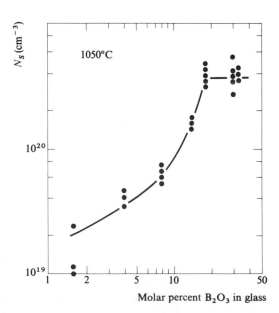

FIGURE 3-7
Boron surface concentration in silicon after diffusion at 1050°C, as a function of the molar percent of B_2O_3 (after Brown and Kennicott[6]).

Table 3-1[6] DIFFUSION RESULTS FROM DOPED OXIDE (2 h at 1100°C, 1 $\Omega \cdot$ cm $\langle 100 \rangle$ silicon)

Molar percent B_2O_3 in glass	1.6	4	8	14	30	100
Sheet resistance ρ'_s, Ω/sq	180	52	33	10	3.6	3.7
Junction depth x_j, μm	1.6	1.9	2.2	2.6	4.0	2.5
Surface concentration N_s, cm^{-3}	1.4×10^{19}	4.8×10^{19}	7.0×10^{19}	2.3×10^{20}	4.2×10^{20}	4.7×10^{20}

Spin-on Sources for Diffusion

Deposition of doped SiO_2 films for use as diffusion sources can be accomplished by using special liquid chemical compounds instead of by the pyrolytic decomposition methods described above. Commercially available chemical solutions for this purpose are made by Emulsitone Co.,† and are available for SiO_2 and SiO_2 doped with various impurities such as phosphorus, boron, arsenic, antimony, gold, etc. These contain an alcohol solution of a polymer, and a compound for the appropriate impurity atom.

The liquid solution is applied to the wafer and the wafer is spun, in a manner similar to that used for photoresist, at about 3000 r/min. A film about 1500 Å thick is formed, which becomes doped oxide when the solvent evaporates. The wafer is then baked at 200°C for 15 min to harden the film, and the diffusion cycle is carried out. A typical cycle for the Emulsitone Emitter Diffusion Source N-250, for example, is 15 min at 1150°C. This produces an impurity profile which approximates a complementary error function with surface concentration 1×10^{21} cm^{-3}. For this case, if the resistivity of the p-type material is 5 Ω · cm, the sheet resistance of the n-type emitter is 4 Ω per square and the junction depth is 1.54 μm.

Solutions are also available for low-surface-concentration diffusions, and are quite useful for such applications as MOS transistor fabrication. Surface concentrations of boron and phosphorus as low as 5×10^{16}, with junction depths of 1.2 μm, have been obtained with these solutions.

Undoped oxides can also be deposited in the same manner. It is therefore possible to avoid the growth of thick thermal oxides for masking against diffusion. One can use the liquid solution to deposit undoped SiO_2, then use photoresist procedures to open diffusion windows, followed by use of liquid solutions to deposit doped oxide, after which the diffusion is carried out. Since deposition of SiO_2 from liquid solution does not require heat treatment at temperatures above 200°C, a final passivating layer can be deposited on the wafer after metallization, and windows can then be opened over the bonding pads.

Ion Implantation[7,8]

While the diffusion technology of the basic process enables the designer to fabricate high-performance devices, this technology also imposes some severe limitations on design flexibility. Each diffusion profile must have a surface concentration higher than that of the material into which it is diffused; this means that large values of sheet resistance are difficult to obtain. Moreover, since the material is compensated several times in, for example, formation of emitter regions, the total number of impurities becomes large with a resulting degradation of mobility and lifetime of carriers. For these reasons, the total number of diffusions in any given region is usually limited to two, as in the basic process, although triple-diffused processes can be designed if careful control of the process parameters is established. Finally,

---

† Emulsitone Co., 41 E. Willow St., Millburn, N.J.

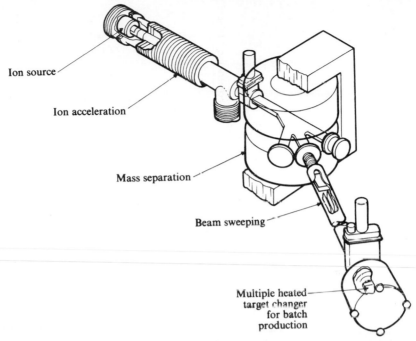

Ion source

Ion acceleration

Mass separation

Beam sweeping

Multiple heated
target changer
for batch
production

FIGURE 3-8
Ion-implantation scheme (after Mayer et al.[7]).

the designer has little control over the shape of the impurity profile resulting from diffusion.   Except for regions very close to the surface where oxide growth affects the redistribution of impurity atoms, the profile will generally approximate a gaussian function or a complementary error function.

Many of the objections to, and limitations from, diffusion can be overcome by ion implantation, an important addition to integrated-circuit processing technology.   The ion-implantation scheme is shown in Fig. 3-8.   Ions such as boron, for example, are emitted by a source and accelerated by an electric field; the ion beam is passed through a magnetic field so that only ions with the desired mass are collected at the exit port of the mass separator.   This ion beam is passed between deflection plates so that it can be swept across the target, in this case a silicon wafer.   The depth of penetration of the ions is controlled by their energy and hence by the accelerating field; the density of ions implanted is controlled by the beam current.   When the ions penetrate the silicon wafer, they produce dislocations; these are removed by annealing the wafer at temperatures of the order of 600°C.

The impurity profile resulting from ion implantation is gaussian, with a peak occurring at a depth R_p, called the *mean range*, and a half-width ΔR_p, called the *straggle*; these are shown in Fig. 3-9.   The mean range depends on the ion mass and energy, while the relative width $\Delta R_p/R_p$ depends on the ratio between ion mass

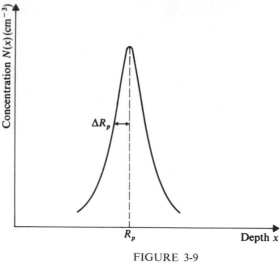

FIGURE 3-9
Distribution of implanted ions.

and silicon atom mass.   Heavier ions produce narrower impurity profiles.   Theoretical and measured profiles for boron implanted in silicon are shown in Fig. 3-10 for an energy of 30 keV.

    Much better control of the impurity profile is possible with implantation techniques than with diffusion methods.   Sheet resistances in excess of 10,000 Ω per square can be obtained, and low surface concentrations which are well controlled can be implanted.   For example, in a MOS transistor requiring precisely controlled threshold voltages, boron-implanted layers with surface concentration controlled to $3.5 \pm 1 \times 10^{15}$ cm^{-3} have been achieved.   Fabrication of shallow devices which would be difficult with diffusion can be achieved with good reproducibility by ion implantation.   Impurity profiles can be tailored to permit optimization of device parameters.

EXAMPLE 3A   *Fabrication of a Microwave Transistor by Ion Implantation.*[8]
Fabrication of a shallow device and the control of the impurity profile are both required in the construction of an integrated microwave transistor.   In this device, a double boron implant is used; the first is carried out with an energy of 100 keV and a dose of 10^{14} ions per cm^2.   This provides a surface concentration of 10^{19} cm^{-3} to permit good ohmic contact and to prevent inversion of the surface.   The second boron implant is at an energy of 200 keV and a dose of 10^{13} ions per cm^2; this produces a controlled number of impurity atoms in the base region at the proper depth, and optimizes the current gain of the device.   The base region formed of this composite implantation has a sheet resistance of 780 Ω per square and a junction depth of 0.8 μm.

    Next, the emitter is implanted by using arsenic ions with 150 keV energy and

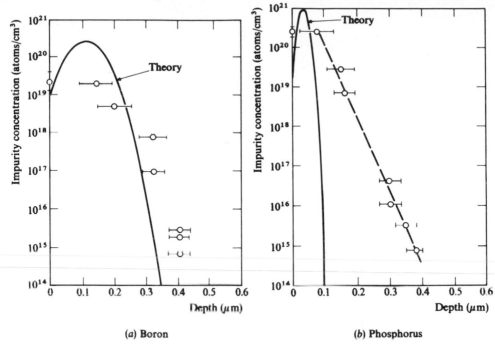

(a) Boron   (b) Phosphorus

FIGURE 3-10
Theoretical and measured profiles for boron (a) and phosphorus (b) in $\langle 111 \rangle$ silicon. Ion energy 30 keV, substrate temperature 625°C, implantation dose 3×10^{15} ions/cm² (after Mayer et al.[7]).

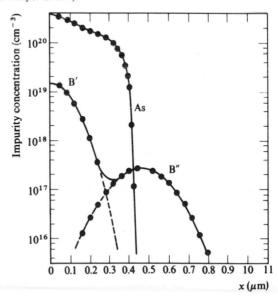

FIGURE 3-11
Impurity profiles for a microwave transistor (after Lepselter[8]).

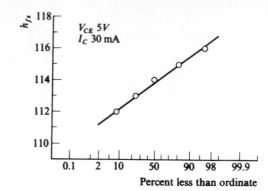

FIGURE 3-12
Distribution of current gain for microwave transistors (after Lepselter[8]).

a dose of 10^{16} ions per cm^2. Finally the entire structure is annealed for 30 min
at 1000°C. The resulting impurity profiles are shown in Fig. 3-11; the emitter
junction depth is 0.4 μm and the base width is 0.4 μm. An indication of the degree
of process control that can be obtained with ion implantation is given by the
distribution of current gain h_{fe} shown in Fig. 3-12.   ////

Implantation in selected regions is accomplished either by using thick oxide
and opening windows by photolithography where implantation is to take place, or
by depositing metal such as aluminum and using photolithography to remove it in
the areas to be implanted. The depth of penetration of ions in SiO$_2$ is comparable
to that in silicon. Metal is used to advantage as an implant mask in the self-
aligning gate MOS transistor, discussed in Chap. 6.

3-3 THIN-FILM COMPONENTS[9]

Thin-film Resistors

Metal deposition on the surface of an oxide-covered chip can be used to perform
several circuit functions in a manner compatible with the rest of the monolithic
circuit construction. Resistors and capacitors are often better realized in thin-
film form than in diffused form in order to achieve particular temperature coefficients, sheet resistance, or parasitic effects different from those of diffused
components. Thin-film elements can be used to augment performance available
from diffused components, thereby increasing circuit design flexibility.
    Perhaps the most widely used thin-film resistors are those made by evapora-
tion of nichrome, a nickel-chromium alloy. As in the case of conventional
aluminum interconnections, resistor dimensions are obtained by etching the
nichrome following a photoresist step. Nichrome resistor sheet resistances range
from 40 to 400 Ω per square, depending upon the film thickness. Temperature
coefficients of nichrome thin-film resistors are generally less than 100 parts per
million per degree centigrade (ppm/°C), whereas diffused silicon resistor tempera-

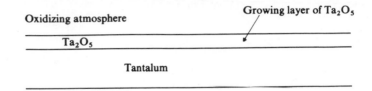

FIGURE 3-13
Tantalum resistor structure.   Note that conversion of the base tantalum metal to Ta_2O_5 by oxidation allows the resistor value to be increased.

ture coefficients are typically 2000 ppm/°C for the standard 200 Ω per square sheet resistance.

Another material widely used for thin-film resistors is deposited tantalum. Here again, various selective etching procedures can be used to define the resistor geometry.   Tantalum, unlike nichrome, has the advantage that the deposited layer of tantalum, once etched, may be varied in thickness by controlled growth of tantalum pentoxide as shown in Fig. 3-13.   By converting some of the tantalum to Ta_2O_5, one can reduce the cross-sectional area of the remaining tantalum. Tantalum resistors made in this way have a range of sheet resistance from about 80 to 4000 Ω per square and a temperature coefficient of resistance from 0 to -150 ppm/°C.   An important aspect of tantalum resistors is the possibility of balancing the negative tantalum resistor temperature drift against the positive temperature drift of diffused silicon resistors in order to achieve very low net temperature variation.

Aluminum contacts are made to both types of thin-film resistors; thus a deposited thin-film resistor appears as shown in Fig. 3-14.   Important parameters of tantalum and nichrome resistors are given in Table 3-2.

Thin-film resistors have the advantage that their values can be adjusted after fabrication, an advantage not shared by diffused resistors.   Such trimming is done by use of a laser beam.   A slot is cut in the resistor structure by a laser beam, with the result shown in Fig. 3-15.   If the resistance is monitored during the trimming, it can be adjusted to within any desired tolerance automatically.   This process requires only a few seconds, and makes possible fabrication of precise resistors. Furthermore, the method can be applied to either nichrome or tantalum resistors.

Table 3-2   CHARACTERISTICS OF THIN-FILM RESISTORS

	NiCr	Ta	Diffused resistor
Range of sheet resistance, Ω/sq	40–400	80–4000	50–500
Temperature coefficient, ppm/°C	<100	0 to -150	2000
Fabrication tolerance, %	5	5	20

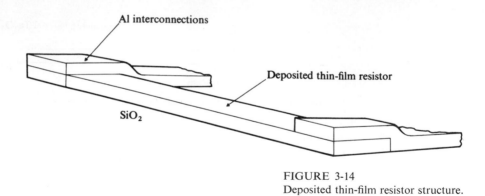

FIGURE 3-14
Deposited thin-film resistor structure.

Thin-film Capacitors

Thin-film techniques can also be used to fabricate capacitors. We have already seen the use of one type of thin-film capacitor, the MOS capacitor, in Chap. 1. The capacitance per unit area of an MOS capacitor is typically of the order of $0.25 \, pF/mil^2$. Of course, a *p-n* junction can also be used as a capacitor, but typical capacitances for junction capacitors are also of the order of $0.25 \, pF/mil^2$. Larger values of capacitance per unit area can be obtained by using a tantalum oxide capacitor.

Since the tantalum oxide layer described in the preceding section is an insulator, a capacitor can easily be fabricated by applying an upper metal plate to this layer. Ordinarily, aluminum would be the first choice for this upper plate, since use of aluminum is compatible with fabrication of the integrated circuit. However, aluminum is soluble in tantalum oxide, even at low temperatures. Therefore a layer of tantalum is sputtered over the tantalum oxide, and this tantalum layer forms the top plate of the capacitor. Typical characteristics of thin-film capacitors are given in Table 3-3.

It is to be emphasized that while thin-film components can be fabricated which are compatible with monolithic integrated circuits in the sense that they can be fabricated on the same substrate, extra processing steps are required for their fabrication. This means not only an increase in the cost of producing the inte-

Table 3-3   CHARACTERISTICS OF THIN-FILM
CAPACITORS

	SiO_2	Ta_2O_5
Capacitance per unit area, pF/mil^2	0.25	2.5
Breakdown voltage	50	20
Fabrication tolerance	20%	20%
Temperature coefficient, ppm/ C		200
Dielectric constant	3.9	21.2

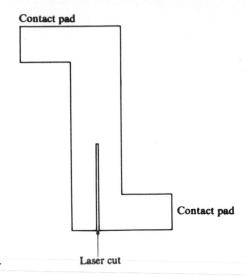

Contact pad

Contact pad

Laser cut

FIGURE 3-15
Laser trimming of thin-film resistors.

grated circuit, but also a decrease in the yield since the yield is an inverse function of the number of fabrication steps. Therefore, thin-film components should be employed only if it has been ascertained that they are absolutely necessary.

EXAMPLE 3B   *Design of a Thin-film Distributed RC Circuit for an Oscillator.*
We now consider the use of thin-film components in the design of a phase-shift oscillator. The basic circuit is shown in Fig. 3-16, with all bias circuits omitted. A distributed RC structure is to be fabricated in thin-film form; it is connected to a transistor as shown. The small-signal collector current passes through the RC structure and reaches the base, diminished in amplitude and shifted in phase. The transistor current gain has a phase shift of $-180°$. If, at the frequency at which the phase shift of the RC structure is $-180°$, the gain of the transistor is equal to the attenuation of the structure, oscillation results. It can be shown that the frequency at which the phase shift of the structure is $-180°$ is given by

$$\omega_0 = \frac{2\pi^2}{RC}$$

where R is the total resistance and C the total capacitance of the structure.
    Suppose it is desired that the frequency of oscillation be $f_0 = 100$ kHz, and that for biasing purposes the resistor have a resistance of 20 KΩ. Then the total capacitance required is

$$C = \frac{2\pi^2}{R\omega_0} = 1570 \text{ pF}$$

    If we use a layer of Ta_2O_5 that is 500 Å thick, the capacitance is 2.5 pF/mil^2, and we require an area of 628 mils^2 to realize 1570 pF. This layer of oxide can be formed on top of the tantalum used for the resistance.

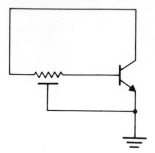

FIGURE 3-16
Basic circuit of a phase-shift oscillator.

For a layer of tantalum 100 Å thick, the sheet resistance is 500 Ω per square.[7] Thus we require 40 squares to realize 20 kΩ. Let the width of the resistor be W and the length L. The requirements which must be satisfied are

$$\frac{L}{W} = 40$$

$$LW = 630 \text{ mils}^2$$

From these we obtain

$$W = 3.96 \text{ mils}$$

$$L = 159 \text{ mils}$$

To complete the design, end pads would have to be added for making contact to metallization stripes. Since it is unlikely that the chip on which the device is being fabricated will be as large as 159 mils on a side, a folded pattern would probably have to be used.　　　　　////

3-4 INTERCONNECTIONS

Interconnections in integrated circuits occur at two levels: the interconnection of the devices on the monolithic chip, and the interconnection of the chip with the system by means of a header or similar substrate. In Chap. 1 the use of aluminum inter-connections on the chip and the connection of the chip to the header by wire bonding were described. We now consider qualitatively some of the problems with conventional aluminum interconnections, and examine alternatives to these and to wire bonding.

Conventional Aluminum Interconnections

As we saw in Chap. 1, aluminum interconnections are accomplished by evaporating aluminum in a vacuum chamber containing the wafers. The entire surface of the wafers is coated with aluminum, and the photoresist procedure is then used to remove the metal from areas where it is not wanted. The evaporation of aluminum from tungsten filaments should not be performed in the fabrication of MOS devices for the following reason. In order to make tungsten sufficiently ductile to be drawn into wire, sodium is added to the tungsten. When tungsten filaments are heated to a temperature sufficiently high to evaporate aluminum, some sodium is also evaporated. Sodium is a major contaminant for MOS devices, since sodium ions apparently are able to migrate along the oxide-silicon interface, even at room temperature, thus drastically altering the device characteristics. Therefore, aluminum deposition for MOS devices is usually done by electron-beam evaporation, a technique which does not require use of tungsten filaments.

After the unwanted metal has been removed, the wafer is sintered for a few minutes at elevated temperature to improve the ohmic contact between the aluminum and silicon and to harden the aluminum. Here care must be taken to keep the temperature below 577°C. At that eutectic temperature a silicon-aluminum alloy forms; this occurs rapidly and the aluminum penetrates the silicon and will destroy the devices. Sintering is therefore carried out at approximately 450°C. Even at this temperature, care must be taken not to sinter for long times. Silicon has a high solubility limit (1.5 percent) in aluminum, and even at temperatures as low as 450°C a solid solution is formed.[10] As time progresses, more silicon is used up in the solution, with the result that shallow junctions can eventually be shorted by this process, as is shown in Fig. 3-17.

In Chap. 1, it was noted that during the fabrication of the integrated circuit, the emitter diffusion must be performed not only in emitter areas but also in any areas where contact must be made to lightly doped n-type material, such as the epitaxial collector region. Aluminum is an acceptor impurity in silicon; hence aluminum diffusing into n-type silicon during sintering may cause a diode to form, rather than an ohmic contact. Fortunately, the solid solubility of aluminum in silicon is only 6×10^{18} cm^{-3}; therefore use of the emitter diffusion in lightly doped areas increases the donor concentration to a level well above that of the aluminum, and no junction can form.

The resistance of evaporated aluminum internal connections is determined by the resistivity of aluminum and the thickness, length, and width of the interconnections according to the familiar resistance relation

$$R = \frac{\rho l}{tW}$$

where ρ = metal resistivity
l = connection length
W = connection width
t = film thickness

Just as with diffused resistors, thickness and resistivity may be combined into a single parameter, sheet resistance:

$$R_s = \frac{\rho}{t}$$

Since for aluminum $\rho = 2.8 \times 10^{-6}$ and t is typically 1500 Å, the sheet resistance of a typical aluminum interconnection is $R_s = 0.187$ Ω per square. While this

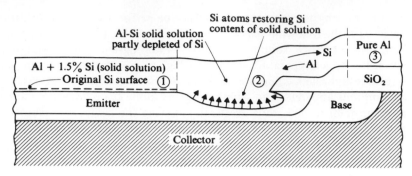

FIGURE 3-17
Penetration of silicon by aluminum at a contact window (after Totta and Sopher[10]).

may seem to be a relatively small value, it can be important in some cases and must be taken into account.   An example of this is given in the design of a 500-mA transistor discussed in Chap. 7.

The above resistance relation holds as long as the film is thick enough that conduction can take place directly along the aluminum film.   For very thin films (hundreds of angstroms), the conduction mechanism changes because of the greater distance between nucleation sites of aluminum composing the film; consequently for thin films the effective resistivity is greater than $2.8 \times 10^{-6} \, \Omega \cdot cm$.

The extensive use of aluminum films for integrated-circuit connections has led to the discovery of new high current density effects in thin metal films.   One such phenomenon, electromigration, is a progressive failure of interconnects during continuous operation.   Consider a typical integrated-circuit interconnection, 1500 Å thick and 1 mil wide.   For a 5-mA current flow in the connection, the current density in the aluminum film is 1.3×10^5 A/cm^2.   At current densities of this order and larger, two effects are believed to contribute to the failure of aluminum interconnections.   The first is the transport of aluminum by momentum exchange with electrons.   It is thought that the thermally activated aluminum ions gain energy from collisions with electrons and are transported away from the contact area.   Since there are no aluminum ions available to fill the vacancies created by the departing ions, these vacancies cluster and form a void which eventually extends across the aluminum stripe, causing a failure.

The second effect which enhances failure is the transport of silicon in aluminum.   At the contact regions, silicon dissolves in aluminum until the solid solubility limit is reached.   Because the activation energy for the diffusion of silicon in aluminum is about 70 percent that of aluminum in aluminum, it is thought that the silicon is readily activated and at the positive end of the stripe is swept away from the contact area by the force resulting from the rate of momentum exchange between activated ions and electrons.   More silicon can now dissolve at the contact.   It is thought that as the silicon is transported down the film, some regions of the film may become supersaturated with silicon, causing the growth of silicon crystallites which weaken the film and lead to failure.

It has been found that the failure rate of aluminum films can be reduced by the deposition of several thousand angstroms of glass over the aluminum. It is thought that the presence of glass reduces the diffusion of aluminum at the aluminum surface.

Theoretical analysis predicts that the mean time to failure for aluminum films is related to the current density by

$$\frac{1}{MTF} = AJ^2 e^{-\phi/kT}$$

where MTF = mean time to failure in hours

J = current density

A = constant relating to the properties of the metal

ϕ = activation energy for aluminum in aluminum

Experiments performed on glassed aluminum films indicate that the mean time to failure is given by

$$\frac{1}{MTF} = J^2 \times 1.88 \times 10^{-3} \times e^{-1.2/kT}$$

It has been shown that aluminum has many desirable properties as a contact and connection metal: it is a good conductor (only silver, copper, and gold are better), it has a relatively high eutectic temperature with silicon (577°C), it adheres well to both silicon and silicon dioxide, and it is easy to evaporate and etch, yet it does not react rapidly with most materials in a sealed header. It has, however, several shortcomings. One major problem is the reaction of aluminum with gold and silicon to form binary and ternary compounds of the three materials. Since gold wires are often used to make connection between aluminum connections on the chip and the header connections, and since the gold-aluminum reaction causes the connections to fail, this can be a troublesome problem. Failures also occur as a result of gold-aluminum-silicon compounds which can form at a wire bond on the chip, since silicon is present in the SiO_2. It is therefore desirable to use aluminum bonding wires.

Beam-lead Interconnections[11]

Beam-lead interconnections provide interdevice connections on the chip and at the same time eliminate the need for bonding wires. Since metals different from aluminum are used, the failure modes described above do not occur. Moreover, the process results in air isolation, eliminating the isolation diffusion. The method used proceeds as follows.

After contact windows have been opened in the oxide, platinum is sputtered onto the wafer, which is then heated to 700°C in an inert atmosphere. This causes platinum silicide (Pt_5Si_2) to form in the windows. Titanium is now sputtered onto the wafer to form good ohmic contacts with the Pt_5Si_2 in the contact windows. Next, a layer of platinum is sputtered onto the wafer. Platinum is inert and does not react with either the titanium or the gold which is to follow.

Interconnection lead patterns are next defined with gold, and the thickness of gold is built up by electroplating. Gold is used because it bonds well, is suitable for electroplating, and has high elongation. The gold leads are built up in this manner to a thickness of 0.5 mil; where they taper in width to contact small-geometry device terminals, they also taper in thickness. It is therefore possible to have thick, wide leads tapering to thin, narrow contacts. Such leads are used not only as interconnections on the chip but as bonding wires and as mechanical support for devices. Thickness-to-width aspect ratios of 1 can be obtained with electroplated gold.

After electroplating is completed, the platinum over the remainder of the wafer is removed by backsputtering, during which the gold leads serve as a mask, protecting the platinum under the gold. The titanium is removed by etching; again the gold protects the titanium under the leads. The wafer is now turned over, and isolation regions are formed by using photoresist on the back of the wafer to delineate the isolation regions and then etching the silicon in these regions through to the front surface. What now remains in each chip is a group of sub-chips connected together electrically and mechanically by the electroplated gold leads. The thickness of the leads makes them strong enough to act as mechanical beam supports, hence the term *beam lead*. Parasitic capacitances between sub-chips are extremely low since no junctions are involved in isolation, and the isolation dielectric is air.

The group of subchips which originally formed the chip is now interconnected with the header as follows. A header of ceramic or similar material is used which has on its surface film-leads in a pattern with terminals designed to match those of the beam leads. The subchip array is now aligned face down on the header so that the outlying beam-lead pads match the header pads. A weld is then performed by ultrasonic or other means. No bonding wires are used. If the header is made large enough, several arrays of devices of various types can be mounted, and are interconnected by the film pattern on the header.

The beam-lead metal system of titanium-platinum-gold adheres well to the SiO_2 on the chip, makes good contact with platinum silicide, and makes good electrical connection to the external circuit. The isolation method which is part of the beam-lead system makes possible fabrication of high-performance devices with very low parasitic capacitance. The obvious disadvantage of the system is that it requires many extra processing steps and is therefore quite expensive.

Solder-bump Interconnections[12]

An alternative to the use of beam leads or wire bonds for interconnection of the chip to the header is the use of small spheres which serve as mechanical support and as electrical interconnection. A series of metal layers is deposited over the contact pad areas as shown in Fig. 3-18a. The particular metals are chosen to provide good electrical contact, good adherence, and masking of the region below the oxide against contamination by the solder. The wafers are now placed in an automated assembly line, and plated copper spheres 5 mils in diameter are dropped

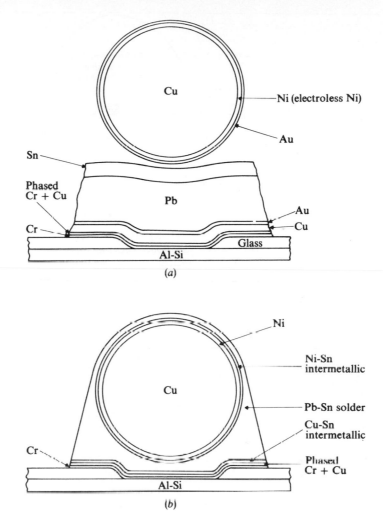

FIGURE 3-18
Solder ball terminal (a) before and (b) after soldering copper ball (after Totta and Sopher[10]).

into the solder wells over the contact pads.   The metals used for plating are chosen to ensure wetting by the solder.   Next the wafers with balls deposited are placed in a furnace and heated until the solder melts.   The solder flows over the balls, forming a solder-coated bump on the contact pad as shown in Fig. 3-18b. Chips can now be separated from the wafers, and mounted face down on headers. As in the beam-lead case, the headers may consist of a ceramic substrate containing film interconnections and contact pads which will align with the bumps on the chip.   The assembly is heated again and the solder reflows, connecting the

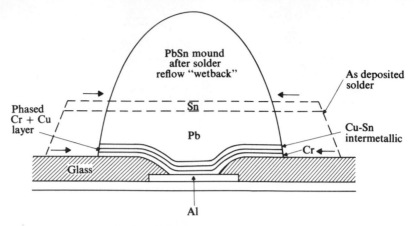

FIGURE 3-19
Controlled collapse solder bump (after Totta and Sopher[10]).

chip to the header. During this procedure the copper balls hold the chip above the header, and also provide a means for containment of the solder around the pads.

An alternative to copper balls is the use of controlled-collapse solder bumps.[13] In this method, some means, such as a glass dam, is used to confine the solder to a given region. Metallic layers are deposited followed by deposition of solder as shown by the broken lines in Fig. 3-19. When the wafer is heated, a reflow "wetback" of the solder occurs, causing a bump to form. The chips are then separated and mounted face down on the header substrate, and the assembly is again heated. If glass dams are used on the substrate pads, the solder cannot flow laterally away from the pads. Instead, a controlled collapse of the solder bump occurs, maintaining the chip elevated above the substrate surface. No wire bonds are used, and the process can be automated for mass production.

PROBLEMS

3-1 A boron-doped oxide is deposited on a silicon wafer having $N_{BC} = 10^{16}$ cm^{-3}. The wafer is then subjected to 1100°C for 2 h. If the boron concentration in the oxide is large enough to maintain solid solubility at the silicon surface during the entire 2 h, find the resulting junction depth.

3-2 Suppose that boron can be ion-implanted in a wafer with $N_{BC} = 10^{16}$ so that a gaussian distribution forms with a peak boron concentration of 10^{20} cm^{-3} at a depth of 5.0 μm. Suppose also that the concentration drops to 10^{19} cm^{-3} 0.2 μm on either side of the peak. Find the location of the junctions if a diffusion is performed for 1 h at 1100°C. To simplify the problem, assume that the wafer is infinite in extent.

3-3 A 2-in-diameter wafer is subjected to a uniformly distributed beam current of boron ions of I A for t s. If the voltage between the ion source and the substrate is 50,000 V, calculate the energy and dose of the beam.

REFERENCES

1 PELTZER, D. L., and W. H. HERNDON: Isolation Method Shrinks Bipolar Cells for Fast, Dense Memories, *Electronics*, pp. 52–55, March 1, 1971.

2 BAKER, W. D., W. H. HERNDON, T. A. LONGO, and D. L. PELTZER: Oxide Isolation Brings High Density to Production Bipolar Memories, *Electronics*, pp. 65–70, March 29, 1973.

3 MUDGE, J., and K. TAFT: V-ATE Memory Scores a New High in Combining Speed and Bit Density, *Electronics*, pp. 65–69, July 17, 1972.

4 FINNE, R. M., and D. L. KLEIN: A Water Amine Complexing Agent System for Etching Silicon, *J. Electrochem. Soc.*, vol. 114, p. 965, 1967.

5 WARNER, R. M., and J. N. FORDEMWALT: "Integrated Circuits: Design Principles and Fabrication," chap. 6, McGraw-Hill Book Company, New York, 1965.

6 BROWN, D. M., and P. R. KENNICOTT: Glass Source B Diffusion in Si and SiO_2, *J. Electrochem. Soc.: Solid-State Science*, pp. 293–300, Feb. 1971.

7 MAYER, J. W., L. ERIKSSON, and J. A. DAVIES: "Ion Implantation in Semiconductors," Academic Press, Inc., New York, 1970.

8 LEPSELTER, M. P.: Ion Implantation—Impact on Device Fabrication, J. Huff and R. Burger (eds.), "Semiconductor Silicon," pp. 842–859, Electrochemical Society, 1973.

9 BERRY, R. W., P. M. HALL, and M. T. HARRIS: "Thin Film Technology," D. Van Nostrand Company, Inc., Princeton, N.J., 1968.

10 TOTTA, P. A., and R. P. SOPHER: SLT Device Metallurgy and Its Monolithic Extension, *IBM J. Res. Dev.*, vol. 13, pp. 226–238, 1969.

11 LEPSELTER, M. P.: Beam Lead Technology, *Bell System Tech. J.*, vol. 45, pp. 233–253, 1966.

12 TOTTA and SOPHER: op. cit.

13 MILLER, L. F.: Controlled Collapse Reflow Chip Joining, *IBM J. Res. Dev.*, vol. 13, pp. 239–250, 1969.

4

PASSIVE COMPONENTS AND THEIR PARASITIC EFFECTS

Now that we have studied basic layout fundamentals and basic processing technology, we can investigate the terminal behavior of integrated devices to see how this behavior is influenced by the constraints imposed by geometry and processing techniques. We begin by considering the simplest passive component, the MOS capacitor, and then continue with interconnection parasitic effects, spiral inductors, thin-film resistors, p-n junctions, diffused resistors, and the "pinch" resistor.

4-1 MOS CAPACITORS[1]

In Chap. 1 we assumed that the MOS capacitor was a parallel-plate capacitor in which both plates were essentially perfect conductors, and we used the formula

$$C_0 = \frac{K_0 \varepsilon_0 A}{l} \qquad (4\text{-}1)$$

to calculate the capacitance. According to (4-1), the MOS capacitor would be a linear circuit element whose capacitance was independent of applied voltage. However, measurements indicate a capacitance-voltage relation similar to that

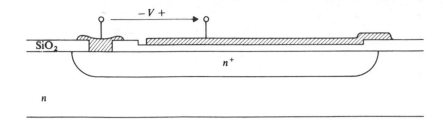

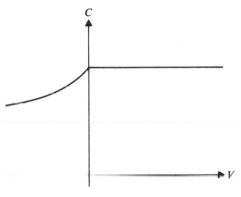

FIGURE 4-1
Typical capacitance-voltage characteristic for an MOS capacitor.

sketched in Fig. 4-1. That the capacitance is a function of voltage is explained as follows.

First consider an MOS capacitor with no applied voltage, and assume that no work-function difference exists between the metal of the top plate and the silicon n^+ layer of the bottom plate. For this case there is no accumulation of charge at the Si SiO₂ interface, and the capacitance is that given by (4-1).

Now assume that a positive voltage is applied as in Fig. 4-2a. A positive charge accumulates on the surface of the upper plate, and a negative charge accumulates at the surface of the silicon n^+ layer. This negative charge is supplied by mobile electrons in the n^+ layer, and because they are accumulated at the surface of the layer, the capacitance is still given by (4-1).

If a negative voltage is applied, a negative charge accumulates on the bottom surface of the metal plate. An equal positive charge must accumulate in the silicon layer. But since there are no mobile holes in the n^+ layer, the positive charge must be created by the formation of a depletion layer. This depletion layer consists of a region of depth w_d below the silicon surface, in which there are essentially no mobile electrons, thereby producing a positive space charge equal to the charge of the bound donor atoms in the depletion region. This is shown in Fig. 4-2b. The thickness of this depletion layer adds to the effective spacing between the plates, reducing the capacitance. Because the donor atoms are bound in the lattice, they

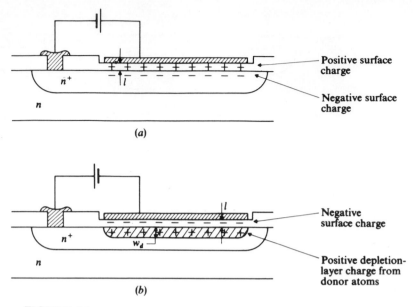

FIGURE 4-2
Charge in the MOS capacitor for (a) positive applied voltage, (b) negative applied voltage.

cannot move to the surface to provide a surface charge. Therefore when the applied voltage is made more negative, requiring more positive charge in the n^+ layer, the increase of charge must be obtained by an increase of the depletion-layer thickness. Thus the capacitance is voltage-dependent for negative applied voltages.

Charge density, electric field, and potential in an MOS capacitor are sketched in Fig. 4-3.

Qualitatively, one can think of the total capacitance for negative applied voltage as resulting from the parallel-plate capacitor whose capacitance C_0 is given by (4-1) in series with a capacitor whose capacitance C_d is that of the depletion region

$$C_d = \frac{K_s \varepsilon_0 A}{w_d}$$

where K_s is the relative permittivity of silicon, as is shown in Fig. 4-4. It should be noted that the dielectric permittivity of SiO_2 is $3.9\varepsilon_0$, while for silicon it is $11.7\varepsilon_0$. Qualitatively, this means that for a particular thickness of depletion region, less voltage is developed across the depletion layer than would be developed across an oxide layer of the same thickness, having the same surface charge density. That the variation of capacitance with voltage is usually negligible is demonstrated by the following example.

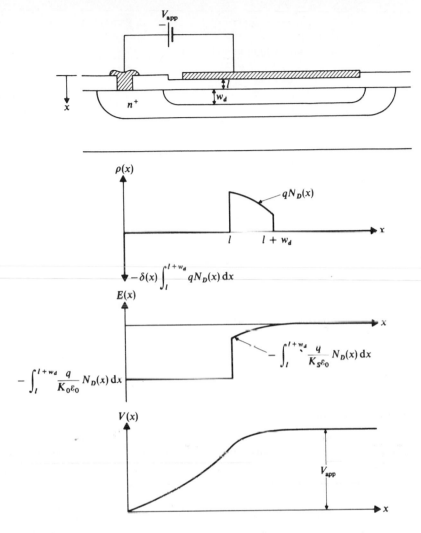

FIGURE 4-3
Charge density, field, and voltage in an MOS capacitor with negative applied voltage.

EXAMPLE 4A *Calculation of Capacitance Variation with Voltage.* We consider the case of an MOS capacitor formed by growing 0.9×10^{-5} cm of oxide over an n^+ layer whose surface concentration is $N_D(0) = 10^{19}$ cm^{-3}. Although the donor concentration decreases with depth into the silicon, to simplify the calculations we assume the donor concentration to be constant.

Let us calculate the applied voltage at which the contribution of the depletion layer results in a decrease of the total capacitance to 90 percent of the value of the

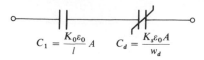

FIGURE 4-4
Equivalent circuit of an MOS capacitor.

$$C_1 = \frac{K_0 \varepsilon_0}{l} A \qquad C_d = \frac{K_s \varepsilon_0 A}{w_d}$$

oxide layer capacitance. Since the two capacitors are in series, we have, for the above condition,

$$C_d = 9C_0$$

The voltage V_0 across the oxide layer is related to the voltage V_d across the deple-tion layer by

$$V_0 = 9V_d$$

The thickness w_d of the depletion layer is

$$w_d = l\frac{K_s}{9K_0} = 3 \times 10^{-8} \text{ m}$$

Now for this value of w_d, the depletion-layer voltage is

$$V_d = \frac{q}{2K_s \varepsilon_0} N_D w_d^2 = 7 \text{ V}$$

The voltage across the oxide layer is therefore

$$V_0 = 63 \text{ V}$$

It is shown in Chap. 6 that a surface inversion layer forms at an even lower voltage.   ////

Breakdown Voltage

Breakdown of an MOS capacitor occurs when the electric field in the oxide exceeds 600×10^6 V/m.   For the case considered in Example 4A, breakdown occurs when $V_0 = 54$ V.   Thus we see that the capacitor would break down before the total capacitance changed by 10 percent.
In practice, imperfections resembling pinholes occur in the oxide when the thickness is reduced. Yield considerations generally impose a lower limit of approximately 10^{-5} cm of oxide thickness.   For this thickness the breakdown voltage is 60 V.   If surface concentrations of the order of 10^{19} or greater are used in the n^+ layer, capacitance variations with voltage will generally be negligible.

Parasitic Effects in MOS Capacitors

Parasitic effects in MOS capacitors arise in the following form:

   1 Leakage conductance of the oxide layer
   2 Series resistance of the contacts
   3 Series resistance of the n^+ layer
   4 Leakage effects resulting from the multiple-layer nature of the silicon substrate

The conductance of the oxide layer is of the order of 10^{-9} $\mho$, and is usually negligible.

For purposes of qualitative discussion, the other parasitic effects may be considered in lumped form as is shown in Fig. 4-5.   The resistances R_C and R_n result from the fact that the capacitor current must flow through the bottom n^+ plate to the contact.   Because the capacitor may be in an isolation region, the effects of the substrate must be considered.   The p-n junction between substrate and epitaxial layer is represented in Fig. 4-5 by the diode and the capacitance C_S.   Since current which flows through the substrate junction must also pass through the epitaxial layer, a resistance R_1 is added to include the epitaxial-layer resistance contribution.   Finally, a resistance R_{SS} is added to account for the resistance of the substrate material.

Probably the most important parasitic resistance contributing to the degradation of capacitor performance is the resistance R_n.   It must be remembered that R_n is distributed over the entire bottom plate, and with the capacitor C_0 it forms a distributed RC structure.   A quick estimate of the value of R_n to use in the first-order equivalent circuit can be made as follows.   Consider the case of a capacitor whose lower plate has a stripe contact, as shown in Fig. 4-6a.   The first-order equivalent circuit is shown in Fig. 4-6b.   We assume that the resistance of the bottom plate is small enough that voltage variations across the bottom plate are negligible in comparison with the voltage between top and bottom plate.   The surface charge on the lower plate is then essentially uniformly distributed over the lower plate.   For a surface charge density ρ, the total displacement current is

$$YX\frac{\partial \rho}{\partial t}$$

At any point x in the lower plate, the current in the n^+ layer must be

$$i(x) = Y(X - x)\frac{\partial \rho}{\partial t}$$

The current density in the n^+ layer is

$$j_x(x) = \frac{i(x)}{l_1 Y}$$

where l_1 is the thickness of the layer.   Now the total voltage along the n^+ layer is given by

$$v = \int_0^X \frac{j_x}{\sigma}\,dx = \frac{\partial \rho}{\partial t}\frac{X^2}{2\sigma l_1}$$

If we define $R \triangle v/i$, we obtain

$$R = \rho_s' \frac{X/Y}{2}$$

where ρ_s' is the sheet resistance of the n^+ layer.

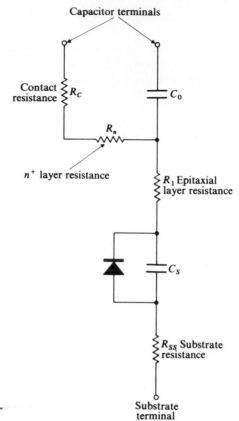

Capacitor terminals

Contact resistance R_C

C_0

R_n

n^+ layer resistance

R_1 Epitaxial layer resistance

C_S

R_{SS} Substrate resistance

Substrate terminal

FIGURE 4-5
First-order lumped equivalent circuit for the MOS capacitor.

A more accurate solution can be obtained by treating the MOS capacitor as a distributed RC structure. If we let c be the capacitance per unit area and ρ'_s the sheet resistance of the n^+ layer, the partial differential equations describing the structure are

$$\frac{\partial i}{\partial x} = -cY\frac{\partial v}{\partial t} \qquad (4\text{-}2)$$

$$\frac{\partial v}{\partial x} = -\frac{\rho'_s i}{Y} \qquad (4\text{-}3)$$

Combining these, we obtain

$$\frac{\partial^2 i}{\partial x^2} = \rho'_s c \frac{\partial i}{\partial t} \qquad (4\text{-}4)$$

Next we take the Laplace transform of (4-2) to (4-4), and employ the boundary conditions

$$I(0, s) = I_0(s)$$

$$I(L, s) = 0$$

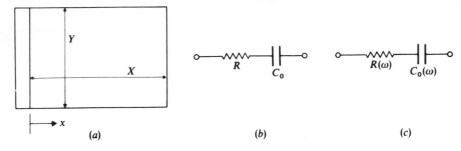

FIGURE 4-6
(a) Bottom plate of an MOS capacitor, with a stripe contact; (b) equivalent circuit obtained by a first-order analysis; (c) equivalent circuit obtained by considering distributed effects.

Solving for $V(x, s)$ and letting $Z(s) \triangleq V(0, s)/I_0(s)$, we obtain

$$Z(s) = \frac{1}{Y} \sqrt{\frac{\rho_s'}{sc}} \coth \sqrt{s\rho_s' c} \qquad (4\text{-}5)$$

If we let $s = j\omega$, we can then resolve $Z(j\omega)$ into a real and an imaginary part. We can thus represent Z by the equivalent circuit of Fig. 4-6c, in which

$$R(\omega) \triangleq \text{Re } Z(j\omega) \qquad (4\text{-}6)$$

$$\omega C_0(\omega) \triangleq \frac{1}{\text{Im } Z(j\omega)} \qquad (4\text{-}7)$$

By using these values one can calculate the dissipation factor of the MOS capacitor.

Exercise 4-1   An MOS capacitor of 100 pF is made with an oxide thickness of 1.5×10^{-5} cm. The sheet resistance of the n^+ diffusion is 50 Ω per square, and the depth is 1 μm. The epitaxial layer has a resistivity of 1 Ω · cm, and a thickness of 10 μm. Estimate the values of R_c, R_n, and R_1 in the equivalent circuit of Fig. 4-5.

4-2   INTERCONNECTIONS

The metallization stripes used as interconnections between devices in an integrated circuit can contribute parasitics resulting from the transmission-line nature of the stripes.

For a straight interconnection stripe such as that shown in Fig. 4-7, the series resistance is

$$R = \frac{L/W}{\sigma_m t_m} = \rho_s' \frac{L}{W}$$

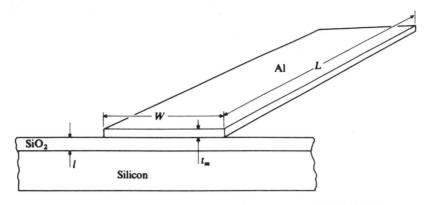

FIGURE 4-7
Interconnection stripe.

where σ_m = conductivity of stripe
    t_m = its thickness, and
    ρ'_s = its sheet resistance
    For an aluminum stripe 1.5×10^{-5} cm thick, the sheet resistance is 0.186 Ω per square.

The capacitance per unit length between stripe and silicon is approximately

$$\frac{C}{L} \approx \frac{K_0 \varepsilon_0 W}{l}$$

if fringing effects are neglected. For a 1-mil-wide stripe over an oxide layer 2.5×10^{-5} cm thick, the capacitance is 0.089 pF/mil.
    The inductance of the stripe can be estimated by assuming that the magnetic field is confined to the region between the metal and the silicon. For a current I, the magnetomotive force H in the oxide is

$$H = \frac{I}{W}$$

The flux density B and flux Φ are

$$B = \mu H = \frac{\mu I}{W}$$

$$\Phi = BLl = \mu I \frac{Ll}{W}$$

where μ is the permeability of the oxide. Since the inductance per unit length is the flux per unit length divided by the current, we have

$$\text{Inductance per unit length} = \frac{\mu l}{W}$$

For the 1-mil stripe over 2.5×10^{-5} cm of oxide, the inductance is 3.14×10^{-13} H/mil.

At high frequencies, there is a penetration of the H field into the silicon, the skin depth for which is

$$\delta = \sqrt{\frac{2}{\omega\mu\sigma}}$$

where σ is the conductivity of silicon. For $1\ \Omega\cdot\text{cm}$ silicon, the skin depth at 1 GHz is 62.6 mils, considerably greater than the typical 10-mil wafer thickness. If we therefore assume H to be uniform in the silicon, we obtain an inductance of 3.14×10^{-10} H/mil. Although this overestimates the inductance, it can be used as a guide in calculating metallization stripe inductance.

Losses also occur as a result of eddy currents produced in the silicon by the magnetic field. From Faraday's law we have

$$\mathbf{J} = \nabla \times \mathbf{H}$$

which for our case yields a magnitude

$$J = \frac{H_0}{\delta}\, e^{-x/\delta}$$

Since at high frequencies δ is much greater than the wafer thickness, we use the approximation

$$J \approx \frac{H_0}{\delta}$$

The power density thus produced is

$$P = \frac{J^2}{\sigma} = \frac{H_0{}^2}{\delta^2\sigma} = \frac{I^2}{W^2\delta^2\sigma}$$

If the volume encompassing the magnetic field is of cross-sectional area A, the power dissipation per unit length is given by

$$\frac{P}{L} = \frac{I^2 A}{W^2\sigma\,\delta^2}$$

This corresponds to the losses which would result in the stripe from an equivalent series resistance per unit length of

$$\frac{R}{L} = \frac{A}{W^2\delta^2\sigma} = \frac{A\omega\mu}{2W^2}$$

For $A = 2 \times 12$ mils, $R/L = 2.41\ \Omega/\text{mil}$ at 1 GHz. Increasing the width of the stripe decreases the series resistance.

The velocity of propagation along the interconnection is sufficiently large that delay associated with interconnections is not generally a factor in integrated-circuit design. Therefore the shunt capacitance and series loss from both bulk resistance and substrate coupling are the primary interconnection parasitic effects.

Parasitic effects associated with a single stripe can be expanded to encompass coupling between adjacent interconnections. There are two coupling mechanisms between adjacent interconnections: inductive coupling and substrate coupling. Direct capacitive coupling between interconnections is small, since the stray coupling fringe capacitances are small.

Substrate coupling occurs when interconnect signals are capacitively coupled

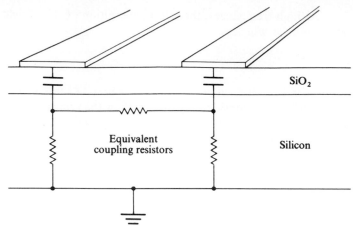

FIGURE 4-8
Substrate resistive coupling between two interconnections.

into the substrate. The substrate resistance to ground appears in an equivalent circuit as shown in Fig. 4-8, and the parasitic substrate signal is then coupled into the second interconnection through the oxide capacitance.

Magnetic coupling between interconnections occurs because of the fringing magnetic fields. The effect of this coupling is to produce a transformer equivalent between the two adjacent leads. For closely spaced stripes, coupling factors as high as 0.3 have been observed.

Interconnection parasitic effects are important in integrated circuits at frequencies above 100 MHz; however, bulk resistance and stray capacitance are the primary effects in most cases.

4-3   HEADER PARASITICS

There are several types of headers which can be used for mounting integrated-circuit chips; they are made of various materials such as plastic, ceramic, and metal alloys. The type of header selected will generally depend upon the system application in which the integrated circuits are being used. As far as parasitic effects are concerned, the metal can-type header is the worst offender. This type of header is, however, quite useful where it is necessary to replace or interchange circuits quickly, since it is easily inserted in a socket which can be mounted on an etched card.

Parasitic effects in headers result from lead inductance, pin capacitance, header losses, and mutual coupling between bonding leads. Many of these effects become important in the metal TO-5 header at lower frequencies than for other headers. This is because the iron-nickel-cobalt alloy, chosen because its thermal coefficient of expansion matches the glass used for the pin seals, is ferromagnetic and therefore contributes to the inductive parasitic effects.

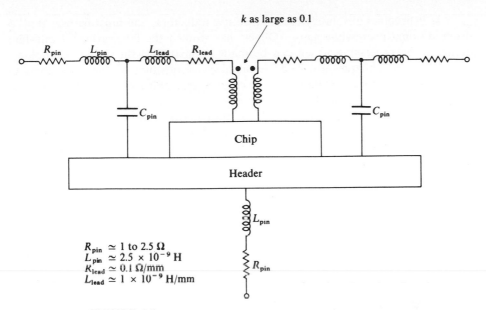

FIGURE 4-9
First-order model for header and lead parasitic effects. Values are for a TO-5 header.

A first-order model for lead and header parasitics is shown in Fig. 4-9 for only two bonding wires and a header connection. This circuit configuration applies for most headers; the values shown are for the TO-5 package.

4-4 SPIRAL INDUCTORS

It is almost impossible to fabricate inductors of any reasonable value of inductance in integrated circuits without resorting to additional fabrication techniques. First, because of the limitations of the processing technology, it is almost mandatory that inductors be planar. This means that some form of spiral geometry must be used. Second, the metal deposition and etching techniques used in standard processing do not lend themselves to the production of thick, closely spaced metal patterns. Consequently the resistance of the metal used to form the inductor is sufficiently large that it is not possible to obtain Q factors much larger than unity. Third, no materials having high permeability are directly available in the standard fabrication procedure. This means that relatively large spirals are required to produce a reasonable fraction of a microhenry. But as the size of the spiral increases, so does the resistance of the metal pattern. The silicon wafer also contributes eddy-current losses, further degrading the Q. The result is that if it is at all possible, one avoids the use of inductors, substituting instead different circuit techniques to produce inductive effects.

If it becomes absolutely essential to use inductors, one must be prepared to use additional processing steps. Olivei[2] has studied the fabrication of thin-film inductors by processing which can be made compatible with planar technology. Basically, the techniques involved are the electrodeposition of thick gold films to yield a low-resistance metal pattern, the electrolytic etching of closely spaced spiral turns, and the deposition of ferrite films over the spiral.

It is interesting to note that there is significant difference in performance among inductors having different geometrical forms of spiral. The three spirals considered by Olivei are the Archimedes, the logarithmic, and the hyperbolic spirals, whose polar equations are, respectively,

$$S_A(r, \theta) = R_0 + a\theta$$

$$S_l(r, \theta) = R_0 \, e^{l\theta}$$

$$\frac{1}{S_h} = \frac{1}{R_0} - h\theta$$

Power dissipation, skin-effect losses, and inductance of air-coil spirals are shown in Fig. 4-10a, b, and c. Here A, B, and C refer to the spirals in the order listed above. R_n is the radius of the outermost turn of the spiral. Inductance of spirals with deposited ferrite is shown in Fig. 4-10d, and Q for spirals with deposited

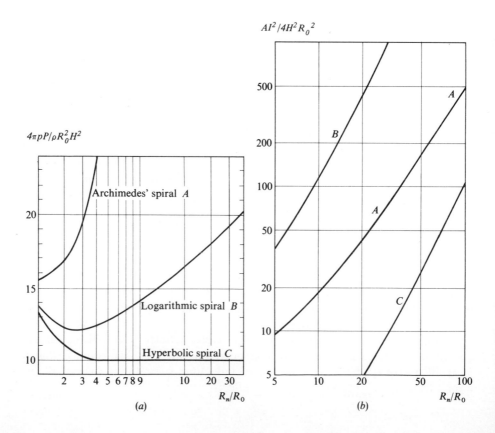

(a)

(b)

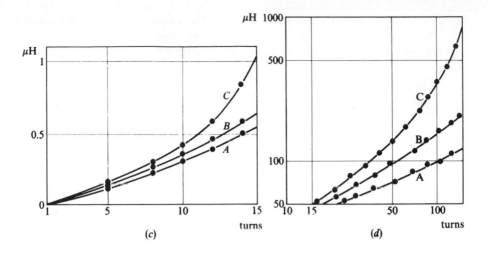

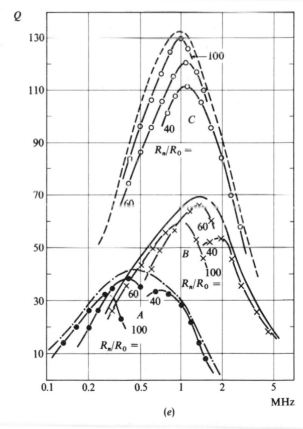

FIGURE 4-10
(*a*) Power dissipation of Archimedes (*A*), logarithmic (*B*), and hyperbolic (*C*) spirals; (*b*) skin-effect losses; (*c*) air-coil inductance; (*d*) inductance of spirals with deposited ferrite; (*e*) *Q* of spirals with deposited ferrite (after Olivei[2]).

ferrite is given in Fig. 4-10e.   It is easily seen by inspection of Fig. 4-10 that the hyperbolic spiral offers significant advantages over the others in all cases.   Olivei reports that inductance values of 1 to 1200 μH with Q's of 100 over the frequency range of 0.5 to 1.5 MHz have been obtained.

The parasitic capacitance between spiral inductors and the silicon wafer can easily be computed from knowledge of the surface area of the spiral and the thickness of the oxide.

4-5   THIN-FILM RESISTORS

In Chap. 3, we saw that resistors can be made by using thin films of various types, such as nichrome and tantalum.   Since these resistors are deposited on the surface of the SiO$_2$ layer, a capacitance is formed between the thin film and the silicon under the oxide.   Thus the thin-film resistor is actually a distributed RC network.

Analysis of the distributed RC network can be handled in a manner similar to that used for distributed effects in MOS capacitors in Sec. 4-1.   We consider a resistor film of length L and width W, as shown in Fig. 4-11a.   Usually the sheet resistance of the film will be larger than that of the silicon, and we therefore make the approximation that the voltage drop along the surface of the silicon is negligible.   For this approximation, a model for a section of length Δx is as shown in Fig. 4-11b, where r and c are the resistance and capacitance per unit length.   Writing the Kirchhoff voltage and current-law equations for the section of Fig. 4-11b, and taking the limit as $\Delta x \rightarrow 0$, we obtain the partial differential equations for the structure.[3]   These can then be solved for the y parameters.   The results are

$$y_{11}(s) = y_{22}(s) = \frac{1}{r}\sqrt{src}\, \coth L\sqrt{src}$$

$$y_{12}(s) = y_{21}(s) = -\frac{1}{r}\sqrt{src}\, \operatorname{csch} L\sqrt{src}$$

These can also be written in terms of ρ_s' and ε, the sheet resistance of the film and dielectric constant of the oxide:

$$y_{11}(s) = y_{22}(s) = \frac{W}{\rho_s'}\sqrt{\frac{s\varepsilon\rho_s'}{l}}\, \coth L\sqrt{\frac{s\varepsilon\rho_s'}{l}}$$

$$y_{12}(s) = y_{21}(s) = \frac{W}{\rho_s'}\sqrt{\frac{s\varepsilon\rho_s'}{l}}\, \operatorname{csch} L\sqrt{\frac{s\varepsilon\rho_s'}{l}}$$

where $\varepsilon = K_0\varepsilon_0$.

For most practical applications it is not necessary to deal with the distributed network; rather, one can make a first-order lumped model which adequately represents the device up to approximately its cutoff frequency $\omega_c = 2.43/RC$. This is done by using a single-pi-section model as shown in Fig. 4-11c, in which

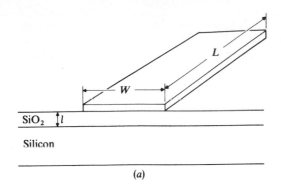

(a)

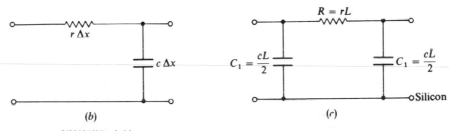

(b)   (c)

FIGURE 4-11
(a) Section of thin-film resistor; (b) approximation used for a section of length Δx; (c) first-order lumped model for the resistor and its parasitic capacitance.

the resistance and capacitance values are chosen so that the steady-state current flow and stored charge of the model match those of the distributed structure. This results in

$$C_1 = \frac{\varepsilon L W}{2l}$$

$$R = \rho'_s \frac{L}{W}$$

The above analysis applies for resistors which are made with either stripe or maze geometry. As we have seen in Chap. 3, laser trimming is often used to obtain precise values for thin-film resistors. For this case, considerably more area will be used than for the stripe geometry. Consequently, for resistors of a given resistance, the surface area will be larger for the trimmed resistor, and the capacitance will be correspondingly larger.

EXAMPLE 4B *Capacitance of a Thin-film Resistor.* We consider the design of a 500-kΩ thin-film resistor for an application which requires a large-value resistor with low-temperature coefficient. The resistor is made by depositing tantalum and oxidizing the film, as described in Chap. 3, to produce a sheet resistance of 1000 Ω per square.   For a 500-kΩ resistor, 500 squares are required.   If we use a line width of $W = 0.5$ mil, the stripe length is 250 mils, and the surface area is 125 mils^2.   (Contact pads are assumed to be negligible.)

If we assume that the resistor is deposited on the layer of SiO$_2$ 6 $\times$ 10^{-5} cm thick, the total capacitance of the resistor is

$$C = 4.64 \text{ pF}$$

The cutoff frequency of the open-circuit voltage transfer ratio of this device would be

$$f_c \approx 167 \text{ kHz}$$

while the lumped model of Fig. 4-11c predicts a cutoff frequency of

$$f_c \approx 137 \text{ kHz} \qquad ////$$

4-6   INTEGRATED JUNCTIONS

While p-n junctions form the basis for the construction of diodes, transistors, field-effect transistors, and other integrated devices, our first concern with junctions at this point is the understanding of their depletion regions and the capacitances that arise therefrom.   Essential to the analysis of integrated juctions is the depletion approximation, without which calculation of, for example, capacitance becomes tedious even for the simplest cases.

We first review the calculation of the locations of the junctions themselves. Sketches of typical impurity profiles for double-diffused integrated devices are shown in Fig. 4-12a.   The net impurity distribution is sketched in Fig. 4-12b.   As we have seen in Chap. 2, the so-called "metallurgical junctions" occur at the points in the material where the net impurity is zero, that is, at the points where the impurity distribution changes from donor to acceptor, or vice versa.   If we assume that in the vicinity of these junctions the impurity profiles resulting from diffusion are well approximated by gaussian functions, then the junction locations are those values of x which are solutions for the equation

$$N_{sD}\, e^{(-x/x_{0D})^2} + N_{BC} - N_{sA}\, e^{-(x/x_{0A})^2} = 0 \qquad (4\text{-}8)$$

where N_{sD} and N_{sA} = surface concentrations after diffusion
$\qquad\qquad N_{BC}$ = background concentration of epitaxial layer
$\quad x_{0D}$ and x_{0A} = diffusion lengths
Equation (4-8) is transcendental, and requires numerical solution for the junction depths.   However, as we have seen in Chap. 2, simplifying approximations can often be made for typical diffusions used in practice.   For the collector junction, one can often assume that the emitter diffusion produces a donor profile which, near the collector junction, is negligible in comparison with the background concentration N_{BC}.   Making this approximation in (4-8), one obtains

$$N_{sA}\, e^{-(x_{jC}/x_{0A})^2} \approx N_{BC} \qquad (4\text{-}9)$$

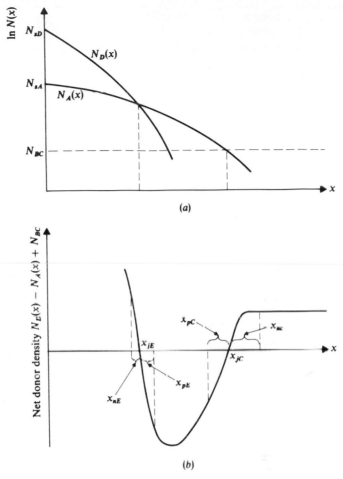

FIGURE 4-17
(a) Impurity profiles for a double-diffused structure; (b) net impurity distribution, showing junction depths and depletion regions.

which can be solved for x_{jC}. Similarly, it often occurs that, near the emitter junction, the background concentration is negligible in comparison with the donor concentration of the emitter diffusion. Making this approximation in (4-8), we obtain

$$N_{sD}\,e^{-(x_{jE}/x_{0D})^2} \approx N_{sA}\,e^{-(x_{jE}/x_{0A})^2} \qquad (4\text{-}10)$$

which can be solved for x_{jE}. These approximations usually hold for structures with junction depths of the order of 1 to 2 μm. For structures with junction depths of 5 to 6 μm, the approximations usually do not hold. If there is some question of the validity of (4-9) and (4-10), the following consistency check can be used:

1   Assume (4-9) to be valid, and calculate x_{jC}.
2   Calculate $N_D(x_{jC}) = N_{sD} e^{-(x_{jC}/x_{0D})^2}$.
3   If $N_D(x_{jC}) \ll N_{BC}$, (4-9) is valid.

A similar consistency check can be used for (4-10).

Depletion Regions[4]

On either side of the metallurgical junctions there will be regions in which there is
net space charge resulting from a difference between the density of impurity atoms
and the density of mobile carriers. The resulting charge dipole which appears
across the junction causes a voltage to appear, and is responsible for the junction
capacitance. In thermal equilibrium, the voltage between the two sides of the
junction is called the built-in voltage ϕ'_C, and is of the order of 0.7 V. When a
voltage is applied to the junction and thermal equilibrium no longer obtains, the
charge distribution in the dipole layer changes to conform to the requirements of
the applied voltage.

The potential distribution—and hence the electric field and the charge
distribution—near the junction are not conveniently solved for, and in order to
expedite the solution a very important approximation, the *depletion approximation*,
is used. The depletion approximation assumes that a region exists on either side
of the metallurgical junction in which the density of mobile carriers is much less
than the density of ionized impurity atoms. These regions are called depletion
regions. In other regions, the depletion approximation assumes space-charge
neutrality; that is, the charge of mobile carriers is exactly balanced by the charge
of impurity ions.

A consequence of the depletion approximation is that the electric field is
zero outside the depletion regions, but large within them. That the depletion
approximation is valid for a typical integrated junction can be seen from Fig.
4-13.[5] Here one can easily define regions outside which the electric field is less
than 10 percent of the peak field at the junction.

Use of the Depletion Approximation

To appreciate the utility of the depletion approximation, we first review the case
of an abrupt junction of Fig. 4-14a in thermal equilibrium. According to the depletion approximation, there will be regions of extent x_n and x_p around the junction
where the density of mobile carriers is much less than the density of impurity ions.
Outside these regions, the net space-charge density is zero. Since the impurity
concentrations on either side of the junction are uniform, the charge dipole which
results will be as shown in Fig. 4-14b. Moreover, the charge dipole requires that

$$N_D x_n = N_A x_p$$

The electric field is $E(x) = (1/\varepsilon) \int \rho(x)\, dx$, which is easily seen from simple
geometrical arguments to have the form shown in Fig. 4-14c. Since the electrostatic potential is related to the field by $-d\phi/dx = E(x)$, the potential can

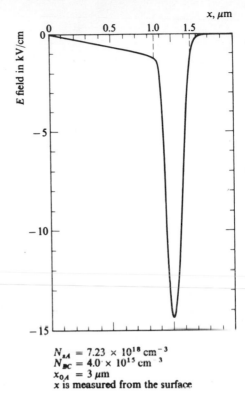

FIGURE 4-13
Thermal equilibrium electric field E versus distance normalized to junction depth for a gaussian-diffused p-n junction (after Hoff and Everhart[5]).

$N_{sA} = 7.23 \times 10^{18}$ cm^{-3}
$N_{sC} = 4.0 \times 10^{15}$ cm^{-3}
$x_{0A} = 3\,\mu$m
x is measured from the surface

easily be calculated. If the potential of the donor side of the junction, outside the depletion region, is taken as zero, the potential difference between the neutral regions on either side of the junction is the built-in voltage ϕ'_C, and the potential is as shown in Fig. 4-14d.

If a voltage V is applied to the junction, thermal equilibrium no longer obtains. The total voltage V_T across the depletion regions is now $V_T = V + \phi'_C$, and the plots of Fig. 4-14 must change accordingly. Since the charge density in each depletion region is fixed by the impurity distribution, the widths of the depletion regions x_n and x_p must change. For example, if V has the same sign as ϕ'_C, $V + \phi'_C > \phi'_C$. The maximum field must then increase, which in turn requires an increase of x_p and x_n, making the depletion region larger.

Clearly, before any calculations of depletion-region sizes can be made, it is necessary to know the built-in voltage ϕ'_C. If one assumes that in thermal equilibrium the net current flow across the junction is zero and that also for each carrier type the drift and diffusion components of current exactly balance, one can show that for the abrupt-junction case

$$\phi'_C = \frac{kT}{q} \ln \frac{N_A N_D}{n_i^2}$$

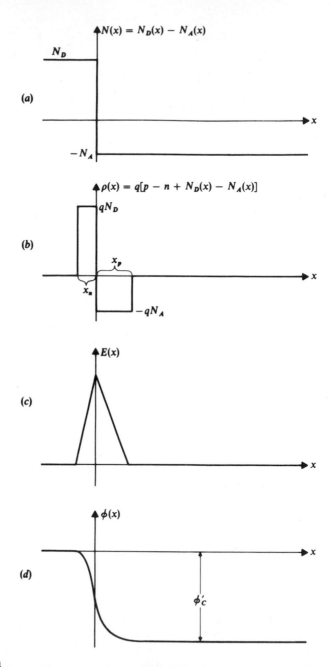

FIGURE 4-14
The abrupt junction in equilibrium; (a) Impurity distribution; (b) net space-charge density according to the depletion approximation; (c) resulting electric field; (d) electrostatic potential.

where n_i is the free electron (or hole) density for intrinsic material. With this, the calculation of all the distributions of Fig. 4-14 in terms of N_A and N_D becomes straightforward and simple.

The charge dipole of the depletion regions also produces a capacitance. Since the width of the regions changes with applied voltage, the capacitance is voltage-dependent. If the capacitance is defined as

$$C \triangleq \frac{dQ}{dV}$$

where Q is the space charge on one side of the junction, it is easily shown by use of the relations which apply to Fig. 4-14 that the capacitance per unit area of the junction is

$$\frac{C}{A} = \frac{C_{0A}}{(1 + V/\phi_c')^{1/2}}$$

where C_{0A} is a function of N_A and N_D.

Exercise 4-2 Find C_{0A} for an abrupt junction.

One other type of junction can be conveniently and easily analyzed. This is the so-called *linear graded junction*, in which the impurity profile changes from donor to acceptor in a linear fashion, as shown in Fig. 4-15. Note that for this case, the charge dipole imposes the requirement that $x_p = x_n$. The capacitance of the linearly graded junction is easily shown to be

$$\frac{C}{A} = \frac{C_{0l}}{(1 + V/\phi_c')^{1/3}}$$

While the abrupt and linear junctions do not exist in practical integrated circuits, they provide at least a qualitative understanding of the behavior of general p-n junctions. Integrated junctions differ in quantitative detail from abrupt and linear junctions, but the first-order qualitative behavior of the former is the same as for the latter. Some cases exist for which abrupt or linear junctions can be used as approximations of actual junctions for purposes of quantitative calculations. However, care must be used in employing these approximations, as we shall later see.

Exercise 4-3 Find C_{0l} for a linear graded junction.

The Lawrence-Warner Curves for Diffused Collector Junctions[6]

The abrupt and linear junctions can be analyzed quantitatively with relative ease because the impurity distributions have very simple analytical form. Unfortunately, the junctions formed by diffusion in an integrated circuit are not easily

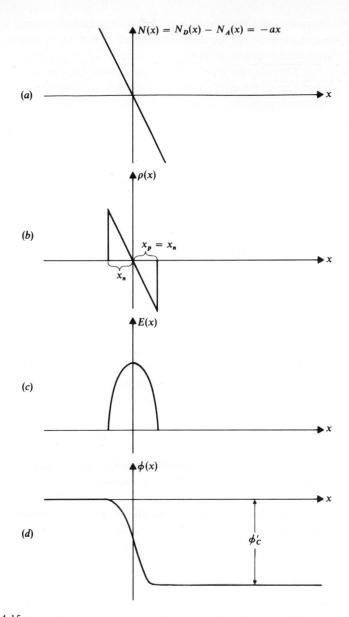

FIGURE 4-15
The linearly graded junction. (*a*) Net impurity profile; (*b*) space-charge density; (*c*) electric field; (*d*) electrostatic potential.

analyzed because the net impurity profile is not a simple form. However, Lawrence and Warner, using computer-aided analysis techniques, have generated nomographs to simplify the calculations for two types of integrated junctions: a gaussian profile diffused into a constant background concentration, and a complementary-error-function profile diffused into a constant background concentration. The latter applies to the junction formed after base predeposit; the former, to the collector junction resulting after subsequent base and emitter drive-in steps.

A set of Lawrence-Warner curves for gaussian profile and constant background concentration is shown in Fig. 4-16. These curves enable the calculation of capacitance, total depletion-layer thickness, and percent of depletion-layer thickness on each side of the junction, in terms of total depletion-layer voltage V_T. Note that since $V_T = V + \phi'_C$, it is necessary to know the built-in voltage for the junction before the curves can be used. Fortunately, the built-in voltage shows rather weak dependence on the impurity profiles, so that one can assume as a reasonable approximation $\phi'_C \approx 0.7$ V for all junctions.

EXAMPLE 4C *Use of the Lawrence-Warner Curves for a Typical Collector Junction.* Consider the case for which the base diffusion results in a boron impurity distribution, after all drive-ins, having surface concentration $N_{sA} = 5 \times 10^{18}$ cm^{-3}, background concentration $N_{BC} = 5.4 \times 10^{15}$ cm^{-3}, and junction depth 3×10^{-4} cm. Let us calculate the zero-bias capacitance and depletion-layer thickness for this junction.

Since $V = 0$, we have $V_T = \phi'_C$, which we assume to be 0.7. Then $V_T/N_{BC} = 0.13 \times 10^{-15}$. We enter the family of curves at this value, and move to the left along a *diagonal* line corresponding to 0.13×10^{-15}. (Interpolation is necessary in order to find this diagonal line.) We continue along the diagonal until we intersect a line corresponding to $x_j = 3 \times 10^{-4}$ cm. (Interpolation between 2×10^{-4} and 3×10^{-4} is necessary.) Now we move *horizontally* left from this intersection and find $C/A \approx 1.5 \times 10^4$ pF/cm^2, and $x_m =$ total depletion-layer thickness $\approx 0.70 \times 10^{-4}$ cm $= 0.70$ μm.

Next we go to the x_1/x_m family of curves corresponding to $N_{BC}/N_s = 10^{-3}$. Here x_1 is the thickness of the depletion region on the *more heavily doped* side; for our case this is the base side of the junction. Entering the family at $V_T/N_{BC} = 0.13 \times 10^{-15}$, we move vertically until we intersect the curve corresponding to $x_j = 3 \times 10^{-4}$ cm. (Interpolation is required.) We then move left horizontally and read

$$\frac{x_1}{x_m} = 0.39$$

Thus the depletion thickness x_{pC} into the base side of the collector junction is

$$x_{pC} = 0.39 \times 0.70 \times 10^{-4} \text{ cm} = 0.273 \text{ μm}$$

and the depletion thickness x_{nC} into the epitaxial collector is

$$x_{nC} = 0.61 \times 0.70 \times 10^{-4} \text{ cm} = 0.427 \text{ μm}$$

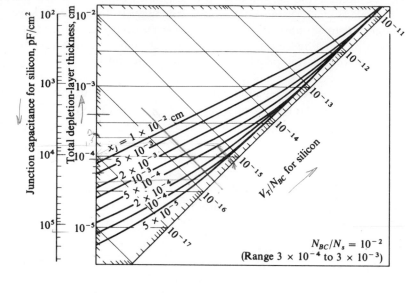

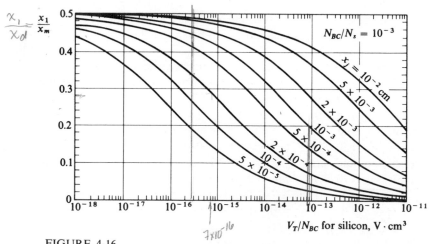

FIGURE 4-16
Lawrence-Warner curves for gaussian impurity distribution into constant background.

The use of the Lawrence-Warner curves for this example is summarized in Fig. 4-17.

Note that if a linear-graded-junction approximation were used, it would yield $x_{pC} = x_{nC}$. From the results obtained with the Lawrence-Warner curves, we conclude that a linear graded approximation would be very poor in this case.

////

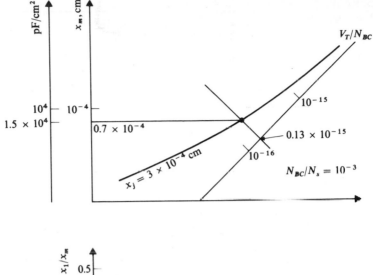

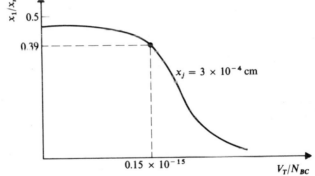

FIGURE 4-17

Summary of the use of the Lawrence-Warner curves for Example 4C.

Since the Lawrence-Warner curves apply to diffusions into constant background concentration, they can be used for calculating the capacitance of collector junctions and isolation junctions. They are not, however, directly applicable to emitter junctions since these involve diffusion into a background consisting of both the epitaxial layer and the base diffusion. For the special cases in which approximation (4-10) is valid for calculating junction depths, the Lawrence-Warner curves can be used as follows to estimate emitter junction capacitance:

1 Use (4-10) to calculate junction depth.
2 With this junction depth, calculate the concentration of *base* diffusion acceptors at the junction.
3 Use this concentration as an effective background concentration.
4 Use the appropriate Lawrence-Warner curves to estimate capacitance

Since the "background" in this case is not constant but decreases with depth, the Lawrence-Warner curves will overestimate the depletion width on the emitter side of the junction and underestimate the depletion width on the base side. The total depletion-layer width, and hence the capacitance, is probably a reasonable estimate.

Design of Junction Capacitors

Our discussion thus far has been confined to planar junctions. If we wish to fabricate a capacitor in an integrated circuit, we could, for example, use a base diffusion into the epitaxial material. The cross section of such a capacitor is shown in Fig. 4-18. As was mentioned in Chap. 2, because diffusion proceeds laterally as well as vertically, and because the epitaxial material has constant background concentration, the location of the junction sidewall will be approximately a quarter circle centered at the edges of the diffusion windows. Since the impurity concentration gradient of the p diffusion is the same along the junction sidewalls as along the bottom, the sidewall capacitance and bottom capacitance (per unit area) are the same. Thus it is only necessary to compute the total effective area of the junction, and multiply by the capacitance per unit area obtained, for example, from the Lawrence-Warner curves.

EXAMPLE 4D   *Design of a 10-pF Junction Capacitor.*   We now consider use of the base diffusion to fabricate a 10-pF capacitor. Although any reasonable surface geometry which results in the correct area is permissible, for simplicity we assume the diffusion window to be a square W mils on a side. The problem then is to calculate W for the particular base diffusion and background concentration. Let the processing parameters be

$$N_{BC} = 10^{16} \text{ cm}^{-3}$$

$$N_{sA} = 5 \times 10^{18} \text{ cm}^{-3}$$

$$x_{jC} = 3.0 \times 10^{-4} \text{ cm}$$

First we calculate $N_{BC}/N_{sA} = 2.00 \times 10^{-3}$, so we use the Lawrence-Warner family of curves corresponding to $N_{BC}/N_s = 10^{-3}$. We wish to calculate the capacitance per unit area at zero bias, so we need the built-in voltage ϕ'_C. Again we assume $\phi'_C = 0.7$ and enter the curves at $V_T/N_{BC} = 0.7 \times 10^{-16}$. We find

$$\frac{C}{A} \approx 2 \times 10^{-4} \text{ pF/cm}^2 = 0.12 \text{ pF/mil}^2$$

Now we must calculate the area of the junction. If we neglect the small fillets that occur at each of the four corners of the diffusion, we can write

$$A \approx W^2 + 4W(\tfrac{1}{4} \times 2\pi x_{jC})$$

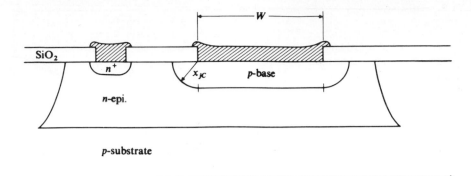

FIGURE 4-18
Cross section of a junction capacitor using the base diffusion.

Then we have

$$C_{total} = (W^2 + 0.742W)\frac{C}{A}$$

Solving for W we obtain

$$W = 8.77 \text{ mils}$$

It is interesting to compare this with the value obtained by neglecting the sidewalls. For that case, $W = 9.13$ mils. ////

Typical values of C/A for base diffusions in most integrated circuits are 0.2 pF/mil^2 or less. This is approximately the same as that which can be obtained with MOS capacitors, so comparable areas are required for either. However, no extra processing steps are required for the junction capacitor, whereas an extra photoresist step is required for the MOS capacitor in order to have oxide of the proper thickness. The junction capacitor will generally exhibit more voltage dependence than the MOS capacitor.

Junction Breakdown Voltage

Breakdown in collector junctions is usually caused by avalanche multiplication, which occurs when the electric field in the silicon reaches a critical value

$$E_C \approx 3 \times 10^5 \text{ V/cm} = 30 \text{ V/}\mu\text{m}$$

As inspection of Figs. 4-13 to 4-15 shows, the maximum field always occurs at the metallurgical junction, and breakdown begins to take place there when the maximum field E_{max} reaches the critical value E_C.

Lawrence and Warner have also computed a nomograph for breakdown calculations; it is shown in Fig. 4-19, and applies for gaussian or complementary-error-function profiles diffused into a constant background.

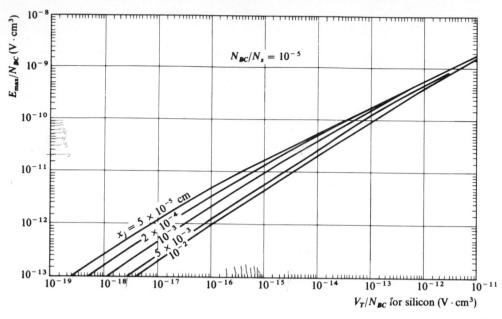

FIGURE 4-19
Lawrence-Warner chart for maximum field.

EXAMPLE 4E *Breakdown Voltage of a Junction Capacitor.* We now calculate the breakdown voltage of the 10-pF capacitor designed in the previous example. First, we note that $N_{BC}/N_s = 2.00 \times 10^{-3}$, but the chart is for $N_{BC}/N_s = 10^{-5}$. Fortunately, the maximum field is only a weak function of N_{BC}/N_s, and so the chart will be used.

The problem is to find what applied voltage will produce $E_{max} = E_C$. Therefore we enter the chart on the ordinate at $E_{max}/N_{BC} = 3 \times 10^5/10^{16} = 0.30 \times 10^{-10}$. Moving horizontally to the right until we intersect the $x_j = 3.0 \times 10^{-4}$ curve (interpolation is necessary), we then move vertically down to read

$$\frac{V_T}{N_{BC}} \approx 5 \times 10^{-15} \text{ V} \cdot \text{cm}^3$$

Thus we obtain

$$V_T \approx 50 \text{ V} = V + \phi_c'$$

For this case the built-in voltage is negligible in comparison with the total depletion-layer voltage, so the applied voltage at which the onset of breakdown occurs is 50 V.   ////

In this example, we have assumed that the maximum field is uniform everywhere along the metallurgical junction. This is true for a planar junction, but sidewall curvature will cause nonuniformities for diffused junctions because the

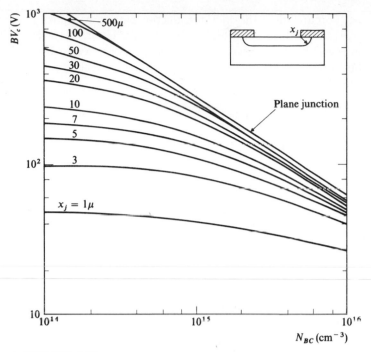

FIGURE 4-20
The effect of curvature on junction breakdown voltage (after Grove[7]).

field lines are concentrated in regions of curvature.   This causes the maximum field to be larger in curved regions than in planar regions.   The result is that breakdown in diffused juctions occurs at somewhat lower voltages than those calculated for planar junctions.   Grove has calculated this effect for one-sided junctions; the results are shown in Fig. 4-20.[7]

Other Junctions

We have thus far confined our discussion to the collector-junction capacitor; we now consider other junctions.   There are two basic factors which influence the choice of junctions for capacitor applications: capacitance per unit area and breakdown voltage.   From Fig. 4-14, we see qualitatively that as the impurity concentration on either side of the junction is increased, the depletion-layer width on that side is decreased.   This means that the capacitance is also increased; furthermore it can be shown that the maximum field also is increased.   Thus the breakdown voltage is less.   Therefore, in general, larger capacitance per unit area is obtained at the expense of lower breakdown voltage.

The emitter junction can be expected to exhibit slightly higher capacitance per unit area than the collector junction since the impurity concentrations on both sides of the former are higher than for the latter. However, the emitter junction does not have a constant background. As we saw in Chap. 2, the concentrations on both sides of the emitter junction are largest near the surface. This means that:

1 The capacitance per unit area is not uniform along the junction, but is largest at the surface.
2 The maximum field is also largest near the surface, and breakdown will occur there.

Finally, as we saw in Chap. 2, the sidewall is not a section of a circle, but has a shape which depends on the parameters of the emitter and base profiles. Thus the area will not be easy to calculate.

Typical breakdown voltages for emitter junctions are in the range 7 to 9 V. Therefore these junctions are used for capacitors only in circuits where the capacitor voltage is very low.

Because it is difficult to make calculations for any but collector and substrate junctions, Table 4-1 is given as a means of estimating capacitance for other junctions.[8] Since it contains a range of background concentrations, it is very useful for quickly estimating capacitances for almost all integrated circuits.

Computer Calculations for General Impurity Profiles

In some cases, notably structures having deep junctions, it is not possible to make approximations (4-9) and (4-10), and the Lawrence-Warner curves cannot be used even for collector junctions. In the general case it is necessary to use a computer solution. This is done as follows.

First, the net impurity profile $N(x)$ is calculated from processing data (times, temperatures, etc.) or obtained by profile measurements. From these calculations, the computer can locate the junction depths.

Next, the built-in voltage at each junction is calculated by an iterative method. It can be shown that

$$\phi_c' = \frac{kT}{q} \ln \frac{N(x_j - x_p)N(x_j + x_n)}{N_i^2} \tag{4-11}$$

where x_p and x_n are the depletion widths on p and n sides of the junction. It can also be shown that

$$V_T = \frac{1}{K_s \varepsilon_0} \int_{x_j - x_p}^{x_j + x_n} x\rho(x) \, dx \tag{4-12}$$

which is valid for both equilibrium and nonequilibrium.

Table 4-1 TYPICAL JUNCTION CAPACITANCE (ZERO BIAS) FOR MONOLITHIC INTEGRATED CIRCUITS*

Junction		Collector resistivity, 0.1 Ω·cm			Collector resistivity, 0.5 Ω·cm			Collector resistivity, 1.2 Ω·cm		
		Capacitance per unit area		Break-down voltage	Capacitance per unit area		Break-down voltage	Capacitance per unit area		Break-down voltage
		pF/mil^2	pF/mm^2		pF/mil^2	pF/mm^2		pF/mil^2	pF/mm^2	
Emitter-base	Sidewall	0.65	1000	7	0.65	1000	7	0.65	1000	7
	Bottom	0.38	600	7	0.29	450	7	0.23	350	7
Base-collector		0.23	350	25	0.13	200	50	0.10	150	70
Collector-substrate	Sidewall	0.16	250	35	0.10	150	70	0.065	100	100
	Bottom	0.065	100		0.065	100		0.065	100	

Emitter sheet resistance = 2–3 Ω/sq
Base sheet resistance = 200 Ω/sq
Substrate resistivity = 10 Ω/cm

Emitter-base junction depth = 2.3×10^{-4} cm
Base-collector junction depth = 2.7×10^{-4} cm
Collector-substrate junction depth = 12.5×10^{-4} cm

* Entries in this table are calculated values and conseqently involve a number of approximations. They are intended for use in estimating junction capacitance.

The following algorithm can now be easily implemented on the computer:

1 Choose a point x_L on the left side of the junction. Calculate

$$Q_L = \int_{x_L}^{x_J} \rho(x)\, dx = \int_{x_L}^{x_J} qN(x)\, dx$$

2 Choose a point x_R on the right side of the junction. Calculate

$$Q_R = \int_{x_J}^{x_R} qN(x)\, dx$$

3 If $Q_R \neq Q_L$, change x_R and repeat (2) until $Q_R = Q_L$.
4 We have now established a charge dipole between x_L and x_R. Now calculate

$$V_T = \frac{1}{K_s \varepsilon_0} \int_{x_L}^{x_R} xqN(x)\, dx$$

5 Calculate

$$\phi_C' = \frac{kT}{q} \ln \frac{N(x_L)N(x_R)}{N_i^2}$$

6 If $\phi_C' \neq V_T$, choose a new x_L and repeat the entire procedure.

This algorithm enables the calculation of a built-in voltage, and the equilibrium depletion width $x_R - x_L$. For nonequilibrium with an applied voltage V, steps (1) through (4) are repeated, but the remainder of the algorithm is

5 Calculate $V_{T_1} = V + \phi_C'$.
6 If $V_{T_1} \neq V_T$, choose a new x_L and repeat the entire procedure.

In this manner one can calculate the built-in voltage, the equilibrium depletion width, and the variation of depletion width with applied voltage for a junction having an arbitrary impurity profile. Once the depletion width has been obtained, the capacitance is simply

$$\frac{K_s \varepsilon_0 A}{x_R - x_L}$$

Moreover, as we shall later see, such a computer routine also forms the basis for calculating the parameters of the double-diffused junction-gate field-effect transistor.

Results of computer calculations for a variety of devices are shown in Table 4-2. For both deep structures, approximations (4-9) and (4-10) could not be used. It is interesting to note, by comparing x_p with x_n, that none of the net impurity profiles can be reasonably approximated by a linear profile in the vicinity of the junctions. It is also interesting to note that the built-in voltages vary from 0.592 to 0.793.

Junction Capacitor Parasitic Effects

The principal parasitic effects associated with a collector junction capacitor are the resistances of the p and n materials, and the capacitance of the substrate junction. The resistance resulting from the p diffusion can be minimized by metallizing over the entire surface area of the p diffusion. A first-order representation of the remaining parasitic effects is shown in Fig. 4-21. By inspection of Fig. 4-21a one notes that all the parasitic effects are distributed. In the first-order lumped model of Fig. 4-21b, R_l represents the lateral resistance of the epitaxial layer between the junction and the n^+ contact, R_v is the vertical resistance of the epitaxial layer between the collector junction and the substrate junction, C_j is the collector junction capacitance, C_s is the substrate junction capacitance including sidewalls, and R_s is the resistance of the substrate between the substrate junction and the substrate contact. R_l can be minimized by using a ring-type contact to the epitaxial layer, and R_v can be minimized by using as thin an epitaxial layer as possible.

It should be noted that the p diffusion, the n-epitaxial layer, and the substrate form a vertical p-n-p transistor. In order that this not become an active parasitic effect, both junctions must always be reverse-biased.

Relation of Depletion Regions to Junction Depth Measurements

We have seen from Example 4C and from Table 4-2B that the depletion-layer width can be an appreciable fraction of the junction depth. As was discussed in Chap. 2, the methods used to measure junction depth typically involve lapping or grooving a sample and then applying a stain which causes p regions to darken. Most stains are somewhat sensitive to electric field, and as a result the area which appears dark after stain is only the region of the p material which is approximately space-charge-neutral. That is, very little, if any, of the depletion region is stained. Therefore, care must be exercised in the use of results from such measurements.

It is usually safe to assume that the edge of the dark region after stain is not the metallurgical junction, but the edge of the depletion region in the p material. Consider, for example, the deep structure "super β" device of Table 4-2B. This device has a p-type base region with actual base width 1.4 μm between metallurgical junctions. However, the undepleted base width w between depletion regions in the p material is only 0.53 μm. It is this base width which would be observed by staining.

For the collector junction, the following procedure can be used to correct for the depletion layer when using stain measurements for junction depth:

1   Assume that the edge of the stained region is the metallurgical junction.
2   Using the Lawrence-Warner curves, calculate the amount of depletion into the p material.
3   Add this amount to the depth measured in (1). The result is approximately the junction depth.

For emitter junctions, corrections would have to be calculated iteratively by computer.

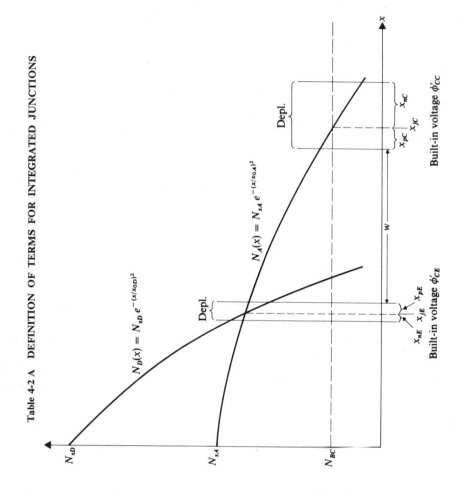

Table 4-2 A   DEFINITION OF TERMS FOR INTEGRATED JUNCTIONS

B COMPUTER CALCULATIONS FOR VARIOUS INTEGRATED DEVICES

Type of device	N_{BC}	N_{sA}	x_{OA}	N_{sD}	x_{OD}	x_{jE}	ϕ'_{CE}	x_{nE}	x_{pE}	x_{jC}	ϕ'_{CC}	x_{pC}	x_{nC}	w
Commercial BJT	1.1×10^{15}	5.00×10^{19}	1.08	7.6×10^{20}	0.883	2.53	0.774	0.10	0.16	3.52	0.620	0.32	0.73	0.51
Medium-depth JFET	3.40×10^{15}	8.00×10^{18}	0.852	4.00×10^{20}	0.601	1.68	0.793	0.09	0.14	2.37	0.668	0.24	0.44	0.31
Deep structure Super-β BJT	9.7×10^{14}	5.00×10^{18}	2.07	7.30×10^{19}	1.66	4.60	0.688	0.22	0.36	6.00	0.592	0.51	0.80	0.5
Deep structure BJT	9.7×10^{14}	5.00×10^{18}	2.04	8.00×10^{19}	1.52	3.80	0.759	0.14	0.18	5.97	0.597	0.48	0.80	1.51
"Textbook" BJT	5.0×10^{15}	4.0×10^{18}	1.16	1.0×10^{20}	0.803	2.00	0.792	0.10	0.134	3.00	0.671	0.262	0.396	0.609

Note: All N in cm^{-3}.
    All x in μm.
    All ϕ in volts.
    w in μm.

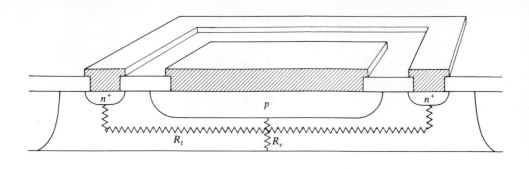

(a)

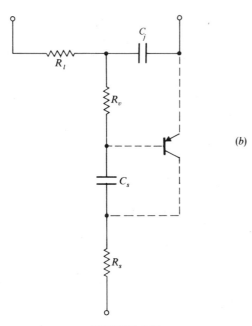

(b)

FIGURE 4-21
(a) Collector junction capacitor; (b) first-order model including parasitic effects.

4-7 DIFFUSED AND EPITAXIAL RESISTORS

Resistors which make use of the various diffused or epitaxial regions common to bipolar transistor fabrication are attractive because they require no extra processing steps. However, the permissible range of resistance values is somewhat limited by the requirements imposed on resistivities by the bipolar transistor. Three such resistors are shown in Fig. 4-22.

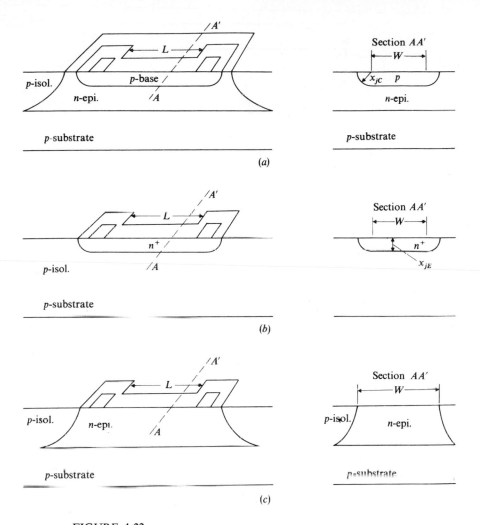

FIGURE 4-22
(a) Base resistor; (b) emitter resistor; (c) epitaxial (collector) resistor.   Sections
A-A′ are vertical sections normal to current flow.

Sheet Resistance

In Chap. 1, we used the concept of sheet resistance in determining the layout of
diffused resistors.   We now wish to relate the sheet resistance to the impurity
profile.   Consider first an idealized rectangular structure in which the current flow
is one-dimensional, as shown in Fig. 4-23.   The current density is

$$J = \sigma(x)E_y = \frac{-\sigma(x)\,d\phi}{dy}$$

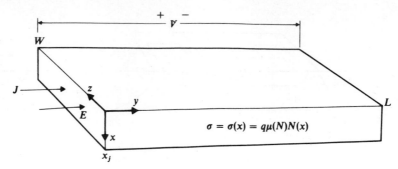

FIGURE 4-23
Idealized model for sheet-resistance calculations.

where E_y = electric field
$\sigma(x)$ = conductivity
ϕ = electrostatic potential
The total current I is given by

$$I = \int_0^W dz \int_0^{x_j} J\, dx = -W \frac{d\phi}{dy} \int_0^{x_j} \sigma(x)\, dx$$

Integrating both sides from $y = 0$ to $y = L$, we obtain

$$IL = W[\phi(0) - \phi(L)] \int_0^{x_j} \sigma(x)\, dx$$

But $\phi(0) - \phi(L)$ is the voltage V across the resistor with polarity as shown in Fig. 4-23. If we define the resistance of the resistor as

$$R \triangleq \frac{V}{I}$$

we have

$$R = \frac{L/W}{\displaystyle\int_0^{x_j} \sigma(x)\, dx}$$

The sheet resistance ρ_s' is now defined by

$$\rho_s' = \frac{1}{\displaystyle\int_0^{x_j} \sigma(x)\, dx}$$

The resistance is

$$R = \rho_s' \frac{L}{W}$$

Note also that the average conductivity $\bar{\sigma}$ of the diffused layer is

$$\bar{\sigma} = \frac{\displaystyle\int_0^{x_j} \sigma(x)\,dx}{\displaystyle\int_0^{x_j} dx} = \frac{1}{x_j}\int_0^{x_j}\sigma(x)\,dx$$

The sheet resistance can therefore be written in terms of the average conductivity as

$$\rho'_s = \frac{1}{x_j\bar{\sigma}}$$

We saw in Chap. 2 how sheet resistance can be conveniently measured and how Irvin's curves can be used to relate measured sheet resistance to assumed profile parameters. Analytically, we can relate $\sigma(x)$ to the impurity profile by

$$\sigma(x) \approx q\mu|N(x)|$$

where $N(x)$ = net impurity concentration
  μ = majority carrier mobility
Unfortunately, μ is a function of impurity concentration

$$\mu = \mu(N)$$

so that the evaluation of ρ'_s from knowledge of the profile is not simple. Figure 4-24 shows the variation of mobility with impurity concentration.[9] It can be seen that for the range of impurity concentration typically encountered in base diffusions, a logarithmic function can be used to approximate the mobility.

Sidewall Conductance

Because of lateral diffusion, the rectangular model Fig. 4-23 does not adequately characterize diffused resistors unless $W \gg x_j$. In principle the correction for lateral diffusion can be handled by dividing the resistor into three parts: the rectangular part corresponding to the model of Fig. 4-23, and the two sidewall regions resulting from lateral diffusion. Since these three regions are effectively connected in parallel, the total conductance of the resistor is

$$G = 2G_{\text{side}} + \frac{1}{\rho'_s(L/W)}$$

where G_{side} is the conductance of one sidewall region. It is convenient to normalize the resistance to the value which would be obtained in the absence of sidewalls; this yields

$$\frac{R}{\rho'_s(L/W)} = \frac{1}{1 + \rho'_s(L/W)2G_{\text{side}}} \qquad (4\text{-}13)$$

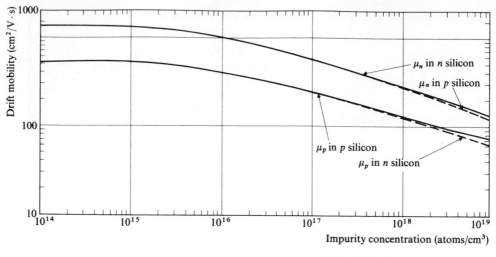

Drift mobility (cm²/V · s)

Impurity concentration (atoms/cm³)

FIGURE 4-24
Mobility in silicon (after Ghandi[9]).

For emitter resistors, G_{side} is difficult to calculate because the analytical expressions for sidewall shape and net impurity profile in the sidewall regions are not simple. However, for base resistors, reasonable approximations can be made to enable analytical treatment of the problem.

EXAMPLE 4F *Effects of Sidewalls on Base Resistors.* If we assume that the lateral diffusion has proceeded radially from the edge of the diffusion window, that the locus of the collector junction is a quarter circle, as shown in Fig. 4-25, and that depletion regions may be neglected, we can write the sidewall conductance as

$$G_{side} = \frac{1}{L} \int_0^{\pi/2} d\phi \int_0^{x_j} r\sigma(r) \, dr$$

We now assume that the p material of the base diffusion is sufficiently extrinsic that the conductivity is determined by majority carriers, that the majority carrier density is equal to the net acceptor impurity concentration, and that in the base diffusion region the net acceptor profile is approximately equal to the base diffusion profile $N_A(x)$. Then

$$\sigma(r) = q\mu[N_A(r)]N_A(r)$$

where $\mu[N_A(r)]$ is the hole mobility in p-type silicon. The sidewall conductance is

$$G_{side} = \frac{\pi}{2L} q \int_0^{x_j} \mu(N_A) r N_A(r) \, dr$$

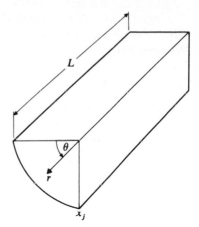

FIGURE 4-25
Model for the calculation of sidewall conductance of a base resistor.

From the plot of μ versus N_A shown in Fig. 4-24, it is observed that between $N_A = 3 \times 10^{15}$ cm^{-3} and $N_A = 3 \times 10^{18}$ cm^{-3}, the mobility is reasonably well approximated by an expression of the form

$$\ln \mu \sim \ln K + m \ln N_A$$

From the curve of Fig. 4-24, we obtain the values

$$N_A = 10^{16} \text{ cm}^{-3} \qquad \mu = 400 \text{ cm}^2/V \cdot s$$
$$N_A = 10^{18} \text{ cm}^{-3} \qquad \mu = 130 \text{ cm}^2/V \cdot s$$

Using these values in the expression for μ, we obtain

$$\ln K = 15.0$$
$$m = -0.246$$

Next we assume that the base diffusion profile results from a standard two-step diffusion cycle, and we neglect any outdiffusion at the surface. Then

$$N_A(r) = N_{sA} e^{-(r/x_{0A})^2}$$

With this expression for $N_A(r)$, the integral in G_{side} can be evaluated, and we obtain

$$G_{side} = \frac{\pi q K N_{sA}^{m+1}}{(m+1)L} \frac{x_{0A}^2}{4} [1 - e^{-(m+1)(x_j/x_{0A})^2}]$$

From approximation (4-9), we can calculate $(x_j/x_{0A})^2$:

$$\frac{x_j^2}{x_{0A}^2} = \ln \frac{N_{sA}}{N_{BC}}$$

Then we have

$$G_{side} = \frac{\pi q K N_{sA}^{m+1}}{(m+1)L} \frac{x_{0A}^2}{4} \left[1 - \left(\frac{N_{BC}}{N_{sA}}\right)^{m+1}\right]$$

For most practical integrated circuits, N_{sA} is at least two orders of magnitude larger than N_{BC}, so the second term within the brackets may be safely neglected. We then have

$$G_{side} \approx \frac{\pi q K N_{sA}^{m+1}}{4(m+1)L} x_{0A}^2 \qquad (4\text{-}14)$$

Now consider a typical base diffusion with $N_{BC} = 10^{15}$, $\rho_s' = 200\ \Omega$ per square, and junction depth 3 μm. In Example 2E we used Irvin's curves to determine the effective surface concentration, with the result that

$$N_{sA} = 3 \times 10^{18}\ \text{cm}^{-3}$$

Inserting these values in (4-14), we obtain

$$G_{side} = \frac{0.726}{L} \times 10^{-6}$$

When this is inserted in (4-13), the result is

$$\frac{R}{\rho_s'(L/W)} = \frac{1}{1 + (290 \times 10^{-6}/W)} \qquad (4\text{-}15)$$

From (4-15) we can determine what error is encountered through neglecting the effects of sidewall conductance. Table 4-3 shows the values of $R/\rho_s'(L/W)$ for several resistor widths. Note that even for 1-mil resistors, the error is approximately 10 percent.

For emitter resistors, computer calculations would be necessary to determine the effects of sidewall conductance.   ////

Collector Resistors

We have seen that typical base diffusions produce sheet resistance of approximately 200 Ω per square. Such a low sheet resistance is disadvantageous when fabricating large-value resistors because large surface area is required. A larger sheet resistance can be obtained by using the n-epitaxial collector material in the

Table 4-3 **EFFECTS OF SIDEWALL CONDUCTANCE FOR BASE RESISTORS***

W, mils	$R/(\rho_s'L/W)$,
1.0	0.898
0.50	0.814
0.20	0.637

* $N_{BC} = 10^{15}/\text{cm}^3$
$N_{SA} = 3 \times 10^{18}$
$\rho_s' = 200\ \Omega/\text{sq}$
$x_J = 3.0\ \mu$m

Note: If sidewall conductance is neglected, $R/(\rho_s'L/W) = 1$.

fabrication of resistors.   In this case, one selects the appropriate surface geometry and surrounds it with an isolation diffusion, as shown in Fig. 4-22c.   Note that the cross section of collector resistors differs significantly from that of base and emitter resistors, and that the sidewall effects cannot be ignored.

To estimate the advantage obtained with collector resistors, note that because the epitaxial material is uniformly doped, the conductivity is constant, and $\sigma = \bar{\sigma}$.   The sheet resistance is

$$\rho'_s = \frac{1}{\bar{\sigma} x_{js}}$$

where x_{js} is the depth of the substrate junction.   Consider the epitaxial layer with $x_{js} = 8$ μm, and $N_{BC} = 5 \times 10^{15}$ cm^{-3}.   This layer has a sheet resistance

$$\rho'_s \approx 1250 \ \Omega/\text{sq.}$$

Note, however, that for a given line width W, the sidewalls of the collector resistor make the effective resistor width larger than W.

EXAMPLE 4G   *Design of a 200 kΩ Collector Resistor.*   For this example, we assume that the end pads may be neglected, and we assume that the isolation diffusion produces isolation junctions whose sidewalls are approximately quarter circles, as shown in Fig. 4-26.   We also neglect depletion regions.   Since the conductivity is constant, and since the geometry of the cross section is not rectangular, it is more convenient to deal with conductance rather than resistance, and with conductivity rather than sheet resistance.   The total conductance of the resistor is

$$G = \frac{\sigma W x_{js}}{L} + 2\frac{\sigma}{L}(x_{js}^2 - \tfrac{1}{4}\pi x_{js}^2) = \frac{\sigma x_{js}}{L}(W + 0.43 x_{js}) \qquad (4\text{-}16)$$

Note that the effective number of squares in the resistor is not L/W but $L/(W + 0.43 x_{js})$, as a result of the sidewall regions.

Now for a 200-kΩ resistor in a typical epitaxial layer with $x_{js} = 8.0$ μm, $N_{BC} = 5 \times 10^{15}$, and $W = 1$ mil, (4-16) yields

$$L = 182 \text{ mils}$$

For our typical base diffusion of 200 Ω per square, we would require (neglecting sidewalls) $L = 200 \times 10^3/200 = 1000$ mils.   A saving of about five times in surface area has been achieved with the collector resistor.   However, to maintain accuracy for collector resistors, the isolation diffusion must be carefully controlled.   Furthermore, depletion resulting from substrate junction bias may significantly affect the resistance of the collector resistor.   ////

Emitter Resistors

Emitter resistors are designed in the same manner as base resistors.   However, because the sheet resistance of the emitter diffusion is typically of the order of 5 Ω per square, emitter resistors are generally used only where small-value resistors are required.

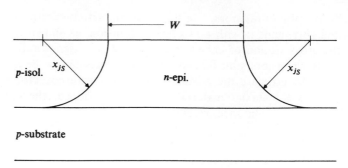

FIGURE 4-26
Cross section of the collector resistor.

Parasitic Capacitances

For base resistors, it is again noted that the base is diffused into the uniformly doped epitaxial material, and the sidewalls are approximately quarter circles. The capacitance is uniform along the junction, and can be obtained either from Table 4-1 or from the Lawrence-Warner curves. Two parasitic capacitance terms arise, one resulting from the body of the resistor, the other from the pad regions which must be designed to accommodate the contact windows. Referring to Fig. 4-22a, we note that the capacitance C_{base} of the base resistor is obtained in terms of the capacitance per unit area C_{bc} as follows:

$$C_{\text{base}} = C_{bc}LW + C_{bc}\left(2\frac{2\pi}{4}x_{jC}L\right) + 2C_{\text{pad}} \qquad (4\text{-}17)$$

where x_{jC} is the collector junction depth and C_{pad} is the capacitance of each contact pad region. The first term of (4-17) is the contribution of the bottom of the resistor, while the second term is the contribution of the sidewalls. If it is assumed that the contact window is a square W on a side, and that the registration clearance is W, then the contact pad is a square $3W$ on a side. If the fillets at the corners of the pad resulting from lateral diffusion are neglected, the pad capacitance is

$$C_{\text{pad}} \approx C_{bc}3W \times 3W + \frac{2\pi}{4}x_{jC} \times 11W \times C_{bc} \qquad (4\text{-}18)$$

Here the first term is the contribution of the bottom of the pad and the second term is the contribution of the sidewalls. Inserting (4-18) in (4-17), we obtain

$$C_{\text{base}} = C_{bc}WL\left(1 + \frac{\pi x_{jC}}{W}\right) + 2C_{bc}W^2\left(9 + \frac{5.5\pi x_{jC}}{W}\right) \qquad (4\text{-}19)$$

For emitter resistors, we assume for simplicity that the sidewalls are approximately vertical. If C_{be} is the capacitance per unit area of the bottom of the emitter and C_{bs} is the sidewall capacitance per unit area, we obtain for C_{emitter}, the capacitance of the emitter resistor,

$$C_{\text{emitter}} = C_{be}LW + 2Lx_{jE}C_{bs} + 2C_{\text{pad}}$$

where x_{jE} is the emitter junction depth. Again assuming a square pad $3W$ on a side, we have

$$C_{pad} = 9W^2 C_{be} + 11Wx_{jE} C_{bs}$$

Thus we obtain

$$C_{emitter} = C_{be} LW\left(1 + 18\frac{W}{L}\right) + 2C_{bs} Lx_{jE}\left(1 + 11\frac{W}{L}\right) \qquad (4\text{-}20)$$

For collector resistors, we refer to the cross section of Fig. 4-26. Let C_s be the capacitance per unit area of the bottom of the substrate junction and C_{ss} the capacitance per unit area of its sidewalls. Note that C_{ss} can be obtained from the Lawrence-Warner curves, since the isolation region is diffused into the epitaxial material. The total capacitance $C_{collector}$ is given by

$$C_{collector} = C_s L(W + 2x_{js}) + 2C_{ss}\frac{\pi}{2} x_{js} l + 2C_{pad}$$

where x_{js} is the substrate junction depth. Again assuming a square pad $3W$ on a side, we find

$$C_{pad} = C_s(3W + 2x_{js})^2 + 11W\frac{\pi}{2} x_{js} C_{ss}$$

Thus we obtain

$$C_{collector} = C_s[L(W + 2x_{js}) + 2(3W + 2x_{js})] + \pi C_{ss} x_{js}(L + 11W) \qquad (4\text{-}21)$$

Effects of Resistor Bias

Two types of bias affect the performance of resistors: the bias voltage between the resistor and the surrounding material, and the bias voltage applied across the resistor terminals. For purposes of illustration, we confine our discussion to base resistors.

Usually the n-epitaxial region surrounding resistors will be connected to the most positive power supply voltage in order to ensure that the p-n junction surrounding the resistor is never forward-biased. It can easily occur that the voltages at the terminals of the resistor are considerably lower than this power supply voltage. This produces a reverse bias of the junction and a corresponding depletion-layer widening.

Consider the case of a resistor in which the voltage between the resistor terminals is negligible, but for which the junction is reverse-biased. Two effects will be observed, both resulting from the increased width of the depletion region: the total parasitic capacitance will decrease, and the total resistance will increase slightly. Qualitatively, we expect the change of resistance to be small, because most of the impurity atoms contributing carriers for current flow are near the surface, and the depletion of a few more near the junction will not appreciably affect the average conductivity. But the change of capacitance should be more significant, since the capacitance depends directly on the depletion width. Calculation of the change of capacitance and resistance provides an instructive exercise in the use of both the Lawrence-Warner curves and Irvin's curves.

EXAMPLE 4H *Change of R and C of a Base Resistor with Junction Bias.* A resistor is made by using a base diffusion in epitaxial material having $N_{BC} = 10^{16}$ cm^{-3}. The base diffusion is calculated to have a sheet resistance of 200 Ω per square with metallurgical junction depth $x_{jc} = 3.0$ μm. The calculation assumed no depletion regions. The resistor is to be 1 mil wide and have a resistance of 2 kΩ. The epitaxial material is 10 V positive relative to the base resistor.

For a sheet resistance of 200 Ω per square, if depletion regions are neglected the average conductivity of the 3.0 μm layer is 16.7 $(\Omega \cdot \text{cm})^{-1}$. This yields $N_{BC}/N_{sA} = 10^{-2}$. From this we find that for a built-in voltage of 0.7, the zero-bias depletion width and capacitance are actually

$$C_{bc} = 2.0 \times 10^4 \text{ pF/cm}^2 = 0.129 \text{ pF/mil}^2$$
$$x_m = 0.55 \text{ } \mu\text{m}$$
$$x_1 = 0.225 \text{ } \mu\text{m}$$

If we use the calculated value of 200 Ω per square for ρ'_s, then the resistor body must contain 8.7 squares, since the end pads contribute 1.3 squares. Computing the total capacitance of the resistor from (4-19), we obtain

$$C_{\text{base}} = 4.39 \text{ pF}$$

at zero bias.

When the bias is increased to 10 V, the new values become

$$C'_{bc} = 7 \times 10^3 \text{ pF/cm}^2 = 0.05 \text{ pF/mil}^2$$
$$x'_m = 1.60 \text{ } \mu\text{m}$$
$$x'_1 = 0.575 \text{ } \mu\text{m}$$

Thus the total capacitance is

$$C'_{\text{base}} = 1.8 \text{ pF}$$

which represents a change of the capacitance by a factor of 3.

To compute the change of resistance, we need to use Irvin's curves. It will be recalled that these curves allow the calculation of the average conductivity of a general layer between x and x_{jc}. The conductance of the resistor can be calculated as follows:

Let G_0 be the conductance for no depletion region

$$G_0 = \frac{\bar{\sigma} x_{jc}}{\Gamma}$$

where Γ is the total number of squares. The conductance subtracted by the depletion region is

$$G_d = \frac{\bar{\sigma}_d x_1}{\Gamma}$$

where x_1 is the depletion width obtained from the Lawrence-Warner curves and $\bar{\sigma}_d$ is the average conductivity of this layer. The conductance of the resistor is then

$$G = G_0 - G_d = \left(1 - \frac{\bar{\sigma}_d x_1}{\bar{\sigma} x_{jc}}\right) \frac{\bar{\sigma} x_{jc}}{\Gamma}$$

The fractional change of conductance due to depletion is

$$\frac{\Delta G}{G} = \frac{\bar{\sigma}_d x_1}{\bar{\sigma} x_{jc}}$$

From Irvin's curves we find

$$\bar{\sigma}_d \approx 0.5 \; (\Omega \cdot cm)^{-1} \qquad \text{for 0 V}$$
$$\bar{\sigma}_d \approx 1.0 \; (\Omega \cdot cm)^{-1} \qquad \text{for 10 V}$$

Thus we obtain

$$\frac{\Delta G}{G} = 0.22 \times 10^{-2} = 0.22 \text{ percent} \qquad \text{for zero bias}$$

$$\frac{\Delta G}{G} = 1.15 \times 10^{-2} = 1.15 \text{ percent} \qquad \text{for 10-V bias}$$

We see that our original qualitative reasoning was correct: the resistance changes by only about 1 percent, while the capacitance changes by a factor of 3.   ////

Let us now consider briefly the case for which a bias voltage is applied between the resistor terminals, but zero bias is applied between one resistor terminal and n material. The applied bias produces a voltage gradient along the resistor, which in turn produces a depletion region of nonuniform width along the resistor. If the applied voltage is sufficiently large, the resistor value will change appreciably, since it now becomes a nonlinear resistor. Moreover, if the applied voltage is large enough, breakdown of the junction will occur.

Actual calculation of the variation of resistance with terminal voltage is difficult. For a quick estimate, one can obtain an upper bound on the variation by assuming the terminal voltage to be applied everywhere as reverse bias across the junction. The method of the preceding example can be used to find $\Delta G/G$, which for the actual case overestimates the change of conductance.

4-8 THE PINCH RESISTOR

We have seen that base resistors typically have sheet resistance of the order of several hundred ohms per square, and that an improvement of about five times is possible by using collector resistors. In some cases, this improvement is not

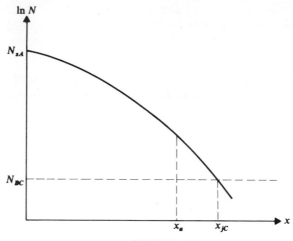

FIGURE 4-27
Impurity profile for base resistor.

sufficient; this is particularly true when resistances of the order of a megohm must be fabricated in reasonably small area.   What is needed is some way to

1   Decrease the impurity concentration so fewer carriers are available
2   Make thinner regions so the conductance will be lower

If we refer to the impurity profile for the base diffusion, sketched in Fig. 4-27, we see that the concentration of impurities is very large near the surface, and much smaller near the junction.   If we could remove the impurities from the surface down to x_a, we could accomplish both (1) and (2) above.   However, it is not necessary that the impurities be removed from the material, only that their associated mobile carriers be prevented from participating in the current flow.   This can be done by forming a junction at x_a so that all current in the resistor is confined between x_a and x_{jC}.   But no additional processing steps are required to do this since the emitter diffusion can be employed to form the junction.   The material between $x = 0$ and $x = x_a$ will then be *n*-type, and between x_a and x_{jC} it will be *p*-type.   Such a resistor is called a *pinch resistor*; its surface geometry is shown in Fig. 4-28.   Note that the n^+ diffusion must everywhere overlap the body of the resistor in order that the resistor current be confined to the narrow channel between x_a and x_{jC}^-.

Effective Sheet Resistance

If we are willing to neglect depletion regions, we can get a quick estimate of the effective sheet resistance of a pinch resistor directly from Irvin's curves.   The depletion-region effects can be estimated by using a combination of Irvin's curves and the Lawrence-Warner curves.

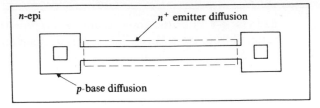

FIGURE 4-28
Surface geometry of a pinch resistor.

It is instructive to estimate the improvement of sheet resistance for the case for which depletion regions are neglected. Let the average conductivity of the base diffusion be $\bar{\sigma}$ and the junction depth be x_{jC}. Let the emitter junction depth be x_{jE}. The sheet resistance ρ'_{sc} of the channel between $x_a = x_{jE}$ and x_{jC} is

$$\rho'_{sc} = \frac{1}{\bar{\sigma}_c(x_{jC} - x_{jE})}$$

where $\bar{\sigma}_c$ is the average conductivity of the channel. Since the sheet resistance ρ'_s of the base diffusion is

$$\rho'_s = \frac{1}{\bar{\sigma}(x_{jC})}$$

the improvement is

$$\frac{\rho'_{sc}}{\rho'_s} = \frac{\bar{\sigma}/\bar{\sigma}_c}{(1 - x_{jE}/x_{jC})}$$

EXAMPLE 4I *Improvement in Sheet Resistance for a Typical Base Diffusion.*
Consider a process yielding

$$\text{Base diffusion:} \quad N_{sA} = 5 \times 10^{18}/\text{cm}^3$$
$$x_{jC} = 3.0 \ \mu\text{m}$$
$$x_{jE} = 2.0 \ \mu\text{m}$$
$$N_{BC} = 10^{15}/\text{cm}^3$$

From Irvin's curves we obtain $\bar{\sigma} = 20 \ (\Omega \cdot \text{cm})^{-1}$ and $\bar{\sigma}_c = 1.3 \ (\Omega \cdot \text{cm})^{-1}$. Then we calculate

$$\frac{\rho'_{sc}}{\rho'_s} = 46$$

Since this neglects depletion regions, the actual improvement will be somewhat larger because the depletion regions cause further narrowing of the channel.

////

Exercise 4-4 For the above example, suppose that the depletion into the p material from the collector junction is 0.23 μm and from the emitter junction is 0.12 μm. Find the improvement in sheet resistance compared with the base resistor.

Parasitic Capacitance

The pinch resistor has two parasitic junction capacitances: one from the collector junction and one from the emitter junction. These two are effectively in parallel. Because the channel of the pinch resistor is so thin compared with its width, one can safely neglect the sidewall capacitance of the channel. However, the pad regions will still be basically the same as those of the base resistor, so their sidewalls must be considered. The capacitance is

$$C_{pinch} = (C_{bc} + C_{be})WL + 2C_{pad}$$

and the pad capacitance is

$$C_{pad} = C_{bc}\left[9W^2 + 11\frac{\pi}{2}Wx_{jc}\right]$$

Thus we obtain

$$C_{pinch} = C_{be}WL + C_{bc}\left[WL + 18W^2 + 11\pi Wx_{jc}\right] \qquad (4\text{-}22)$$

Effects of Bias

The effects of bias are much more serious for pinch resistors than for base resistors. This is because the channel is very thin even at zero bias, and small changes of depletion-region width can have a large effect on channel thickness. It is therefore important to avoid bias voltages of more than a few volts, either between the terminals or between p and n materials.

The breakdown voltage of pinch resistors is that of the emitter junction, typically 7 to 9 V. This means that bias voltages either across the resistor or between the resistor and the n material must be kept below the emitter breakdown voltage. However, for most pinch resistors the variation of channel thickness and its attendant variation of the resistance of the resistor will impose limitations even before the breakdown voltage is reached.

The Voltage-Variable Resistor

The variation of channel thickness with reverse junction bias can in some cases be used to advantage. Consider an AGC application in which it is desired to produce a resistor whose resistance can be varied by a control voltage. If the circuit is properly designed, it can be arranged so that the resistor has no bias voltage between its terminals, and also so that the applied signal is sufficiently small that it produces negligible change of the depletion region.

A pinch resistor is ideally suited to this application. The signal is applied to the terminals of the p material, while the control voltage is applied as a reverse bias between n material and one terminal of the p material. Control voltage variations thus change the value of the resistor, but signal variations do not. It should be noted that the parasitic capacitance of the resistor also varies with the control voltage. Furthermore the resistor is actually a distributed RC structure in which R and C are varied. Such a structure can be combined with a gain device to produce a voltage-controlled oscillator.

4-9 COMPARISON OF MONOLITHIC RESISTORS

It is instructive to compare the surface area required, and the parasitic capacitance encountered, for the various types of monolithic resistors. This is best done by a numerical example. Values chosen for the diffusions conform approximately to those of Table 4-1.

EXAMPLE 4J *Comparison of Resistor Types for a 200-kΩ Resistor*

ASSUMED PROCESSING DATA

$$N_{BC} = 5 \times 10^{15} \text{ cm}^{-3}$$

$$x_{js} = 8.0 \ \mu\text{m}$$

Base diffusion: $\quad N_s = 5 \times 10^{18} \text{ cm}^{-3}$

$$x_{jC} = 3.0 \ \mu\text{m}$$

$$\rho'_s = 200 \ \Omega/\text{sq}$$

Emitter diffusion: $\quad x_{jE} = 2.0 \ \mu\text{m}$

$$\rho'_s = 3.0 \ \Omega/\text{sq}$$

ESTIMATED CAPACITANCES

$$C_{be} = 0.23 \text{ pF/mil}^2$$

$$C_{se} = 0.65 \text{ pF/mil}^2$$

$$C_{bc} = 0.12 \text{ pF/mil}^2$$

$$C_{bs} = 0.065 \text{ pF/mil}^2$$

$$C_{ss} = 0.065 \text{ pF/mil}^2$$

RESISTOR PARAMETERS FOR $R = 200$ kΩ

Type	W, mils	L, mils	C_{body}, pF	C_{pad}, pF	C_{total}, pF
Base resistor	1.0	998.6	164.29	1.33	166.9
Collector resistor	1.0	180.35	30.71	1.21	33.13
Pinch resistor	1.0	21.71	7.60	1.33	10.26

Although the pinch resistor has two parasitic junctions, its length is so much less than for the other two types that its total capacitance is the lowest of the three.   ////

4-10 MODELS FOR DIFFUSED RESISTORS

An exact characterization of diffused resistors is unwieldy because the resistors are multiple-layer distributed devices. For most practical purposes, two types of first-order models are generally sufficient. The first is a model to be used primarily to determine how the bias voltages influence the choice of isolation regions. For this purpose we are interested primarily in the junctions. Models incorporating only the salient aspects of the junctions are shown in Fig. 4-29; in these models lumped diodes are used to represent the distributed junctions. For the pinch resistor only two diodes, representing one *p-n* junction, are shown connected to the resistor, although there are actually two *p-n* junctions distributed about the resistor. As was noted in the preceding section, the n^+ diffusion must overlap the p diffusion; therefore the n^+ emitter and the n-epitaxial layer form, at least for electrical purposes, a single n-type layer. Emitter and collector junctions are thus electrically in parallel and provide the same effect as a single junction.

The second type of model required is the first-order small-signal model. As we have seen for thin-film resistors, the equations for the distributed RC structure can be written, but the solution will involve transcendental functions. For tractability, we therefore approximate the distributed structures by the single-pi-section models of Fig. 4-30 in which the steady-state current and the steady-state stored charge match those of the distributed network. Such a quasi-static model is generally adequate for most purposes.

Two models are shown for the base resistor. For relatively small-value resistors (several thousand ohms or less) the resistance R_{epi} of the epitaxial material may be comparable to that of the base resistor; it is included in Fig. 4-30a. The value of R_{epi} depends on the thickness of the epitaxial layer, and the geometry of the base resistor embedded in it. R_{epi} can be calculated by using methods similar to those employed for collector resistors. For medium- and large-value base resistors, folded geometry will probably be used, and for this case R_{epi} will be small compared with R; R_{epi} can then be ignored, as shown in Fig. 4-30b.

For the pinch resistor, the overlapping n^+ diffusion together with the relatively large resistance and small size of the p channel make R_{epi} again negligible.

In Fig. 4-30a, b, and e the capacitance C_{subs} is the total capacitance of the junction between substrate and epitaxial layer. It can be computed from knowledge of C_{bs} and C_{ss} and the geometry of the isolation region. In all cases the substrate resistance has been ignored; this is usually a reasonable approximation because the area of the substrate is so large.

It should also be noted that for base and pinch resistors, the contact to the n region will almost always be connected to a power supply voltage. Since this is also the case for the substrate contact for all resistors, substrate capacitance will have little or no effect for base and pinch resistors.

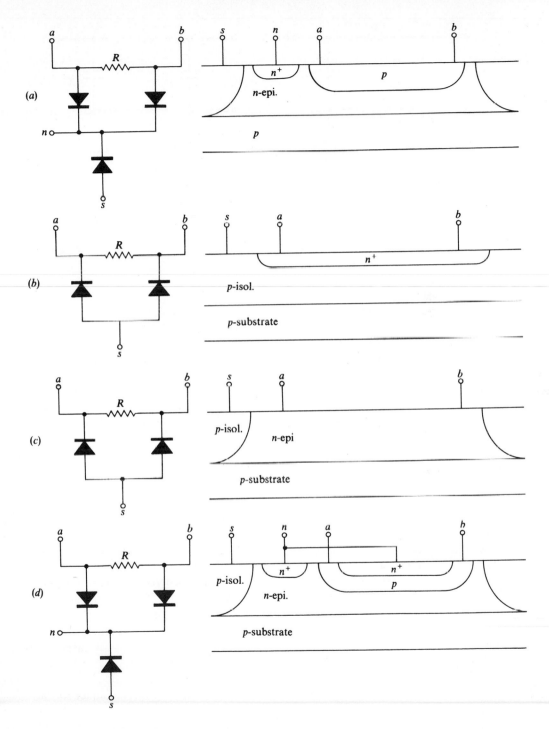

FIGURE 4-29
First-order resistor models for biasing purposes.   (*a*) Base resistor; (*b*) emitter
resistor; (*c*) collector resistor; (*d*) pinch resistor.

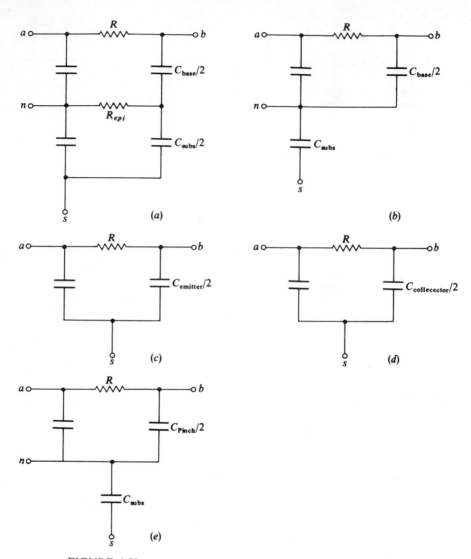

FIGURE 4-30
First-order resistor models for small-signal behavior. (*a*) Small-value base resistor; (*b*) medium- or large-value base resistor; (*c*) emitter resistor; (*d*) collector resistor; (*e*) pinch resistor.

PROBLEMS

4-1 An integrated transistor is fabricated with the following results: $N_{BC} = 10^{16}$ cm^{-3}, substrate resistivity 10 $\Omega \cdot$ cm, isolation diffusion surface concentration 5×10^{17} cm^{-3}, isolation sidewall junction radius 10 μm, base diffusion $N_{sA} = 5 \times 10^{18}$ cm^{-3}, emitter diffusion $N_{sD} = 10^{20}$ cm^{-3}, $x_{JC} = 2$ μm, $x_{JE} = 1$ μm. Use the Lawrence-Warner curves to estimate the capacitance of the bottom of the emitter junction and to calculate the collector isolation sidewall capacitance. Use a step-junction approximation to calculate the substrate bottom capacitance. Find the breakdown voltage of the isolation junction.

4-2 Use the conditions of the depletion approximation to derive Eq. (4-12).

4-3 For the epitaxial resistor of Example 4G, calculate the total parasitic capacitance if the substrate resistivity is 1 $\Omega \cdot$ cm and the surface concentration of the isolation diffusion is 10^{17} cm^{-3}.

4-4 For the resistor of Prob. 4-3, calculate the resistance of the resistor when depletion regions are included.

4-5 For the resistor of Prob. 4-4, calculate the resistance and total capacitance if 10-V reverse bias is applied between substrate and epitaxial layer.

4-6 Make a plot of the total capacitance of emitter and base resistors as a function of resistance. Assume an epitaxial resistivity of 0.5 $\Omega \cdot$ cm and use Table 4-1 for capacitance values. Include the effects of end pads, and assume $W = 0.5$ mil.

4-7 Design a 50-kΩ collector resistor and compare its capacitance with a 50-kΩ base resistor. Assume 1.2 $\Omega \cdot$ cm epitaxial material 12.5 μm thick, and use Table 4-1 for capacitance values.

4-8 An integrated-circuit process has $N_{BC} = 5 \times 10^{15}$ cm^{-3}; the epitaxial-layer thickness is 8 μm. It is desired to make n-type pinch resistors by adding an extra p-type diffusion step. This extra diffusion produces a junction at $x_J = 6$ μm, and has surface concentration 10^{18} cm^{-3}. The substrate resistivity is 1 $\Omega \cdot$ cm. Make a step-junction approximation for the substrate junction, use the Lawrence-Warner curves to calculate the depletion at the top junction, and calculate the sheet resistance of this pinch resistor.

4-9 In the process of Prob. 4-8, base and emitter diffusions are added which produce $x_{JC} = 3.0$ μm and $x_{JE} = 2.0$ μm, with $N_{sA} = 5 \times 10^{18}$. Depletion regions into the p material are 0.12 μm and 0.23 μm. Compare the total capacitance of a 200-kΩ pinch resistor with the total capacitance of a 200-kΩ resistor made as in Prob. 4-8. Assume the emitter capacitance to be 0.25 pF/mil^2.

4-10 A boron diffusion is performed into an epitaxial layer having $N_{BC} = 10^{16}$ cm^{-3}. The actual metallurgical junction depth is 1.0 μm and the surface concentration is $N_s = 10^{19}$ cm^{-3}.

A resistor is made from this diffusion. If a reverse bias is applied to the junction and the reverse voltage is equal to the breakdown voltage, what is the sheet resistance of the resistor and what is the capacitance per unit area?

4-11 An n-type collector resistor is made by leaving an oxide mask *over* the epitaxial layer and opening a window *around* it for the isolation diffusion. Let the epitaxial-layer thickness for 5 μm and the concentration be $N_{BC} = 10^{15}$ cm^{-3}. The oxide mask, excluding end pads, is 20 μm wide and 1000 μm long.

The resistor is to be represented by a pi-section RC network. Find R and C. Do not use Table 4-1; assume that the bottom capacitance per unit area is the same as that of the sidewalls. Neglect end pads.

REFERENCES

1 GROVE, A. S.: "Physics and Technology of Semiconductor Devices," chap. 9, John Wiley & Sons, Inc., New York, 1967.

2 OLIVEI, A.: Optimized Miniature Thin-film Planar Inductors, Compatible with Integrated Circuits, *IEEE Trans. Parts, Materials Packaging*, vol. PMP-5, pp. 71–88, 1969.

3 GHAUSI, M. S., and J. J. KELLY: "Introduction to Distributed Parameter Networks," pp. 18–23, Holt, Rinehart and Winston, Inc., New York, 1968.

4 HAMILTON, D. J., F. A. LINDHOLM, and A. H. MARSHAK: "Principles and Applications of Semiconductor Device Modeling," chap. 3, Holt, Rinehart and Winston, Inc., New York, 1971.

5 HOFF, P., and T. E. EVERHART: Carrier Profiles and Collection Efficiency in Gaussian *p-n* Junctions under Electron Beam Bombardment, *IEEE Trans. Electron Devices*, vol. ED-17, pp. 458–465, 1970.

6 LAWRENCE, H., and R. M. WARNER, JR.: Diffused Junction Depletion Layer Capacitance Calculations, *Bell System Tech. J.*, vol. 34, pp. 105–128, 1955.

7 GROVE: op. cit., p. 197.

8 MEYER, C. S., D. K. LYNN, and D. J. HAMILTON: "Analysis and Design of Integrated Circuits," p. 19, McGraw-Hill Book Company, New York, 1968.

9 GHANDI, S. K.: "The Theory and Practice of Microelectronics," p. 236, John Wiley & Sons, Inc., New York, 1968.

INTEGRATED JUNCTION-GATE FIELD-EFFECT TRANSISTORS

5-1 INTRODUCTION

In some circumstances in integrated circuits, notably micropower applications, devices are required which have both gain and large input impedances. The junction-gate field-effect transistor (JFET) is such a device, as is also the insulated-gate field-effect transistor (IGFET or MOSFET). Both devices approximate the behavior of a voltage-controlled current source.

The JFET closely resembles the pinch resistor in cross section. For simplicity we consider first the hypothetical JFET structure of Fig. 5-1,[1] for which we assume uniform doping in all regions, and $N_D \gg N_A$. To see qualitatively how the device operates we imagine the gate connected to the source, and a voltage V_{DS} applied between drain and source, as shown in Fig. 5-2. As V_{DS} is made negative, a current flows from source to drain, causing the voltage along the channel to be more negative than the source. Thus the gate junctions are reverse-biased, and negligible current flows in the gates. Moreover, the reverse junction voltage varies along the channel, being largest at the drain; therefore the depletion width into the channel is largest there. As V_{DS} is made more negative, the drain current increases and so does the depletion width all along the channel. When $V_{DS} = -V_p$, the pinchoff

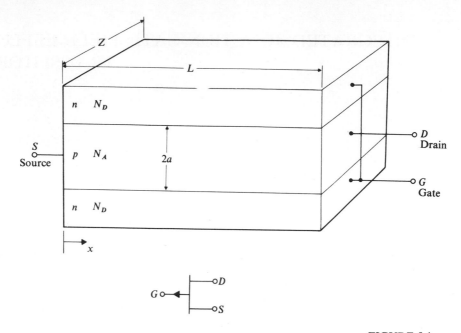

FIGURE 5-1
p-channel JFET.

voltage, all of the channel at the drain end has been depleted. For more negative voltages, no further increase of drain current is observed until breakdown occurs between drain and gate.

If a positive voltage V_{GS} is now applied between gate and source, the result is that even for $V_{DS} = 0$ there will be some depletion caused by V_{GS}, and therefore less drain voltage is required to cause pinchoff of the channel at the drain end. Breakdown will also occur at a smaller value of V_{DS}.

The *VI* characteristics for a *p*-channel JFET are shown in Fig. 5-3*a*. Clearly, a JFET could also be constructed with an *n* channel and *p* gates; all of the above remarks apply if all polarities are reversed. Characteristics for the *n*-channel device are shown in Fig. 5-3*b*. That the JFET can exhibit gain can also be seen from Fig. 5-3. If a load resistor is added so that operation is at point Q, a change of gate voltage shifts operation to Q_1, producing therefore a change of both drain voltage and drain current. Since the gate junctions are always reverse-biased, the only gate current which flows, other than the reverse leakage current, is that required to charge the depletion-layer capacitance.

The important low-frequency parameters of the JFET can be deduced from inspection of Fig. 5-3. For biasing purposes, I_{DSS} will be important. For distortion-free operation, Q must lie between V_p and V_B, and the difference $V_B - V_p$ determines the dynamic range. Finally, the transconductance $g_{fs} \triangleq -(\partial I_D / \partial V_{GS})$ will be important in determining the gain.

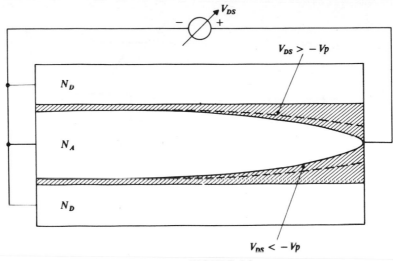

FIGURE 5-2
Channel shape resulting from applied bias.

For $V_{DS} < V_p$, the JFET behaves like a nonlinear resistor.   The small-signal conductance between source and drain in this region is

$$g_{ds} \triangleq \frac{\partial I_D}{\partial V_{DS}}$$

It can be shown that for a JFET with uniformly doped channel, the depletion region voltage at pinchoff is

$$V_{T_p} - \frac{a^2 q N_A}{2\varepsilon} \qquad (5\text{-}1)$$

where $\varepsilon = K_v \varepsilon_0$
$\quad K_s = $ relative dielectric constant of silicon
The voltage which must be applied to the gate-source terminals to completely pinch off the channel is then

$$V_{GS} + \phi'_c = \frac{a^2 q N_A}{2\varepsilon}$$

Since the built-in voltage produces some depletion of the channel even for $V_{GS} = 0$, the drain-source voltage necessary to pinch off the channel at the drain end when $V_{GS} = 0$ is

$$V_{DS} = \frac{a^2 q N_A}{2\varepsilon} - \phi'_c \triangleq V_p \qquad (5\text{-}2)$$

It can also be shown that the drain current I_{DSS} at pinchoff is given by

$$I_{DSS} = \frac{2}{3} V_p a q \mu N_A \frac{Z}{L} \qquad (5\text{-}3)$$

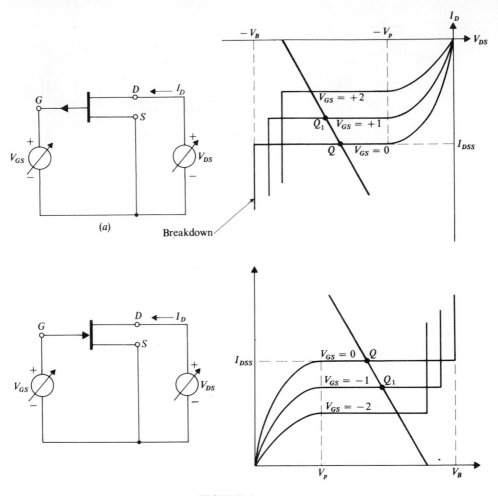

FIGURE 5-3
Characteristics of (a) p-channel JFET; (b) n-channel JFET.

where μ is the majority carrier mobility.   The transconductance is a function of the gate bias, and can be shown to be

$$g_{fs} = g_{fso}\left(1 - \sqrt{\frac{V_{GS} + \phi'_C}{V_{T_p}}}\right) \tag{5-4}$$

where

$$g_{fso} = 2aq\mu N_A \frac{Z}{L} \tag{5-5}$$

Here, of course, the square-root dependence on V_{GS} arises because we have assumed step junctions.   The variation of drain current in the voltage range $V_p < V_{DS} < V_B$

is usually fairly well approximated by assuming the following dependence on V_{GS}:[2]

$$I_D \approx I_{DSS}\left(1 - \frac{V_{GS} + \phi_C'}{V_{T_p}}\right)^2 \qquad (5\text{-}6)$$

It is emphasized that (5-6) is an approximation even for step-junction devices, as can be seen by calculating $\partial I_D/\partial V_G$ from (5-6) and comparing the result with (5-4).

Unfortunately, for any but abrupt or linear graded junctions, the analysis of the JFET becomes tedious and generally requires computer solution. We shall later show, however, that the depletion-voltage algorithm described in Chap. 4 for junctions, together with an integration routine, is basically all that is required for computer-aided analysis of the JFET.

Some qualitative aspects pertaining to JFET design can be noted from the foregoing analysis. In terms of the device geometry, we can control a by choice of junction depth, and Z and L by surface geometry configuration. At first glance it appears that by proper choice of these three parameters we can design for whatever V_p, I_{DSS}, and g_{fso} we choose. Such is not the case. Note from (5-2) that the pinchoff voltage depends on a but not on the surface geometry. Thus for a given N_A, we can select a to give the desired V_p. However, note from (5-3) and (5-5) that both g_{fso} and I_{DSS} depend in the same way on Z/L. Thus we can independently choose I_{DSS} or g_{fso}, but not both. These remarks apply not only to abrupt-junction uniform-channel JFETS but to general JFET structures as well. The procedure generally followed in JFET design is:

1   Select a to produce the desired V_p.
2   Choose Z/L to give either the required I_{DSS} or the required g_{fso}.
3   Choose L equal to the minimum line width that can be produced, in order to minimize the amount of area required by the JFET.

Exercise 5-1 Verify Eq. (5-1) for the abrupt-junction case.

Exercise 5-2 Show that a JFET with channel thickness $2a$ but having only a single gate will have a depletion-layer pinchoff voltage four times that given by Eq. (5-1).

5-2   THE n-CHANNEL INTEGRATED JFET

The cross section of an n-channel JFET is shown in Fig. 5-4. As we shall later see, the requirements for epitaxial-layer thickness x_{js} for this JFET may not be compatible with the requirements of the n-p-n bipolar transistor, so it may be necessary to add extra diffusion steps to the processing in order to achieve the desired compatibility.

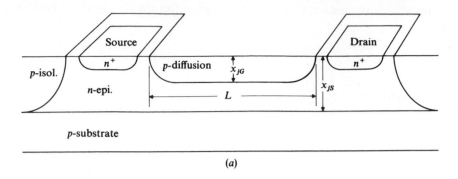

(a)

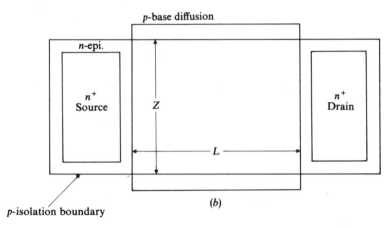

(b)

FIGURE 5-4
The n-channel JFET (a) cross section; (b) surface geometry (open).

The surface geometry in Fig. 5-4 is called open geometry, and for this case the p diffusion must completely cover the n-epitaxial material between source and drain. If it does not, there will be conducting regions between source and drain which are not influenced by the gate; this will have the effect of connecting constant conductances in parallel with the source-drain terminals, and the device will not exhibit JFET characteristics.   But this requirement in turn forces the p diffusion to overlap the isolation regions; therefore the top gate structure is electrically connected to the substrate.   This device must always operate with $V_{GS} = 0$, and is therefore useful only as a bias current source for some other device in the circuit. Since the substrate is connected to a power supply voltage, even the utility of the device for biasing is somewhat limited.   The JFET in which the gates and source are connected in this manner is called a *current limiter* and has the *VI* characteristic shown in Fig. 5-5.

In order to fabricate the device so that the two gates are electrically independent, it is only necessary to change the surface geometry to make it similar

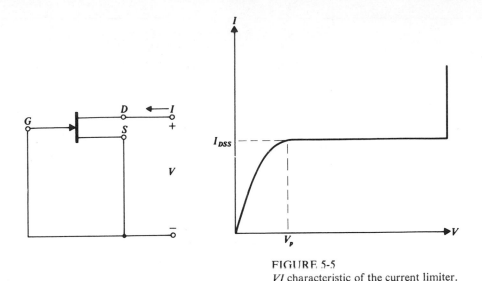

FIGURE 5-5
VI characteristic of the current limiter.

to that shown in Fig. 5-6; this is called closed geometry. For this case the top gate can be biased separately and can also have a signal applied to it.

Note that the closed-geometry device is a four-terminal device. The substrate will probably be connected to the most negative power supply voltage. However, for most integrated circuits the impurity concentration in the substrate is much less than that of the epitaxial layer; therefore the substrate gate will not have much influence on the behavior of the device.

EXAMPLE 5A *Pinchoff Voltage of an n-Channel JFET.* Consider a **JFET** made in epitaxial material having $N_{BC} = 5 \times 10^{15}$ cm^{-3}, and substrate junction depth 6.0 μm. A p-type gate diffusion is performed with $N_{sA} = 2 \times 10^{18}$ cm^{-3} and $x_{JG} = 5.0$ μm. To estimate the pinchoff voltage for the device, we assume the depletion from the substrate into the epitaxial material to be negligible. We also assume the built-in voltage of the top gate to be $\phi'_C = 0.7$. We now need to know what total depletion-layer voltage produces a depletion thickness of 1.0 μm on the n side of the top gate. This can be done by using the Lawrence-Warner curves in iterative fashion: first assume a voltage, then calculate the depletion width, then repeat the process until 1.0 μm is obtained.

Using this procedure, we find that for a depletion-layer voltage V_{T_p} of 4.0 V, the total depletion-layer width is 1.6 μm, of which 1.0 μ is on the n side of the junction. Since we have assumed $\phi'_C = 0.7$ V, the terminal voltage required for pinchoff is

$$V_p = V_{T_p} - \phi'_C = 3.3 \text{ V}$$

The actual pinchoff voltage will be slightly lower, since there will be some depletion from the substrate junction.

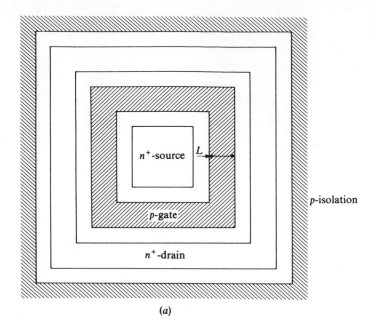

(*a*)

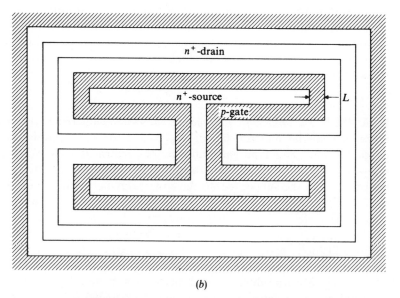

(*b*)

FIGURE 5-6
(*a*) Basic closed-geometry JFET; (*b*) method used to increase Z/L.

It is also of interest to note that for zero applied terminal voltage, the depletion into the channel is 0.465 μm; that is, the channel is almost half pinched off by the built-in voltage.

To estimate the output dynamic range for this device, we recall that breakdown between gate and channel limits the maximum drain voltage. From the Lawrence-Warner curves for breakdown voltage, we find

$$V_B = 60 \text{ V}$$

The output dynamic range is approximately 56.7 V.

This same JFET has a g_{fso} of approximately 53.5×10^{-6} ℧ per unit Z/L. If, for example, a JFET with $g_{fso} = 1000 \times 10^{-6}$ ℧ is required, and if 0.5-mil line width is used, we require the average perimeter of the p diffusion to be $Z = 9.85$ mils.   ////

Exercise 5-3 If the JFET of Example 5A is used as a pinch resistor, find its resistance for $V_{GS} = 0$.

Exercise 5-4 Suppose that for the JFET of Example 5A, the resistivity of the p-type substrate is 10 $\Omega \cdot$cm. If the built-in voltage $\phi'_c = 0.65$ V, calculate the depletion into the channel cased by ϕ'_c.

5-3 THE DOUBLE-DIFFUSED JFET

A p-channel JFET can be fabricated in integrated form in a manner similar to that used for p-type pinch resistors. The epitaxial material forms the lower gate, the base diffusion the channel, and the emitter diffusion the top gate. As was the case for the n-channel device, if open geometry is used, the gate diffusion must extend beyond the channel limits and overlap the epitaxial material, therefore both top and bottom gates are unavoidably connected electrically. Closed geometry can be employed if two separate gates are required.

A sketch of the surface geometry for the p-channel device is shown in Fig. 5-7a; the VI characteristics are shown in Fig. 5-7b. Note that the drain-gate breakdown voltage is now the same as the emitter junction breakdown voltage of the n-p-n transistor: about 7–9 V. This means that for a reasonable output dynamic range, the pinchoff voltage should not exceed 2 or 3 V.

Since two diffusions are involved in the fabrication of the p-channel JFET, the impurity distribution in the channel will not be uniform, and the gate junctions will not be abrupt. The Lawrence-Warner curves are not applicable to this device for the same reason that they could not be used to calculate emitter junction depletion widths. Several approximations have been suggested for the channel impurity profile in order to make analysis tractable.[3,4] Such approximations are generally not very useful because they do not enable the calculation of the built-in voltage. We saw in Chap. 4 that built-in voltages can range from less than 0.60 V

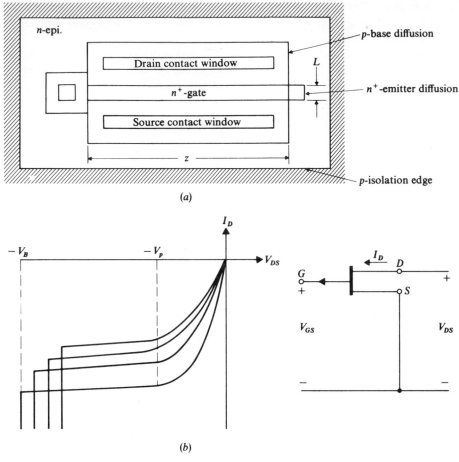

(a)

(b)

FIGURE 5-7
(a) Open surface geometry for a p-channel JFET; (b) VI characteristic.

to almost 0.80 V. Since pinchoff voltages of approximately 1 to 3 V are required, it is clear that the built-in voltage cannot be neglected. We also saw in Chap. 4 that depletion into the p material can be several tenths μm as a result of the built-in voltage alone. Since channel thickness of the order of 1 μm is required for pinchoff voltages of 1 to 3 V, it is clear that the depletion caused by the built-in voltage cannot be neglected.

In Chap. 4 we suggested an algorithm for use in calculating both the built-in voltage and the depletion versus voltage for a general junction. We now demonstrate that this algorithm can also be used to calculate V_p, I_{DSS}, and g_{fso} for a JFET with arbitrary impurity distribution in channel and gates. In Fig. 5-8 is shown the cross section of a channel of such a general JFET. The device is assumed to have two electrically independent gates, and the depletion regions shown are for

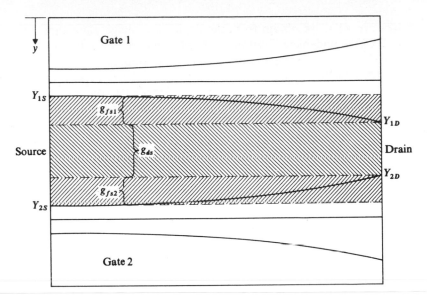

FIGURE 5-8
Geometrical interpretation of g_{fs1}, g_{fs2}, g_{ds} for the general JFET.

operation below pinchoff.   It is a rather remarkable fact that for such a bias condition the values of g_{fs1}, g_{fs2}, and g_{ds} are simply the *conductances* of the material
between the dotted lines as shown in the figure.[5,6]   That is,

$$g_{fs1} = \frac{Z}{L} \int_{Y_{1S}}^{Y_{1D}} q\mu(N)N(y)\, dy \qquad (5\text{-}7)$$

$$g_{ds} = \frac{Z}{L} \int_{Y_{1D}}^{Y_{2D}} q\mu(N)N(y)\, dy \qquad (5\text{-}8)$$

$$g_{fs2} = \frac{Z}{L} \int_{Y_{2D}}^{Y_{2S}} q\mu(N)N(y)\, dy \qquad (5\text{-}9)$$

If both gates are connected together, the total transconductance is

$$g_{fs} = g_{fs1} + g_{fs2} \qquad (5\text{-}10)$$

For operation at pinchoff and beyond, pinchoff occurs at the drain end at some
depth Y_p, and $Y_{2D} = Y_{1D} = Y_p$.   For this condition we see that

$$g_{ds} = 0$$

and
$$g_{fs} = \frac{Z}{L} \int_{Y_{2D}}^{Y_{2S}} q\mu(N)N(y)\, dy \qquad (5\text{-}11)$$

If $N(y)$ is known or can be calculated, and if $\mu(N)$ is known, then in principle we only require Y_{1S}, Y_{2S}, Y_{1D}, and Y_{2D} to complete the analysis.   But these

values are what we obtain from the depletion-voltage algorithm. Note that the details of the undepleted portion of the channel are not important; only the depletion widths at the source and drain ends are of importance.

Note further that

$$I_{DSS} = \int_0^{V_P} g_{ds}(V_{DS}) \, dV_{DS} \qquad (5\text{-}12)$$

But the depletion-voltage routine enables us to calculate Y_{1D} and Y_{2D} for any values of V_{GS} and V_{DS}. Once these are calculated, we can calculate g_{ds} from (5-8). This can be done for V_{DS} from 0 to V_p, and (5-12) can easily be evaluated numerically. The procedure for computer analysis of the device at and beyond pinchoff is therefore as follows:

1 Select the values of gate bias V_{GS1} and V_{GS2} which are of interest.
2 Using the algorithm for built-in voltage, calculate ϕ'_{C1} and ϕ'_{C2}.
3 Use $\phi'_{C1} + V_{GS1}$ and $\phi'_{C2} + V_{GS2}$ in the depletion-voltage routine to find Y_{1S} and Y_{2S}.
4 Use the depletion-voltage routine to calculate and store Y_{1D} and Y_{2D} for values of V_{DS} from 0 to well beyond the anticipated V_p.
5 Search the results of (4) for $Y_{1D} = Y_{2D} \triangleq Y_p$. This is the location of the pinchoff point, and V_{DS} for this Y_{1D} is the drain-source pinchoff voltage V_{DSp}. Note that if $V_{GS1} = V_{GS2} = 0$, $V_{DSp} = V_p$.
6 Use the results of (4) in (5-8) to calculate g_{ds}; integrate this from $V_{DS} = 0$ to $V_{DS} = V_{DSp}$. The result is I_{DSS}.
7 Calculate

$$g_{fs1} = \frac{Z}{L} \int_{Y_{1S}}^{Y_P} q\mu(N)N(y) \, dy$$

and

$$g_{fs2} = \frac{Z}{L} \int_{Y_P}^{Y_{2S}} q\mu(N)N(y) \, dy$$

For both gates connected together,

$$V_{GS1} = V_{GS2}$$

and

$$g_{fs} = g_{fs1} + g_{fs2}$$

This technique has been used to calculate the parameters of double-diffused JFET parameters which were fabricated in the laboratory; good agreement has been obtained between calculated and measured values. Calculated values for several devices are shown in Table 5-1.[7] Calculated and measured parameters for the laboratory device are given in Table 5-2.

It will be noted that the hypothetical "textbook" process produces a JFET whose pinchoff voltage is greater than 10 V. This process would be unacceptable for JFET use because the breakdown voltage, which is that of the emitter junction, is approximately 7 V. This device would therefore exhibit breakdown before pinchoff is reached, and would be unusable as a JFET. Such a problem often occurs in shallow structures because the impurity concentration in the channel is relatively large; this in turn produces large pinchoff voltages. If shallow structures are to be used for JFETs, the channel thickness must therefore be less than for deep structures. For example, the undepleted channel thickness in equilibrium is 0.69 μm for the laboratory device of Table 5-1, while for the deep structure it is 1.30 μm. Both devices have the same pinchoff voltage.

Table 5-1

A   CALCULATED JFET PARAMETERS FOR SEVERAL DEVICES

Device	x_{JE}, μm	x_{JC}, μm	ϕ'_{CE}	ϕ'_{CC}	Z/L	g_{fs1}, $\mu\mho$	g_{fs2}, $\mu\mho$	I_{DSS}, μA	V_p, V
Laboratory	1.80	2.50	0.782	0.663	4.05	97.3	40.2	154	2.45
Medium depth	2.61	3.53	0.758	0.619	1	24.1	7.4	37.7	2.65
Deep structure	4.51	5.99	0.698	0.595	30	454	185	698	2.44
"Textbook"	2.00	3.00	0.792	0.671	10	583	278	3726	10.8

B   PROCESS PARAMETERS USED IN THE CALCULATIONS

Device	N_{BC}, cm^{-3}	Base diffusion N_{sA}, cm^{-3}	x_{OA}, μm	Emitter diffusion N_{sD}, cm^{-3}	x_{OD}, μm
Laboratory	3.4×10^{15}	5.7×10^{18}	0.919	4.8×10^{20}	0.625
Medium depth	1.1×10^{15}	5×10^{19}	1.08	7.5×10^{20}	0.895
Deep structure	9.7×10^{14}	5×10^{18}	2.07	7.4×10^{19}	1.65
"Textbook"	5×10^{15}	4×10^{18}	1.16	1.0×10^{20}	0.803

Table 5.2   COMPARISON OF CALCULATED AND MEASURED RESULTS FOR THE LABORATORY DEVICE

Parameter	Calculated	Measured
V_p	2.43	
V_{GS1} to pinchoff channel with $V_{GS2} = 0$	3.7	3.5
g_{fs1} ($V_{GS1} = V_{GS2} = 0$)	97.3×10^{-6} $\mho$	95×10^{-6} $\mho$
g_{fs2} ($V_{GS1} = V_{GS2} = 0$)	40.2×10^{-6} $\mho$	
g_{fs0} ($V_{GS1} = V_{GS2} = 0$)	137×10^{-6} $\mho$	120×10^{-6} $\mho$
I_{DSS} ($V_{GS1} = V_{GS2} = 0$)	150×10^{-6} $\mho$	150×10^{-6} A
I_{DZ}[†] ($V_{GS1} = V_{GS2} = V_{GZ}$)	26.2×10^{-6} A	25×10^{-6} A

† Zero-temperature-coefficient drain current in pinchoff operation.

5-4   DESIGN OF THE INTEGRATED JFET

We have seen in the preceding section that the process used for the bipolar *n-p-n* transistor may not yield an acceptable double-diffused JFET.   If this is the case, additional processing steps must be added.   This means that the need for JFETs in a particular circuit must be weighed carefully.   If it is determined that the JFET is absolutely essential, the design usually proceeds as follows.

The most frequently occurring reason for unacceptable JFETs is that the pinchoff voltage is too large, principally because the channel thickness is too large for the level of impurity concentration present in the channel.   Since we assumed that the process has been chosen to yield suitable bipolar transistors, we do not wish to tamper with the diffusion schedule used to produce the bipolar device. Instead we add an extra diffusion step.   While there are several choices for how this can be performed, an extra *n* diffusion is generally added.   This is done as follows.   Windows are opened in the oxide for the gates but not the emitters.   An *n*-type predeposit is performed for a time t_1.   Windows are now opened for the emitters, and a predeposit is performed for a time t_2.   The total predeposit time for gates is now $t_1 + t_2$.   A drive-in is next performed for t_3.   We can choose t_3 and t_2 to give the desired emitter junction depth, and t_1 to give the desired gate junction depth.

Many integrated-circuit processes include a so-called *super-β* process which is similar to the procedure described above and which yields, in addition to the normal *n-p-n*, an *n-p-n* device with very narrow base and large current gain.   If such a process is available, it will usually lead to JFET devices with acceptable pinchoff voltages.

To determine what value should be used for t_1, we begin with the final base diffusion profile which will result for the *n-p-n*.   We select a gate junction depth and use the computer routine described in Sec. 5-3 to calculate the pinchoff voltage.   The procedure is iterated until an acceptable result is obtained.   Once the junction depth is known, we can solve for the predeposit required to produce this junction depth for a drive-in time t_3.   The time $t_1 + t_2$ can then be calculated, and since t_2 is known for the bipolar process, t_1 can be calculated.

Whether we use the above procedure or accept an available super-β process, the computer routine has given us the V_p to be expected, together with the g_{fs} and I_{DSS} per unit Z/L that will result.   We can then select Z/L to produce the desired values.   The minimum line width can be used for L, and the remaining surface geometry can be laid out.

EXAMPLE 5B   *Design of Double-diffused JFET Layout.*   Consider a case for which the profile design for an acceptable V_p has been completed, and the results of the computer analysis indicate $g_{fso} = 21 \times 10^{-6}$ ℧ and $I_{DSS} = 23.3 \times 10^{-6}$, both per unit Z/L.   Suppose that $I_{DSS} = 700 \times 10^{-6}$ A is required for the circuit being fabricated.   Then the required Z/L is 30.   If the process being used has 0.5-mil line widths and spaces, and 2.0-mil isolation clearances, the resulting layout

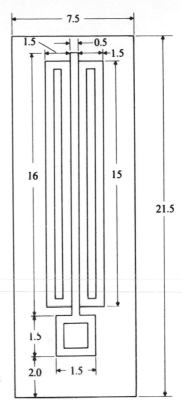

FIGURE 5-9
Layout of the JFET of Example 5B
(dimension in mils).

is as shown in Fig. 5-9.   It is interesting to note that for this example, the surface area required for the active channel is 7.5 mils², but the total surface area required for the isolated device is 161 mils².   ////

Relation of the JFET to the Pinch Resistor

It will be noted that the cross section of the JFET is the same as for the pinch resistor.   It is therefore worthwhile to point out the differences between the two devices.   As far as operation is concerned:

> *1* The pinch resistor is operated with small bias current to prevent non-linearities and pinchoff from occurring.   The JFET requires operation in the pinchoff condition and thus has much larger bias voltages and currents.
> *2* The pinch resistor is basically a two-terminal device, with the *n* material being connected to a power supply voltage, unless the device is being used as a voltage-controlled resistor.   The JFET makes use of the *n* material as a third terminal, and by this means derives its gain property.

As far as layout is concerned:

1 Open geometry is generally used for pinch resistors. For the JFET, the decision to use open or closed geometry is based on whether the circuit requires one or two independent gates.

2 One of the principal reasons for using the pinch resistor is to obtain large resistance in small space. Therefore the minimum line width is used for W, and L is chosen to produce the required L/W ratio. Thus the pinch resistor will appear long and narrow on the surface. The opposite is true for the JFET, for which the minimum line width is used for L, and Z is chosen to give the required Z/L ratio. The pinchoff current of the JFET is increased by increasing Z/L, the JFET channel will usually appear short and wide on the surface, as is evidenced by Fig. 5-9.

5-5 SMALL-SIGNAL MODELS FOR THE JFET

We consider now small-signal models for the JFET, and we restrict our attention to the portrayal of first-order behavior. In preceding sections we have seen that in pinchoff operation, the drain current is first-order-independent of V_{DS}, and the gate current is negligible. At low frequencies, the JFET is reasonably well approximated by a voltage-controlled current source whose transconductance is g_{fs}. The primary parasitic contributions to the degradation of performance with frequency are the capacitances associated with the various contact regions and with the substrate. The device has an intrinsic cutoff frequency[8] because the channel is basically a distributed RC structure, but usually the parasitic capacitances and circuit resistances limit the frequency response to well below the intrinsic cutoff frequency.

In Fig. 5-10, small-signal models are shown for the n-channel JFET with open geometry and the double-diffused p-channel device with open geometry. The "contact" capacitances arise because the source and drain regions to which contact must be made must be enlarged to accommodate the contact windows. This in turn makes the area of the junctions surrounding the contact areas larger than the area of the active channel.

EXAMPLE 5C *Calculation of Parasitic Capacitances.* We return to the layout of the JFET of Example 5B, shown in Fig. 5-9, and we assume for simplicity that the capacitances of the various junctions can be obtained from Table 4-1. We also assume an 8.0-μm epitaxial thickness, and we neglect all sidewalls except those of the isolation diffusion.

For the channel capacitance, we arbitrarily use half the value of the zero-bias capacitance of the junctions involved to approximate the average capacitance. We obtain the following results:

$$C_{gs(\text{channel})} \approx 1.24 \text{ pF}$$
$$C_{gd(\text{contact})} \approx 5.25 \text{ pF}$$
$$C_{gs(\text{contact})} \approx 5.25 \text{ pF}$$
$$C_{g(\text{subs})} \approx 18.7 \text{ pF}$$

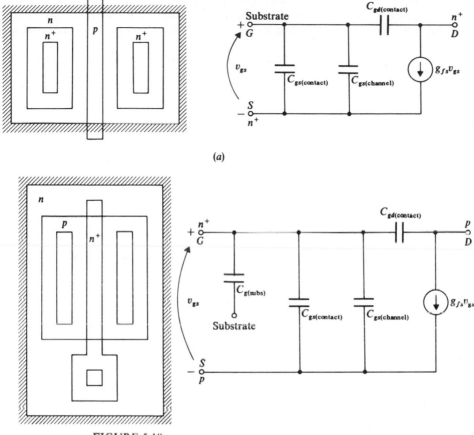

FIGURE 5-10
First-order JFET models. (a) n-channel open geometry; (b) p-channel open geometry.

One can include second-order behavior in the model by means of series resistances in the source and drain leads, and conductance between drain and source.　　　　　////

PROBLEMS

5-1 Derive an expression for the pinchoff voltage for a device for which the impurity profile in and around the channel can be approximated by a linear function.

5-2 Design the layout for a double-diffused JFET with closed geometry and $Z/L = 100$. Assume 0.5-mil line widths, spaces, and registration tolerances, and 1.0-mil isolation clearance.

5-3 Calculate the capacitances for a small-signal model of the JFET of Prob. 5-2. Use capacitance values of Table 4-1.

5-4   A wafer has p-type substrate with impurity concentration $N_{sA} = 10^{16}$ cm^{-3}. An n-type epitaxial layer is grown with linear impurity profile. The layer is 5 μm thick and has a surface concentration of 10^{17} cm^{-3}. The impurity concentration at a depth of 5 μm is 10^{15} cm^{-3}.

A JFET is to be designed with an epitaxial channel. The device is to be used in the common source configuration with the signal applied to the top gate. Top gate bias voltage V_{GS1} is zero. Design the JFET, including surface geometry, to have a pinchoff voltage of 3 V and a pinchoff current of 2 mA. Assume that a p^+ duffusion is to be used to form the top gate, but the depletion into the top gate is negligible. Thus the only parameter of the diffusion that is necessary is the junction depth.

Built-in voltages are 0.7 for the top gate and 0.6 for the bottom gate. Channel mobility is constant, and the value $\mu = 10^3$ cm^2/V · s. The substrate is grounded.

5-5   A hypothetical processing schedule produces emitter junction depths of 2.0 μm and collector junction depth of 3.0 μm. This schedule does not produce an acceptable p-channel JFET. Suppose that the diffusion times and temperatures are those of Exercises 2-1 and 2-2. An extra emitter diffusion step is to be added in order to produce an acceptable JFET without altering the bipolar transistors. Suppose that computer-aided analysis indicates that the JFET channel thickness should be 0.5 μm between metallurgical junctions. Design the required process schedule.

5-6   An n-channel JFET is fabricated in epitaxial material with $N_{BC} = 5 \times 10^{15}$, $x_{jS} = 6.0$ μm, and $x_{jG} = 5.0$ μm. Closed surface geometry is used. Assume that the substrate gate has no effect on the channel, and that the top gate is approximately an abrupt junction with $N_A \gg N_{BC}$.

If the source is connected to ground, and $V_{GS} = -V_{T_p}/4$ and $V_{DS} = +V_{T_p}/4$, calculate g_{fs}. Neglect ϕ'_c.

REFERENCES

1   HAMILTON, D. J., F. A. LINDHOLM, and A. H. MARSHAK: "Principles and Applications of Semiconductor Device Modeling," chap. 4, Holt, Rinehart and Winston, Inc., New York, 1971.

2   WALLMARK, J., and H. JOHNSON (eds.): "Field-effect Transistors: Physics, Technology, and Applications," chap. 5, Prentice-Hall, Inc., Englewood Cliffs, N.J., 1966.

3   BURGER, R. M., and R. P. DONOVAN: "Fundamentals of Silicon Integrated Device Technology," pp. 279–299, Prentice-Hall, Inc., Englewood Cliffs, N.J., 1968.

4   LIN, H. C.: "Integrated Electronics," pp. 250-263, Holden-Day, Inc., Publisher, San Francisco, 1967.

5   RICHER, I.: The Equivalent Circuit of an Arbitrarily-doped Field-effect Transistor, *Solid State Electron.*, vol. 8, pp. 381–393, May 1965.

6   LINDHOLM, F. A., and P. R. GRAY: Large-signal and Small-signal Models for Arbitrarily-doped Four-terminal Field-effect Transistors, *IEEE Trans. Electron. Devices*, vol. ED-13, pp. 819–829, December 1966.

7   HAMILTON, D. J., and J. G. FOSSUM: Final Report: Computer Aided Design of Integrated Junction Field-effect Transistors, Burr-Brown Research Corp., Tucson, Arizona, July 28, 1970.

8   LINDHOLM, F. A.: Unified Modeling of Field-effect Devices, *IEEE J. Solid-State Circuits*, vol. SC-6, pp. 250–259, 1971.

6

MOS FIELD-EFFECT TRANSISTORS

Metal-oxide-semiconductor field-effect transistors (MOSFET) play an important role in integrated circuits, particularly in large-scale integration of digital circuits. Indeed, the use of high-density MOSFET digital circuits has made possible the production of the popular pocket-size electronic calculators offered by several companies. Other acronyms for the MOSFET are IGFET (insulated-gate field-effect transistor) and MISFET (metal-insulator-semiconductor transistor). Like the JFET, the MOSFET depends for its operation upon the modulation of a channel conductance by a gate structure. In the JFET, this is accomplished by using a p-n junction for the gate and channel regions, and using the depletion properties of the junction to modulate the channel conductance. In the MOSFET, no such junction is used. Rather, a metal plate, separated from the semiconductor by an insulator, is used as a gate. The electric field established in the insulator by the gate voltage causes a mobile charge to appear at the surface of the semiconductor; this surface charge forms the conducting channel between source and drain.

In this chapter, we examine the terminal behavior of the MOSFET, and its relation to integrated-circuit processing variables; in a later chapter, MOSFET circuits are discussed. Since the MOSFET channel is formed at the semiconductor surface, we begin by considering surface space-charge regions. The influence of the surface space-charge regions on MOS capacitor behavior is next considered; this is followed by a development of the dc characteristics of the MOSFET. Finally, different methods of fabrication are described. Our treatment closely follows that of Grove[1] and that of Carr and Mize.[2]

6-1 SURFACE SPACE-CHARGE REGIONS

We consider first a region of a p-type material with acceptor concentration N_A, having on its surface an oxide layer of thickness d_0, with a layer of metal on the oxide surface, as shown in Fig. 6-1a. Assume for the moment that both metal and semiconductor have the same work function, and that there are no charges in the oxide layer. With no voltage applied to the gate, the electron energy will be as shown in Fig. 6-1b. Because the structure is in equilibrium, the Fermi level E_F will be the same in metal, insulator, and semiconductor.

   If a small positive voltage is applied to the gate, the Fermi level of the metal will be displaced from that of the semiconductor by an amount of qV_G, where q is the electronic charge. The electric field established across the oxide must terminate on charges in the semiconductor; these charges result from depletion of mobile carriers at the semiconductor surface, leaving bound, negatively charged acceptor ions. Thus a depletion region is formed at the surface, as is shown in Fig. 6-1c, even though there is no p-n junction. Note that this means that not all of the voltage V_G appears across the oxide layer: some voltage must appear between the semiconductor surface and the region deep in the bulk material. Because there is a potential variation between the deep bulk and the surface, the energy levels E_C, E_V, and E_i bend in the semiconductor. In fact, if a large V_G is applied as shown in Fig. 6-1d, the bending will be so large that E_i will fall below E_F at the *surface*. The significance of this is best understood by recalling how the mobile carrier densities for a semiconductor are related to the energy levels:

$$n = n_i e^{(E_F - E_i)/kT} = n_i e^{(-q\phi_F/kT)}$$

$$p = n_i e^{(E_i - E_F)/kT} = n_i e^{(q\phi_F/kT)}$$

where $q\phi_F \triangle E_i - E_F$.

   In the bulk material, $E_i > E_F$, and n is very small; however, we know that deep in the bulk material $p \approx N_A$. Thus we find

$$q\phi_{F(\text{bulk})} = kT \ln \frac{N_A}{n_i}$$

For n-type material,

$$q\phi_{F(\text{bulk})} = -kT \ln \frac{N_D}{n_i}$$

   If V_G is large enough to cause E_i to fall below E_F at the surface, ϕ_F at the surface will be negative and n will be large while p will be small. Thus large V_G has the effect of making the surface of the semiconductor *appear* to be of impurity

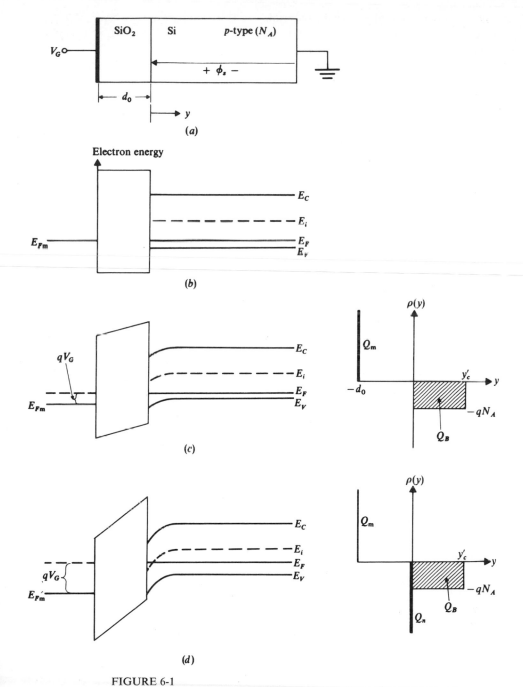

FIGURE 6-1
(a) The MOS structure; (b) electron energy for the equilibrium case $V_G = 0$;
(c) band bending with small positive V_G; (d) positive V_G, large enough to cause
the formation of an inversion layer.

type opposite that of the bulk material.` It is as though a thin layer of n-type material, with mobile electrons, exists at the surface. This layer is called an inversion layer, or a field-induced channel.

We now define *strong inversion* as the condition existing when the band bending at the surface is sufficient to cause E_i to fall below E_F by an amount $q\phi_{F(bulk)}$. Since E_i is above E_F by an amount $q\phi_{F(bulk)}$ deep in the bulk material, the total band bending that occurs from deep bulk to surface is $2q\phi_{F(bulk)}$ for strong inversion. Note that this corresponds to an inversion channel in which the mobile carrier density is

$$n \approx N_A$$

When an inversion channel forms, the mobile carrier density depends exponentially on ϕ_F. Therefore, a further increase of V_G will cause an enormous increase of n. Thus we see that the charges required to terminate the field established by V_G will henceforth be provided primarily by the inversion channel, not by depletion into the bulk material. We therefore make the approximation that depletion reaches its maximum extent $y'_{C(max)}$ when strong inversion occurs. To calculate y'_C, we define a surface potential ϕ_s measured relative to the deep bulk material. Then the depletion depth, for uniformly doped material, is

$$y'_C = \left(\frac{2K_s \varepsilon_0 \phi_s}{qN_A}\right)^{1/2}$$

where K_s is the relative dielectric permittivity of silicon and ε_0 is the permittivity of free space. At strong inversion, ϕ_s is defined to be $2\phi_{F(bulk)}$; by the above approximation, $y'_{C(max)}$ is therefore

$$y'_{C(max)} = \left(\frac{2K_s \varepsilon_0 2\phi_{F(bulk)}}{qN_A}\right)^{1/2}$$

If we let Q_B be the total charge per unit surface area of the depletion region, we have

$$Q_B = -qN_A y'_C$$

and

$$Q_{B(max)} = -qN_A y'_{C(max)} = -(2K_s \varepsilon_0 2\phi_{F(bulk)} qN_A)^{1/2}$$

6-2   THE MOS CAPACITOR

In Chap. 4, we discussed the MOS capacitor and qualitatively described the effects of bias on capacitance as a result of the formation of a depletion layer. We are now in a position to include quantitatively the additional effect of the inversion channel. Let

$Q_n \triangle$ mobile charge, per unit surface area, of inversion channel
$Q_B \triangle$ charge per unit surface area of depletion region
$Q_s \triangle$ total net charge per unit surface area in silicon

The applied gate voltage is

$$V_G = V_0 + \phi_s$$

where V_0 is the voltage across the oxide layer. Since the electric displacement vector must be continuous across the SiO_2-Si interface, it is required that

$$K_0 \varepsilon_0 E_0 = K_s \varepsilon_0 E_s$$

where K_0 = relative permittivity of the oxide
  E_0 = electric field in oxide
  E_s = electric field at surface in silicon
The surface field in the silicon is

$$E_s = \frac{-Q_s}{K_s \varepsilon_0}$$

Then we have

$$E_0 = \frac{K_s \varepsilon_0 E_s}{K_0 \varepsilon_0} = \frac{-Q_s}{K_0 \varepsilon_0}$$

Since the field in the oxide is constant, V_0 is given by

$$V_0 = d_0 E_0 = \frac{-d_0 Q_s}{K_0 \varepsilon_0} = \frac{-Q_s}{C_0}$$

where C_0 is the oxide capacitance per unit surface area, and

$$C_0 = \frac{K_0 \varepsilon_0}{d_0}$$

The gate voltage is thus

$$V_G = \frac{-Q_s}{C_0} + \phi_s$$

One can visualize the total capacitance as resulting from the series combination of the oxide capacitance and the semiconductor capacitance, as we saw in Chap. 4:

$$C = \frac{1}{1/C_0 + 1/C_s}$$

where $C_s = K_s \varepsilon_0 / y_c'$.
  Thus we obtain

$$\frac{C}{C_0} = \left(1 + \frac{2K_0^2 \varepsilon_0 V_G}{qN_A K_s d_0^2}\right)^{-1/2}$$

as long as V_G is positive but less than the value required to produce strong inversion. When V_G reaches the value required to produce strong inversion, no further increase of depletion region occurs, and therefore no further change of capacitance can occur.

The Threshold Voltage V_T

Let V_T be the gate voltage necessary to produce strong inversion, and assume that prior to strong inversion no mobile charge exists near the surface; the only net charge in the semiconductor is that of the depletion region. Just before the onset of strong inversion, the total charge Q_s is

$$Q_s = Q_{B(max)}$$

The surface potential for this condition is

$$\phi_s = 2\phi_{F(bulk)}$$

Therefore the gate voltage at strong inversion is

$$V_G \triangleq V_T = \frac{-Q_{Bmax}}{C_0} + 2\phi_{F(bulk)}$$

The minimum value of capacitance occurs at the onset of strong inversion, and is given by

$$C_{min} = C_0 \left(1 + \frac{2K_0{}^2 \varepsilon_0 V_T}{qN_A K_s d_0{}^2}\right)^{-1/2}$$

For our assumption of no work-function difference and no charge in the oxide layer, the variation of C with applied voltage is as shown in Fig. 6-2.

As was explained in Chap. 4, no variation of C occurs for negative V_G, since an accumulation of positive mobile carriers occurs at the silicon surface.

All of our discussion thus far has assumed p-type bulk material. It is clear that the same development could be made for n-type material, and the results would be the same if N_A were replaced by N_D and appropriate sign changes were made.

EXAMPLE 6A *Threshold Voltage for an MOS Capacitor.* In Chap. 4 we calculated the applied voltage for which the depletion region of an MOS capacitor caused a 10 percent change of capacitance. In that example, it was assumed that only the depletion region contributed to the total charge in the silicon. We now investigate the threshold voltage for the capacitor. We assume, as in Example 4A, that the bulk material is n-type, and that

$$d_0 = 900 \text{ Å} = 0.09 \ \mu\text{m}$$
$$N_D = 10^{19} \text{ cm}^{-3}$$

For this donor level, the potential $\phi_{F(bulk)}$ is

$$\phi_{F(bulk)} = -0.54 \text{ V}$$

while the depletion-layer charge is

$$Q_{B(max)} = 18.5 \times 10^{-7} \text{ C/cm}^2$$

The oxide capacitance is

$$C_0 = 38 \times 10^{-9} \text{ F/cm}^2$$

and the threshold voltage is

$$V_T = \frac{-Q_{B(max)}}{C_0} + 2\phi_{F(bulk)} = -50 \text{ V}$$

It will be recalled from Example 4A that if only the depletion region is considered, an applied voltage of -70 V is required to change the capacitance by 10

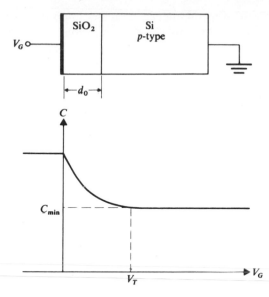

FIGURE 6-2
Variation of C with applied voltage.

percent.   Since $V_T = -50$ V, an inversion channel would form, and variation of C with V_G would cease before C had changed by 10 percent.   Indeed, the inversion layer forms just prior to breakdown of the oxide for this example.

It is interesting to note that if N_D is reduced to 10^{15} cm^{-3} for this capacitor, the threshold voltage becomes $V_T = -0.94$ V.   ////

When the gate voltage is made larger than the threshold voltage, no further change of surface potential ϕ_s results because negligible additional depletion occurs. The semiconductor charge must then satisfy the relation

$$V_G - \frac{-Q_s}{C_0} + 2\phi_{F(bulk)} = \frac{-Q_n - Q_{B(max)}}{C_0} + 2\phi_{F(bulk)}$$

or

$$V_G - V_T = \frac{-Q_n}{C_0}$$

Exercise 6-1   Calculate the threshold voltage for strong inversion for p material with $N_A = 10^{15}$ cm^{-3} and $d_0 = 10^{-5}$ cm.

6-3   EFFECTS OF WORK-FUNCTION DIFFERENCE AND OXIDE CHARGE

The work function of a material is defined as the energy required to remove an electron, with energy equal to the Fermi energy, from the material to vacuum. Work-function values for several metals are given in Table 6-1.   For silicon, the work function depends on the impurity concentration, since the Fermi level is a function of impurity concentration. At room temperature the work function for silicon is

$$\phi_S = 3.8 + \phi_{F(\text{bulk})}$$

For an MOS capacitor using aluminum as the metal, the metal work function ϕ_M will generally be different from that of silicon, as can be seen from Table 6-1[3] and Fig. 6-3a. This means that there will be some band bending in thermal equilibrium, since the Fermi level must be the same in all parts of the system, as is shown in Fig. 6-3b. From Fig. 6-3b it is clear that if the work-function difference ϕ_{MS} is $\phi_M - \phi_S$, then a gate voltage $V_G = \phi_{MS}$ would have to be applied to restore the bands to the flat-band condition. Note that if ϕ_{MS} is negative, as is the case for Fig. 6-3b, the result, for thermal equilibrium, is for the bands to bend *away* from the inversion condition in *n*-type silicon but *toward* the inversion condition for *p*-type silicon. In the general case, the threshold voltage is given by

$$V_T = \frac{-Q_{B(\text{max})}}{C_0} + 2\phi_{F(\text{bulk})} + \phi_{MS}$$

EXAMPLE 6B   *Threshold Voltage for n-Type Silicon.*   Consider an MOS capacitor with $N_D = 10^{15}$, $d_0 = 1000$ Å, and aluminum gate. The threshold voltage if ϕ_{MS} were zero would be

$$V_T = -0.98$$

Since $\phi_M = 3.2$ for aluminum and $\phi_F = -0.29$ for $N_A = 10^{15}$ cm^{-3}, we find

$$\phi_{MS} = -0.3 \text{ V}$$

The threshold voltage is then

$$V_T = -0.98 - 0.3 = -1.28 \text{ V} \qquad ////$$

Exercise 6-2   Calculate V_T for the MOS capacitor of Example 6B if *p*-type silicon with $N_A = 10^{15}$ cm^{-3} is used.

Table 6-1   WORK FUNCTION FOR SEVERAL METALS

Metal	Work function ϕ_M, eV
Al	3.2
Au	4.0
Ag	4.0
Cu	3.7
Mg	2.25
Ni	3.7

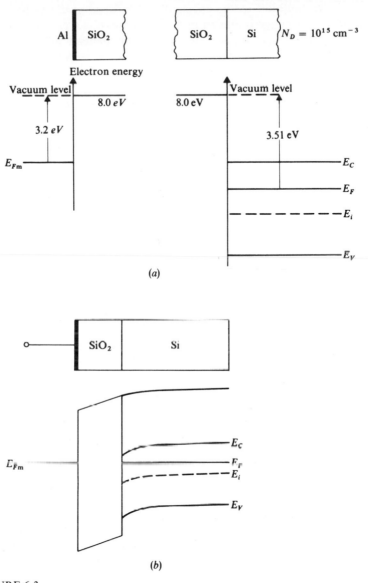

(a)

(b)

FIGURE 6-3
(a) The metal-oxide and oxide-silicon systems considered separately; (b) the MOS system in thermal equilibrium.

In addition to the effects of work-function difference, it is found that the threshold voltage is altered by the presence of charged surface states at the Si-SiO$_2$ interface. These surface charges do not depend upon the impurity level, but rather on the crystal orientation of the silicon surface. Measurements show that the charges are always positive and have the densities shown in Table 6-2. That this affects the threshold voltage can be explained as follows. Assume that the surface charge is at the interface as shown in Fig. 6-4. The total charge within the silicon is still Q_s, so

$$\varepsilon_0 K_s E_s = -Q_s$$

but now

$$K_0 \varepsilon_0 E_0 = -Q_s - Q_{ss}$$

Since the field is still uniform within the oxide,

$$E_0 = \frac{V_0}{d_0}$$

Thus we have

$$V_0 = -\frac{d_0 Q_s}{\varepsilon_0 K_0} - \frac{d_0 Q_{ss}}{\varepsilon_0 K_0}$$

and

$$V_G = V_0 + \phi_s = \frac{-Q_s}{C_0} - \frac{Q_{ss}}{C_0} + \phi_s$$

If we now consider both ϕ_{MS} and Q_{ss} simultaneously, we find

$$V_G = -\frac{Q_s}{C_0} - \frac{Q_{ss}}{C_0} + \phi_s + \phi_{MS}$$

The new threshold voltage is found by setting

$$Q_s = Q_{B(max)}$$

and

$$\phi_s = 2\phi_{F(bulk)}$$

This yields

$$V_T = 2\phi_{F(bulk)} + \phi_{MS} - \frac{Q_{ss}}{C_0} - \frac{Q_{B(max)}}{C_0}$$

Table 6-2   DENSITY OF SURFACE CHARGES Q_{ss} (after Carr and Mize[2])

Silicon surface orientation	Q_{ss}/q, cm^2
111	5×10^{11}
110	2×10^{11}
100	9×10^{10}

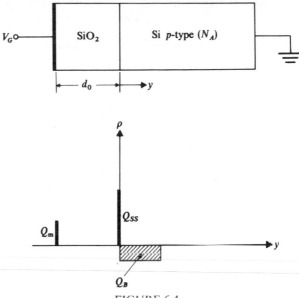

FIGURE 6-4
Surface charge at the Si-SiO$_2$ interface.

EXAMPLE 6C *Threshold Voltage with* ϕ_{MS} *and* Q_{SS}. The MOS capacitor of Example 6B is fabricated with *n*-type silicon having $\langle 111 \rangle$ surface orientation and $N_D = 10^{15}$ cm^{-3}. The threshold voltage for this device is now

$$V_T = -3.6$$

With $Q_{SS} = 0$:

$$V_T = -1.28$$

With $Q_{SS} = 0$ and $\phi_{MS} = 0$:

$$V_T = -0.98 \qquad ////$$

Exercise 6-3 Calculate the threshold voltage for the capacitor of Exercise 6-2 if $\langle 111 \rangle$ silicon is used.

6-4 THE MOS TRANSISTOR

The inversion layer which forms in the MOS capacitor can be used as the channel of an MOS transistor. We consider first the idealized structure having the cross section shown in Fig. 6-5; later we discuss the fabrication of MOSFET devices by integrated-circuit planar technology. The device of Fig. 6-5 is called an *n*-channel MOSFET since the carriers in the inversion layer are electrons. Contacts to the ends of this layer which forms the channel for the MOSFET are made through the n^+ regions; these are called source and drain. If a sufficiently large gate voltage

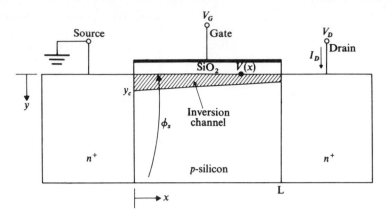

FIGURE 6-5
Idealized n-channel MOSFET.

V_G is applied, an inversion channel forms; if a drain voltage V_D is now applied between source and drain, a channel current I_D will flow. Let the voltage at the silicon surface, measured relative to the source, be $V(x)$, and assume no potential variation in the y direction in the inversion channel. Note that for positive V_D, $V(x)$ increases with x, and the voltage across the oxide decreases with x. Therefore there will be less induced charge in the inversion layer near the drain than near the source. By the same reasoning, the surface potential ϕ_s will be larger near the drain than it is near the source.

Let the total charge in the semiconductor be

$$Q_s = Q_n + Q_B$$

as before, where Q_n is the mobile charge per unit surface area and Q_B is the depletion layer charge per unit surface area. The channel current density is

$$J_D = q\mu n(x, y)\,\frac{-dV}{dx}$$

If the width of the MOSFET is Z, the channel current is

$$I_D = -Z \int_0^{y_c(x)} q\mu n(x, y)\,\frac{-dV}{dx}\,dy$$

where μ is the mobility.

But the charge Q_n is simply

$$Q_n = \int_0^{y_c(x)} -qn(x, y)\,dy$$

The current can therefore be written in terms of Q_n as

$$I_D = -Z\mu Q_n \frac{dV}{dx} \qquad (6\text{-}1)$$

From the analysis of the MOS capacitor, we know that

$$V_G = -\frac{Q_s}{C_0} + \phi_s$$

$$= \frac{-Q_n - Q_B}{C_0} + \phi_s(0) + V(x) \qquad (6\text{-}2)$$

if we temporarily neglect ϕ_{MS} and Q_{SS}.

We now assume that strong inversion exists at the source end of the channel and we make the *important simplifying approximation* that Q_B is constant in the p material and has everywhere its value at the source end. Then we have

$$\phi_s(0) = 2\phi_{F(\text{bulk})}$$

and

$$Q_B \approx Q_{B(\text{max})}$$

Substituting these in (6-2), we obtain

$$\frac{Q_n}{C_0} = -V_G - \frac{Q_{B(\text{max})}}{C_0} + 2\phi_{F(\text{bulk})} + V(x) \qquad (6\text{-}3)$$

Inserting this result in (6-1), integrating both sides from $x = 0$ to $x = L$, and noting that $V(L) = V_D$, we find

$$I_D = \mu \frac{Z}{L} C_0 \left[(V_G - V_t)V_D - \frac{V_D^2}{2} \right] \qquad (6\text{-}4)$$

where V_T is the previously defined threshold voltage

$$V_T = \frac{-Q_{B(\text{max})}}{C_0} + 2\phi_{F(\text{bulk})}$$

If ϕ_{MS} and Q_{SS} are included, (6-4) remains unchanged but the threshold voltage is

$$V_T = \frac{Q_{B(\text{max})}}{C_0} + 2\phi_{F(\text{bulk})} + \phi_{MS} - \frac{Q_{SS}}{C_0}$$

Note that (6-4) only applies if an inversion channel exists; this implies

$$V_G - V_T > 0$$

and

$$V_G - V_T > V_D$$

Because $V(x)$ increases along the channel, inversion is less near the drain, as we have seen. Suppose that V_D is now increased so that $V_D > (V_G - V_T)$. Then no inversion can occur at the drain end of the channel; this situation is similar to pinchoff in the JFET. But current still flows in the channel if $V_G > V_T$. This condition of operation is known as saturation, because the drain voltage no longer can influence the drain current. The drain voltage V_{DS} at which saturation occurs is clearly

$$V_{DS} = V_G - V_T$$

and the drain current for saturation is found by using V_{DS} in (6-4):

$$I_{DS} = \frac{\mu Z}{2L} C_0(V_G - V_T)^2 \qquad (6\text{-}5)$$

Sketches of I_D versus V_D are shown in Fig. 6-6.

It is clear that p-channel devices can be analyzed in the same way, and the results of (6-4) and (6-5) will apply if appropriate sign changes are made.

When the MOSFET is used as a small-signal amplifier, it is operated in the saturation region, and its transconductance is of interest. The transconductance g_m is defined by

$$g_m \triangleq \frac{\partial I_{DS}}{\partial V_G}$$

From (6-5) we obtain

$$g_m = \frac{\mu Z}{L} C_0(V_G - V_T) \qquad (6\text{-}6)$$

As was the case for the JFET, g_m is directly proportional to Z/L.

EXAMPLE 6D *Transconductance of a p-Channel Device.* Consider a MOSFET fabricated in a substrate having $N_D = 10^{15}$ cm^{-3}, and aluminum gate separated from the substrate by 10^{-5} cm of SiO$_2$. From Example 6C, the threshold voltage is

$$V_T = -3.6 \text{ V}$$

if $Q_{ss}/q = 5 \times 10^{11}$ cm^{-2}. The hole mobility for inversion layers on $\langle 111 \rangle$ silicon is 190 cm^2/V $\cdot$ s. If the device has $Z = 1$ mil and $L = 0.2$ mil, the transconductance is

$$g_m = -33.3 \times 10^{-6}(V_G + 3.6) \ \mho$$

Suppose the device is operated with $V_G = -10$ V; then if operation is to be in the saturation region, we must have $V_D \le -6.4$ V. For this condition we find

$$g_m = 213 \times 10^{-6} \ \mho \qquad \qquad ////$$

EXAMPLE 6E *Use of the MOSFET as a Voltage-variable Resistor.* When $|V_D| < |V_G - V_T|$ the MOSFET is said to be operating in the "triode" region, and in this region I_D is a function of both V_D and V_G. Suppose it is desired to use the MOSFET for a small-signal application, such as an automatic gain control circuit, requiring a voltage-variable resistor. The conductance of the channel is given by

$$g_{ds} \triangleq \frac{\partial I_D}{\partial V_D}$$

which, from (6-4), is

$$g_{ds} = -\mu \frac{Z}{L} C_0[(V_G - V_T) - V_D]$$

for a p-channel device.

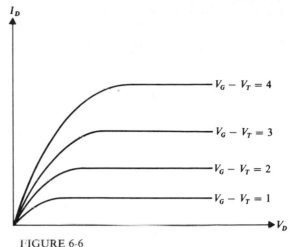

FIGURE 6-6
IV characteristics for the idealized *n*-channel MOSFET.

If we operate the device of Example 6D with $V_D = I_D = 0$, the small-signal resistance r_{ds} between source and drain terminals is

$$r_{ds} = \frac{1}{g_{ds}} = \frac{-1}{\mu(Z/L)C_0(V_G - V_T)}$$

If $V_G = -10$, we obtain

$$r_{ds} = 4.7\text{K }\Omega \qquad \text{////}$$

Conductance in the Saturation Region

Although the sketch of Fig. 6-6 shows zero slope for the $I_D V_D$ characteristics in the saturation region, such is not the case in practical devices. This is because an increase of V_D beyond V_{DS} causes a small reduction of the actual length of the inversion layer, thus causing an increase of the drain current. The effect becomes important in devices with very small L.

Consequences of the Constant Q_B Approximation

The square-law nature of the $I_D V_D$ relationship arose because of our approximation

$$Q_B \approx Q_{B(\text{max})}$$

everywhere in the channel. If the analysis is refined by including the dependence of Q_B on $V(x)$, the result is, for the *n*-channel device,

$$Q_B = -(2K_s \varepsilon_0 q N_A)^{1/2}[V(x) + 2\phi_{F(\text{bulk})}]^{1/2}$$

Use of this result in (6-2) leads to

$$I_D = \mu \frac{Z}{L} C_0 \left\{ (V_G - V_T)V_D - \frac{V_D{}^2}{2} - \frac{2}{3} \frac{1}{C_0} (2K_s \varepsilon_0 qN_A)^{1/2} (2\phi_{F(\text{bulk})})^{3/2} \right.$$

$$\left. \left[1 + \frac{3V_D}{4\phi_{F(\text{bulk})}} - \left(1 + \frac{V_D}{2\phi_{F(\text{bulk})}} \right)^{3/2} \right] \right\} \qquad (6\text{-}7)$$

Effects of Substrate Bias

Thus far we have implicitly assumed that the p-type substrate of the device of Fig. 6-5 is connected to ground. If, instead, it is connected to a voltage V_S, a change of the depletion layer will result. If V_S is negative, the size of the depletion region will increase, and the magnitude of Q_B will increase. At strong inversion we then find the depletion charge $Q'_{B(\text{max})}$ to be

$$Q'_{B(\text{max})} = -\{2K_s \varepsilon_0 qN_A[2\phi_{F(\text{bulk})} - V_S]\}^{1/2}$$

$$= Q_{B(\text{max})} \left[1 - \frac{V_S}{2\phi_{F(\text{bulk})}} \right]^{1/2}$$

The new threshold voltage V'_T is therefore

$$V'_T = \phi_{MS} + 2\phi_{F(\text{bulk})} - \frac{Q_{SS}}{C_0} - \frac{Q_{B(\text{max})}}{C_0} \left[1 - \frac{V_S}{2\phi_{F(\text{bulk})}} \right]^{1/2}$$

Exercise 6-4 The MOSFET of Example 6E is operated with a substrate bias $V_S = +10$ V. Find the threshold voltage.

6-5 INTEGRATED MOSFET DEVICES

MOSFET devices can be fabricated by standard integrated-circuit processing techniques as is shown by the cross-section sketches of Fig. 6-7. For a p-channel device, an n-type substrate is used. Diffusion of source and drain regions is performed. Next oxide is removed over the channel region, and a thin oxide is regrown. While it would be desirable to have very thin oxide to yield low threshold voltage devices, in production oxide thicknesses less than 10^{-5} cm are generally not used because of reliability and yield problems. Following the thin-oxide regrowth, windows are opened in the thick oxide over source and drain regions, and aluminum is deposited for contacts and for gate metal. For an n-channel device, a similar procedure is used.

Note that for the procedure outlined above, it is necessary that the thin oxide and metal gate overlap both source and drain regions, in order to account for registration alignment errors. This means that there will be gate-source and gate-drain overlap capacitances. The latter is the more objectionable in applications

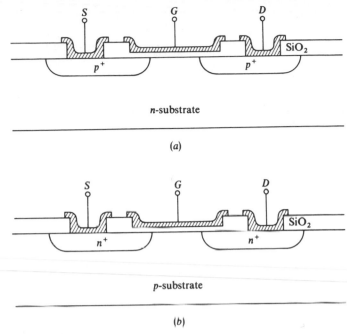

FIGURE 6-7
(a) Integrated p-channel MOSFET; (b) integrated n-channel MOSFET.

where the device is used in the common-source connection, since Miller-effect multiplication of the gate-drain capacitance can often occur.

The surface geometry for a p-channel MOSFET is shown in Fig. 6-8.

Exercise 6-5   Suppose a standard process has geometry constraints as follows:

$$\text{Minimum line width} = 0.4 \text{ mil}$$
$$\text{Minimum spacing} = 0.3 \text{ mil}$$
$$\text{Minimum contact window} = 0.5 \times 0.5 \text{ mil}$$

Lay out a p-channel MOSFET with $Z/L = 10$.

In an integrated circuit, metallization stripes will be used for interconnections and will be deposited on the surface of the thick oxide.   Because the interconnections will at times be above the n-type substrate, there is a possibility that an inversion channel may form in the substrate if large enough voltages are applied between substrate and metal interconnections.   Since the power supply voltages are usually the largest voltages in the circuit, power supply voltages must always be less than that threshold voltage V_{TF} which will produce an inversion of the substrate under

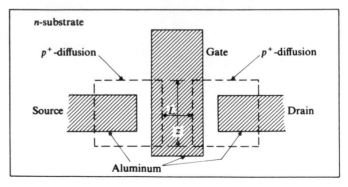

FIGURE 6-8
Surface geometry of an integrated *p*-channel MOSFET.

the thick oxide.   Should such an inversion occur, the channels formed could connect several devices together, causing circuit malfunction.

EXAMPLE 6F   *A Standard-Process p-Channel MOSFET on* ⟨111⟩ *Silicon.*   Consider the device of Example 6D fabricated on ⟨111⟩ silicon with 10^{-5}-cm gate oxide.   As we have seen, the threshold voltage is

$$V_T = -3.6 \text{ V}$$

For the thick oxide, 1.35×10^{-4} cm is approximately the maximum thickness that can be used because of processing constraints.   For this value, the threshold voltage V_{TF} is

$$V_{TF} = -37 \text{ V}$$

Power supply voltages less than 37 V would thus be required.   ////

EXAMPLE 6G   *A Standard-process p-Channel MOSFET on* ⟨100⟩ *Silicon.*   If the MOSFET is fabricated on ⟨100⟩ silicon, Q_{SS} will be lower than for ⟨111⟩ silicon.   Referring to Table 6-2, we find $Q_{SS}/q = 9 \times 10^{10}$ for ⟨100⟩ silicon.   This reduces the threshold voltage to

$$V_T = -1.7 \text{ V}$$

but at the same time the thick-oxide threshold voltage is reduced to

$$V_{TF} = -13 \text{ V}   ////$$

EXAMPLE 6H   *A Standard-process n-Channel Device on* ⟨100⟩ *Silicon.*   Consider a device made on ⟨100⟩ silicon with $N_A = 10^{15}$ cm^{-3} and $d_0 = 10^{-5}$ cm.   For this value of N_A, $\phi_{F(bulk)} = 0.29$ and $\phi_{MS} = -0.90$.   The threshold voltage is

$$V_T = -0.39 \text{ V}$$

Note that a negative threshold voltage for an n-channel device indicates that there will be an inversion layer when $V_G = 0$. Such a device is called a depletion-mode MOSFET. This would be a disadvantage in digital circuits, where it is desirable to have devices cut off with zero gate voltage. ·////

Normally, n-channel devices have a speed advantage, because electron surface mobility is about three times hole mobility. In order to take advantage of the larger electron mobility, it would seem advisable to use n-channel devices and employ a larger impurity concentration in the substrate in order to make V_T positive. Unfortunately, the mobility is an inverse function of impurity level, with the result that the standard-process n-channel devices with positive threshold voltages do not have great speed advantage over p-channel devices.

Complementary MOS Devices (CMOS)

In some applications it is desirable to have both n-channel and p-channel enhancement-mode devices. This is particularly true in micropower circuits, where low standby power is a requirement. If careful process control can be maintained, it is possible to fabricate both types of devices on the same chip. The fabrication sequence is shown in Fig. 6-9. Note that because a p-type diffusion is required for electrical isolation between devices, more surface area will be required for a given number of devices in CMOS than in circuits using only a single device type.

6-6 OTHER PROCESSING METHODS[4]

We have seen that oxide thickness, ϕ_{MS}, and Q_{SS} of the standard process impose limitations on the threshold voltage of MOSFETs, while the standard photolithography is responsible for the overlap capacitance, which limits the speed of operation. It is possible to mitigate these limitations at the expense of process complexity.

Metal-Nitride-Oxide Semiconductor Devices (MNOS)[5]

It will be recalled that the threshold voltage is given by

$$V_T = 2\phi_{F(\text{bulk})} + \phi_{MS} - \frac{Q_{SS}}{C_0} - \frac{Q_{B(\text{max})}}{C_0}$$

If C_0 could be made larger, V_T would decrease. This could be accomplished in two ways: by reducing the oxide thickness and by using a dielectric material with larger K_0. Of course, the dielectric material must be compatible with the processing. One such material is silicon nitride (Si_3N_4) which can be deposited by chemical vapor deposition techniques. Unfortunately, the silicon nitride–silicon interface produces a larger Q_{SS} than SiO_2-Si, and this nullifies the benefit derived from the $K_0 = 7.5$ of Si_3N_4. However, a compromise can be effected by growing

1. *P*-well mask

2. 1st boron deposition
   and diffusion-oxidation

3. *P*-channel source
   and drain mask

4. 2nd boron deposition
   and diffusion-oxidation

5. *N*-channel source
   and drain mask

6. Phosphorus deposition
   and diffusion-oxidation

7. Gate mask

8. Gate oxidation
   and contact mask

9. Metalization
   and metal mask

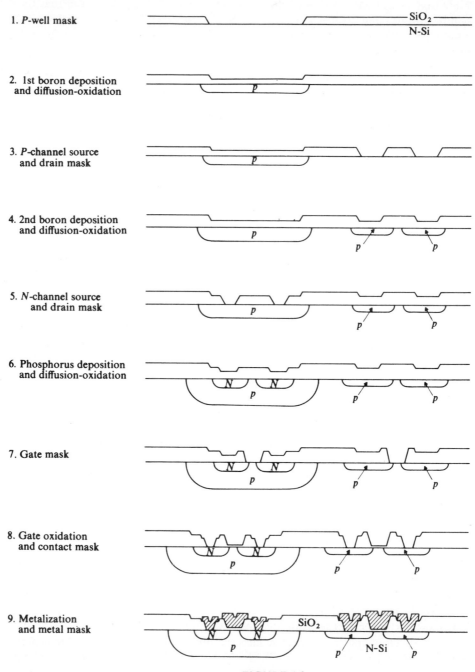

FIGURE 6-9
CMOS fabrication sequence.   (After Penney.[4])

a thin layer (2×10^{-6} cm) of SiO_2, and then depositing a thick enough layer (8×10^{-6}) of Si_3N_4 to reduce the pinhole density.   This provides a two-layer dielectric, as shown in Fig. 6-10, with an effective K_0 larger than that of the same thickness of SiO_2.   Typical effective values of K_0 are approximately 6.8.   For the p-channel device, this reduces the threshold voltage to about -2 V.

Exercise 6-6   A p-channel device is made by using 2×10^{-6} cm of SiO_2, 8×10^{-6} cm of Si_3N_4, and an aluminum gate over $\langle 111 \rangle$ silicon with $N_D = 10^{15}$.   Calculate the threshold voltage.

Self-aligning Gate Structures

To overcome the capacitance problems associated with gate-drain overlap, several techniques have been devised in which the gate-channel registration is automatic, making unnecessary the usual error resulting from alignment tolerances.   These techniques use ion implantation, silicon-gate structures, refractory metal devices, and thick-oxide structures.

Ion Implantation[6]

The ion implantation method is shown in Fig. 6-11.   Standard processing techniques are first used to diffuse source and drain regions, regrow thin oxide, and deposit metal.   However, the p diffusions are spread somewhat farther apart than the final desired channel length.   The gate metal thus falls between, but does not overlap, these regions.   A source of boron ions is used in the implantation chamber, and these ions are accelerated to have sufficient energy to penetrate the thin oxide and the silicon surface.   Note, however, that the gate metal prevents them from penetrating the region below it.   Thus source and drain regions now extend to the edge of, but not under, the gate metal.   Imperfect registration of gate metal with source and drain diffusions is unimportant.

Silicon-Gate Structures[7]

Both overlap capacitance and threshold voltage can be improved by using a gate of doped silicon rather than metal.   The process sequence is shown in Fig. 6-12.   After the thin oxide is regrown, a layer of polycrystalline silicon is deposited over the entire wafer by chemical vapor deposition techniques.   This silicon is etched away from all areas except that where the gate is to be located.   Then windows are opened in the oxide layer where source and drain diffusions are to take place.   A boron diffusion is now performed, and the gate silicon is also doped p-type during this diffusion.   Oxide is next deposited over the wafer, and windows are opened for metal contacts.   Conventional metal deposition follows.

Since the silicon gate determines where the p diffusion will not occur, no

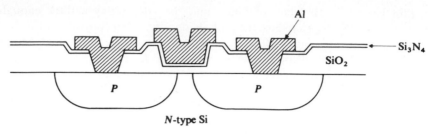

FIGURE 6-10
The MNOS device.   (After Penney.[4])

registration problem exists.   The only overlap capacitance is that resulting from lateral diffusion under the gate oxide.

EXAMPLE 6I   *Improvement of Threshold Voltage.*   To see how the threshold voltage is improved by the silicon gate, we consider once again the *p*-channel device of Example 6C, and we assume that the silicon gate has $N_A = 10^{15}$ cm^{-3}. Although the gate material is polycrystalline, we estimate its work function by

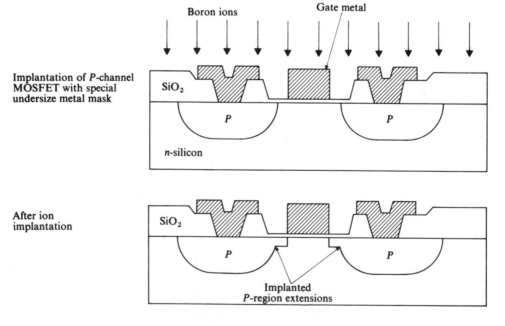

Implantation of *P*-channel MOSFET with special undersize metal mask

After ion implantation

FIGURE 6-11
Self-aligning gate made by ion implantation.   (After Penney.[4])

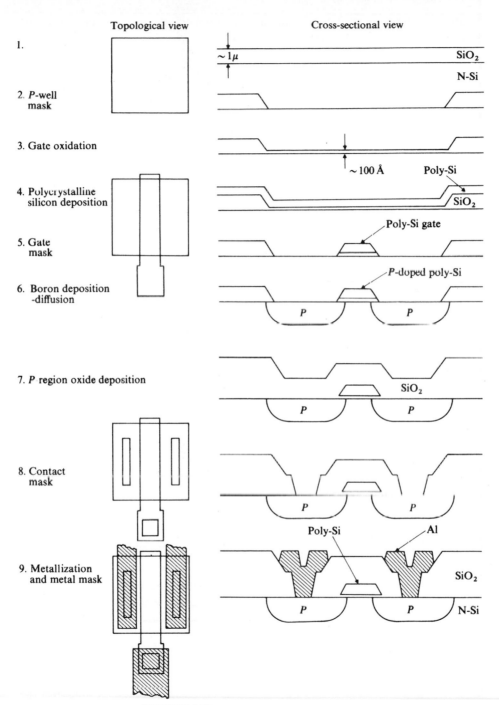

Topological view   **Cross-sectional view**

1.

$\sim 1\mu$ SiO$_2$

N-Si

2. *P*-well mask

3. Gate oxidation

~ 100 Å   Poly-Si

4. Polycrystalline silicon deposition

SiO$_2$

Poly-Si gate

5. Gate mask

P-doped poly-Si

6. Boron deposition -diffusion

P   *P*

7. *P* region oxide deposition

SiO$_2$

P   *P*

8. Contact mask

P   *P*

Poly-Si   Al

9. Metallization and metal mask

SiO$_2$

P   *P*   N-Si

FIGURE 6-12
Self-aligning gate made by doped silicon deposition. (After Penney.[4])

assuming it to be a perfect crystal. For a p-doped silicon gate and the n-type material of the device of Example 6C, we find

$$\phi_{MS} \approx 0.60 \text{ V}$$

Note that ϕ_{MS} is positive, whereas it was negative for an aluminum gate. For $\langle 111 \rangle$ silicon in the substrate the threshold voltage is

$$V_T = -2.5 \text{ V}$$

It will be recalled that for an aluminum gate, $V_T = -3.6$ V.

One more advantage results from the silicon gate. The polycrystalline silicon can also be used for interconnections, with the restriction that such interconnections must not cross any p regions. This restriction arises because the polycrystalline silicon is doped by the same predeposition as the p-type regions; if it crossed them it would mask them against the boron predeposition. Since the silicon gate is covered by oxide before metal deposition, the silicon-gate interconnections are at a different level than the metal.

The work-function difference ϕ_{MS} does not contribute significantly to V_{TF}; thus this device also has $V_{TF} = -37$ V.

The disadvantage of the silicon-gate structure is that it requires several extra processing steps, one of which is the deposition of polycrystalline silicon.   ////

MOSFETs Using Refractory Metal (RMOS)[8]

In the silicon structure and in the ion-implanted MOSFET, the gate was deposited before completion of the source and drain formation. In the latter case aluminum could be used for the gate because no further high-temperature processing was required. In the former case, use of silicon permitted high-temperature diffusions to be performed after the gate was in place. Clearly if a metal with high enough melting point could be found, the silicon-gate process could be used with the deposition of the metal replacing the deposition of polycrystalline silicon. One such metal is molybdenum. The molybdenum process also has the advantage that it can be used for interconnection, with the same restriction applying as for the silicon gate. It also offers the advantage of an improved threshold voltage.

Exercise 6-7 Molybdenum has a ϕ_M of approximately 4.0 V. Calculate the threshold voltage of the device of Example 6C with a molybdenum gate.

Self-aligning Thick-Oxide Structure (SATO)[9]

In the SATO process, silicon nitride is used both as a diffusion mask and to locate the gate region, as is shown in Fig. 6-13. Silicon nitride is first deposited over the entire wafer, and windows are opened for source and drain diffusion. Next the silicon nitride is removed everywhere but over the gate region, and a thick oxide

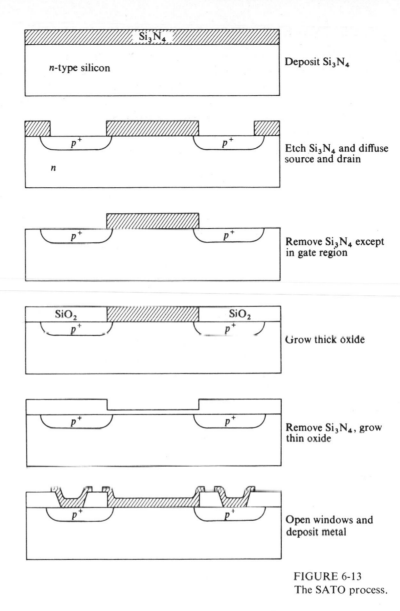

FIGURE 6-13
The SATO process.

is grown. The remaining nitride prevents growth over the gate region. This nitride is then removed and thin oxide is grown. Metal is then deposited as in the standard process. Note that because of alignment tolerances, metal will overlap the source and drain, but where it occurs the overlap will be over thick oxide and will thus produce negligible overlap capacitance. The only overlap on the thin oxide is that resulting from lateral diffusion under the original silicon nitride.

6-7 EQUIVALENT CIRCUITS FOR THE MOSFET[10]

In large-signal applications such as logic circuits, accurate calculation of switching times requires that the nonlinearities of the device be taken into account. This means that computer-aided analysis will be necessary. An equivalent circuit which portrays the first-order device behavior is shown in Fig. 6-14; this model includes the nonlinear gate-source and gate-drain capacitances C_{GS} and C_{DS}, the capacitances of the source and drain diffused regions C_1 and C_2, the nonlinear conductance G_{DS} which is observed in saturated operation, and the $I_D(V_D, V_G)$-dependent current source given by (6-7). It does not include parasitic series source and drain and substrate resistance. The analysis of a particular large-signal circuit can often be simplified, at the expense of accuracy, by averaging all of the nonlinear elements over the operating range. This technique results in a linear equivalent circuit which can be analyzed more rapidly.

For small-signal applications the parameters of the equivalent circuit become linear; a first-order equivalent circuit for saturated operation is shown in Fig. 6-15. Here the elements C_c, g_c, and the dependent source $g_m v_{gs}$ represent the "intrinsic" device. The distributed channel conductance and capacitance are represented in lumped form by C_c and g_c; these are responsible for the intrinsic cutoff frequency of the device. However, in most applications, extrinsic elements will usually dominate, and C_c and g_c can be neglected.

The gate-drain and gate-source overlap capacitances are represented by C_{gd} and C_{gs}, while the capacitance associated with the depletion charge around source and drain diffusions is represented by C_1 and C_2. The conductance g_{ds} represents the slope of the $I_D V_D$ characteristics that results from the shortening of the channel with increasing V_D.

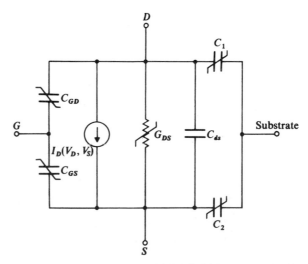

FIGURE 6-14
Large-signal equivalent circuit.

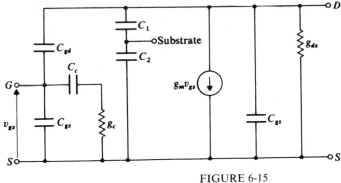

FIGURE 6-15
Small-signal equivalent circuit.

PROBLEMS

6-1 An MOS capacitor has n-type silicon with $N_D = 10^{15}$ and $d_0 = 1000$ Å. Find the charge per unit surface area of the holes in the inversion channel if $V_G = -5$ V.

6-2 Use the exact form for Q_B, and derive (6-7).

6-3 Use (6-7) to find I_{DS} and V_{DS}, the saturation drain voltage and current.

6-4 A p-channel enhancement-mode MOSFET is made on $\langle 111 \rangle$ silicon with an aluminum gate. The device has

$$N_D = 10^{15} \text{ cm}^{-3}$$

$$\phi_F = -0.29 \text{ V}$$

$$\phi_{MS} = 0.30 \text{ V}$$

$$\frac{Q_{ss}}{q} \quad 5 \times 10^{11} \text{ cm}^{-2}$$

$$d_0 = 800 \times 10^{-8} \text{ cm}$$

$$\frac{Z}{L} = 10$$

$$L = 0.5 \text{ mil}$$

Compare I_{DS} and V_{DS} obtained from the square-law formulation with I_{DS} and V_{DS} obtained from (6-7).

6-5 A p-channel enhancement-mode device has gate connected to drain. Find $I_D(V_D)$.

6-6 The device of Prob. 6-4 is to be used as a voltage-variable resistor. Assume that the square-law formulation is valid, and that the drain bias current is zero. Find the small-signal resistance $r(V_G)$ between source and drain terminals.

6-7 Two identical MOSFETs with the parameters of the device of Prob. 6-4 are used in the circuit of Fig. P6-7, in which one MOSFET acts as a small-signal amplifier and the other as a two-terminal small-signal load resistance. Find the dc level of the output and the small-signal gain v_{out}/v_{in}. What do the results indicate about the layout of MOSFETs which are to be used as load resistors when large gain is required?

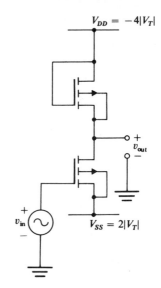

FIGURE P6-7

6-8   An *n*-channel MOSFET has $\phi_{MS} = 0$, $Q_{ss} = 0$, $\mu(Z/L)C_0 = 100 \times 10^{-6}$ A/V, $d_0 = 10^{-5}$ cm.   The device has threshold voltage $V_T = 1.42$ V, and is to be operated with $V_G = 3$ V.   Suppose processing variations cause d_0 to change to $0.5 - 10^{-5}$ cm. By what percentage does the saturation drain current I_{DS} change?

6-9   The device of Prob. 6-8 with $d_0 = 10^{-5}$ cm is used in the circuit of Fig. P6-9.   For what range of gate bias voltage V_G is operation in the saturated mode?

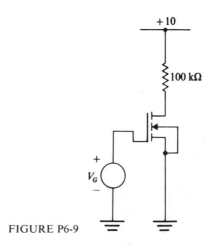

FIGURE P6-9

REFERENCES

1 GROVE, A. S.: "Physics and Technology of Semiconductor Devices," chaps. 9 and 11, John Wiley & Sons, Inc., New York, 1967.

2 CARR, W. N., and J. P. MIZE: "MOS/LSI Design and Application," chaps. 1 and 2, McGraw-Hill Book Company, New York, 1972.

3 DEAL, B. E., E. H. SNOW, and C. A. MEAD: Barrier Energies in Metal-Silicon Dioxide–Silicon Structures, *J. Phys. Chem. Solids*, vol. 27, pp. 1873–1879, December, 1966.

4 PENNEY, W. M. (ed.): "MOS Integrated Circuits," chap. 3, Van Nostrand Reinhold Company, New York, 1972.

5 DEAL, B. E., P. J. FLEMING, and P. L. CASTRO: Electrical Properties of Vapor-deposited Silicon Nitride and Silicon Oxide Films on Silicon, *J. Electrochem. Soc.*, vol. 115, p. 300, March 1968.

6 MAYER, J., L. ERIKSSON, and J. DAVIES: "Ion Implantation in Semiconductors Silicon and Germanium," Academic Press, Inc., New York, 1970.

7 FAGGIN, F., and T. KLEIN: Silicon Gate Technology, *Solid-State Electron.*, vol. 13, p. 1125, 1970.

8 WAKEFIELD, R. H., E. R. WARD, and J. A. CUNNINGHAM: Another Self-aligning MOS Process Has Interconnecting Advantages, *Electronics*, pp. 89–92, Jan. 3, 1972.

9 SCHMID, H.: Making LSI Circuits: A Comparison of Processing Techniques, *IEEE Trans. Manufac. Technol.*, pp. 19–31, December 1972.

10 HAMILTON, D. J., F. A. LINDHOLM, and A. H. MARSHAK: "Principles and Applications of Semiconductor Device Modeling," chap. 5, Holt, Rinehart and Winston, Inc., New York, 1971.

7

BIPOLAR TRANSISTORS AND DIODES

7-1 INTRODUCTION

The *n-p-n* bipolar transistor is the most widely used integrated device. It consumes less space than most resistors or capacitors, and for this reason the integrated-circuit designer attempts to design circuits which use transistors exclusively rather than combinations of resistors and transistors. In this chapter we discuss the behavior of the *n-p-n* transistor, emphasizing first-order effects. We consider first the dc behavior, and base our analysis on an idealized one-dimensional structure. We then investigate parasitic effects, and the relationship of device terminal behavior to physical processes. Next we consider various nonlinear phenomena such as high-level injection, current crowding, and base-width modulation. Finally, we discuss *p-n-p* transistors and integrated diodes.

7-2 THE IDEALIZED INTRINSIC STRUCTURE

Detailed analysis of the bipolar transistor is difficult because multidimensional effects must be considered, as must a multitude of nonlinear effects.[1] However, reasonable results, at least for dc behavior, can be obtained by approximating the

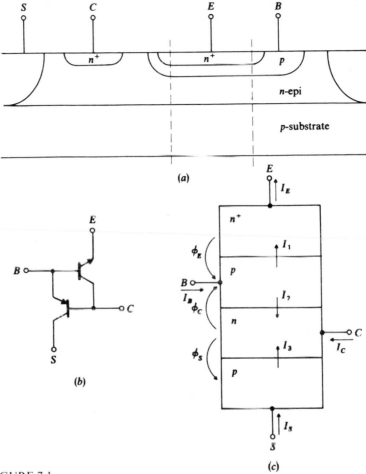

FIGURE 7-1
The *n-p-n* transistor. (*a*) Cross section, showing one-dimensional idealization; (*b*) symbolic representation of the four-layer structure; (*c*) definitions of polarities for the four-layer model.

device by an idealized one-dimensional structure. The cross section of the transistor is shown in Fig. 7-1*a*, in which the dashed lines indicate the region to be treated as a one-dimensional structure. Symbolically, the transistor can be represented as shown in Fig. 7-1*b*, where the *p-n-p* transistor represents the parasitic transistor made up of base, collector, and substrate.

It will be recalled that the dc behavior of a discrete three-terminal three-layer transistor, including the nonlinearities of the junction behavior, can be reasonably well represented by the Ebers-Moll equations,[2] which relate the *terminal* voltages to the *terminal* currents. A similar set of equations can be written for a general multiple-layer device, if the variables chosen are *junction voltages* and *junction* currents rather than terminal voltages and currents. The integrated transistor is a four-layer structure, and the set of equations relating its junction currents to its junction voltages is called *the four-layer nonlinear model*.[3, 4]

In order to be precise in our discussion of the integrated transistor we must define polarities for the junction voltages and currents. So that our model may be specialized to three-layer devices of either *n-p-n* or *p-n-p* type, we adopt the following convention:

> *1*   Junction voltage ϕ is the voltage of *p* material relative to *n* material. Therefore, ϕ is positive for forward-biased junctions and negative for reverse-biased junctions.
>
> *2*   Positive junction currents flow from *p* material to *n* material, that is, in the direction current would flow if the junction were a forward-biased diode.

If this convention is always observed, it will not be necessary to change signs for different types of transistors, or for multiple-layer structures with different order of layers. The polarity convention is shown in the symbol of Fig. 7-1c.

For the device represented by Fig. 7-1c, the four-layer model equations are

$$
\begin{bmatrix} I_1 \\ I_2 \\ I_3 \end{bmatrix} = \begin{bmatrix} 1 & -\alpha_I & 0 \\ -\alpha_N & 1 & -\alpha_{SI} \\ 0 & -\alpha_S & 1 \end{bmatrix} \begin{bmatrix} I_{s1}(e^{q\phi_E/mkT} - 1) \\ I_{s2}(e^{q\phi_C/mkT} - 1) \\ I_{s3}(e^{q\phi_S/mkT} - 1) \end{bmatrix} \tag{7-1}
$$

In (7-1) the α's are current gains for certain explicit connections of the junctions; for example, α_S is the ratio $I_3/-I_2$ for $\phi_E = \phi_S = 0$. The I_s's are junction saturation currents; for example, I_{s2} is the current $-I_2$ when $\phi_E = \phi_S = 0$ and $\phi_C \ll -mkT/q$. The parameter m is an artifice used to account for certain nonlinear phenomena, and ranges from $m = 1$ to $m = 2$.

While (7-1) provides a very convenient and systematic way of representing the relationship between junction voltages and junction currents, it does not directly yield the terminal behavior. This can easily be obtained with the use of (7-1) and Kirchhoff's laws. Having adopted the above convention for junction variables, we are at liberty to assign any polarities we wish to the terminal variables. A particularly convenient set is shown in Fig. 7-1c. For this set, the terminal variables are easily seen to be related to the junction variables by the following equations:

$$
\begin{bmatrix} I_E \\ I_B \\ I_C \\ I_S \end{bmatrix} = \begin{bmatrix} 1 & 0 & 0 \\ 1 & 1 & 0 \\ 0 & -1 & -1 \\ 0 & 0 & 1 \end{bmatrix} \begin{bmatrix} I_1 \\ I_2 \\ I_3 \end{bmatrix} \tag{7-2}
$$

$$
\begin{bmatrix} V_{BE} \\ V_{BC} \\ V_{SC} \end{bmatrix} = \begin{bmatrix} 1 & 0 & 0 \\ 0 & 1 & 0 \\ 0 & 0 & 1 \end{bmatrix} \begin{bmatrix} \phi_E \\ \phi_C \\ \phi_S \end{bmatrix} \tag{7-3}
$$

It is important to note that (7-1) to (7-3) can easily be specialized to represent the three-layer transistor. (For the *n-p-n* transistor, there is no fourth layer; by deleting row 3 and column 3 of the square matrices in (7-1) and (7-3), and row 4 and column 3 of the rectangular matrix in (7-2), we obtain the Ebers-Moll equations for the *n-p-n* transistor. Similarly for the *p-n-p* transistor there is no first *n*-type layer; we delete row 1 and column 1 of the square and rectangular matrices to obtain the Ebers-Moll equations for the *p-n-p* transistor. No sign changes are necessary since the polarities are explicit in Fig. 7-1c.

While the four-layer model provides a very convenient means for analyzing device terminal behavior, considerable care must be exercised in its application. This is attributable to the fact that the equations are nonlinear in the ϕ's. As an example of the troubles that can arise, consider the case where a junction current, for example I_1, is determined by an external voltage $-V$ and resistor R connected to the emitter terminal. For this case the left side of (7-1) will contain a term involving $(V - \phi_E)/R$ while the right side contains a term in $e^{\phi_E/mkT}$. The equations now become transcendental, and some means of obtaining an approximate solution is required.

Fortunately the nonlinearities in (7-1) are quite extreme, and some very simple approximations can be used. Note that at room temperature

$$0.026 \lesssim \frac{mkT}{q} \lesssim 0.052$$

The function $[e^{(q\phi/mkT)} - 1]$ can be approximated to within 10 percent as follows:

$$\text{For } \phi > 2.3 \frac{mkT}{q}: \quad e^{(q\phi/mkT)} - 1 \approx e^{q\phi/mkT}$$

$$\text{For } \phi < -2.3 \frac{mkT}{q}: \quad e^{(q\phi/mkT)} - 1 \approx -1$$

Physically, this means that for forward-biased junctions the -1 term can usually be neglected, and for reverse-biased junctions, the exponential term can usually be neglected.

Further simplifications can also be made. As the sketch of Fig. 7-2 shows, for sufficiently large forward current, very little variation of forward voltage occurs for variations of forward current. In fact, it is easily shown by using the Ebers-Moll equations for a diode that a tenfold increase of current causes only about 10 percent increase of the junction voltage. Thus in many cases the voltage of a forward-biased junction can be considered to be approximately a constant value V_D. For silicon diodes, a reasonable choice for V_D is 0.7 volt. The employment of these approximations in the use of the four-layer model is best illustrated by an example.

EXAMPLE 7A   *Saturation Voltage of the Integrated Inverter.*   The circuit to be considered is shown in Fig. 7-3a.   It is assumed that R_C and R_B are chosen so that

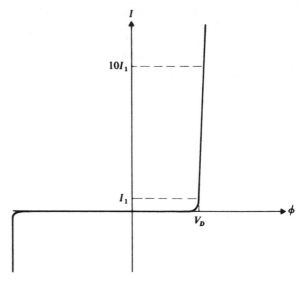

FIGURE 7-2
IV characteristic for a single junction.

the transistor is saturated; that is, both collector and emitter junctions are forward-biased. The substrate is also assumed to be grounded.

In order to determine the relationships among the various voltages, and to aid in determining which junctions are forward-biased, it is often convenient to replace the transistor by the symbol of Fig. 7-1b. We are interested in calculating the saturation voltage V_{CE}. From Fig. 7-3b we see by inspection that $V_{CE} = V_{CS} = \phi_S$. If the transistor is saturated, V_{CE} will be a positive voltage; this will reverse-bias the substrate junction. We therefore make the approximation that $\phi_S < -(mkT/q)$, and that

$$e^{q\phi_S/mkT} - 1 \approx -1$$

Since emitter and collector junctions are forward-biased, we assume their voltages to be approximately V_D, and we can then write

$$I_B \approx \frac{V_{CC} - V_D}{R_B} \qquad (7\text{-}4)$$

$$I_C \approx \frac{V_{CC}}{R_C} \qquad (7\text{-}5)$$

The four-layer model with all polarities specified is shown in Fig. 7-3c.

By use of approximations (7-4) and (7-5) we have essentially assumed that the terminal currents are independent of any variations of junction voltages; this is a reasonable approximation if V_{CC} is of the order of several volts.

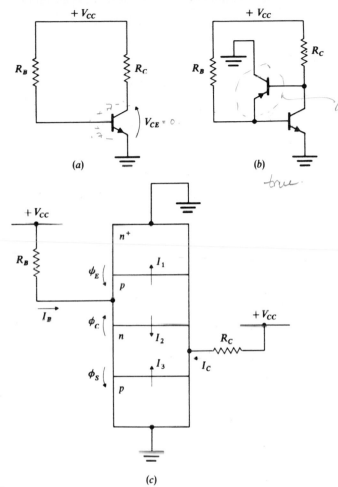

FIGURE 7-3
The saturating inverter. (*a*) The inverter circuit; (*b*) the circuit with *n-p-n* and parasitic *p-n-p*; (*c*) the four-layer model for the transistor.

Since both emitter and collector junctions are foward-biased, we make use of the approximation

$$e^{(q\phi/mkT)} - 1 \approx e^{q\phi/mkT}$$

for those junctions. The equations for the four-layer model are now written as

$$I_B = I_1 + I_2 = (1 - \alpha_N)I_{s1}\, e^{q\phi_E/mkT} + (1 - \alpha_I)I_{s2}\, e^{q\phi_C/mkT} + \alpha_{SI}\, I_{s3} \qquad (7\text{-}6)$$

$$-I_C = (I_2 + I_3) = -\alpha_N I_{s1}\, e^{q\phi_E/mkT} + (1 - \alpha_S)I_{s2}\, e^{q\phi_C/mkT} - (1 - \alpha_{SI})I_{s3} \qquad (7\text{-}7)$$

These equations can now be solved for ϕ_E and ϕ_C. Then by application of Kirchhoff's voltage law to Fig. 7-3c, we have

$$V_{CE} = \phi_E - \phi_C$$

Substitution of the values for ϕ_E and ϕ_C yields

$$V_{CE} = \frac{mkT}{q} \ln\left[\frac{I_{s2}}{I_{s1}} \frac{(I_B - \alpha_I I_{s3})(1 - \alpha_S) - (1 - \alpha_I)((1 - \alpha_{SI})I_{s3} - I_C)}{(1 - \alpha_N)((1 - \alpha_{SI})I_{s3} - I_C) + \alpha_N(I_B - \alpha_{SI} I_{s3})}\right]$$

Typical values for the parameters of an integrated transistor are

$$\alpha_N = 0.99 \qquad I_{s1} = 10^{-16} \text{ A}$$
$$\alpha_I = 0.20 \qquad I_{s2} = 10^{-15} \text{ A}$$
$$\alpha_S = 0.70 \qquad I_{s3} = 10^{-13} \text{ A}$$
$$\alpha_{SI} = 0.10$$

If the terminal currents are microamperes or larger, we see that all terms involving I_{s3} will be negligible, and we can approximate V_{CE} by

$$V_{CE} = \frac{mkT}{q} \ln\left[\frac{I_{s2}}{I_{s1}} \frac{(1 - \alpha_S)\alpha_N + (I_C/I_B)(1 - \alpha_I)}{\alpha_N - (1 - \alpha_N)(I_C/I_B)}\right] \qquad (7\text{-}8)$$

This indicates, as we intuitively expected, that V_{CE} is a function of the ratio I_C/I_B. We could also ask what ratio I_C/I_B is required to produce a given value of V_{CE}; (7-8) can be rearranged to yield

$$\frac{I_C}{I_B} = \frac{(I_{s1}/I_{s2}) e^{qV_{CE}/mkT} - (1 - \alpha_S)}{(1 - \alpha_I) + (I_{s1}/I_{s2})(1 - \alpha_N)e^{qV_{CE}/mkT}} \qquad (7\text{-}9)$$

////

Using the foregoing example as a guide, we can propose a set of general procedures to be followed in the use of the four-layer model:

1 By inspecting the circuit, determine which junctions are forward-biased. Typical voltage across a *junction* which is forward-biased is about 0.7 V. If external circuit voltages are large compared with this, some of the terminal currents can probably be considered to be *circuit-limited*; that is, the current is determined essentially by the external voltages and resistances. Knowledge such as this is extremely useful as it simplifies the nonlinear equations which will have to be solved.

2 Note that if a junction is reverse-biased by several times mkT/q, the term $e^{q\phi/mkT}$ is negligible compared with -1 and the device performance is essentially independent of that junction voltage. This knowledge is useful in eliminating some of the variables in the problem.

3 Using Kirchhoff's current law, write the terminal currents of interest in terms of the junction currents.

4 Using the model equations, write the junction currents in terms of the junction voltages.

5 Solve for the variables of interest.

Exercise 7-1   Find the substrate current for the transistor of Example 7A.

7-3 PARASITIC EFFECTS

Parasitic effects associated with integrated transistors fall into two broad categories: active and passive.   Active parasitic effects are the result of unwanted transistors or diodes which occur because of the additional junction arising from the necessity for isolation, or because of the proximity of several devices in a given isolation region.   These effects can be analyzed by use of the four-layer nonlinear model. In cases where both substrate junction and adjacent junctions contribute parasitic transistors, the three-dimensional problem can be treated as a composite of one-dimensional problems, and each one-dimensional problem can be analyzed with the four-layer model.   An example of this is the lateral *p-n-p* transistor, which is discussed later in this chapter.

Passive parasitic effects result from junction depletion-region capacitances and from resistances which occur between the contacts and the "active" regions of the device.   They can be treated with the techniques similar to those used for diffused resistors.   When calculating parasitic resistance, one should keep in mind that the thickness of the epitaxial material is only a few micrometers, while the surface geometry of the transistor will be several mils in extent.   Thus it is reasonable to approximate current flow in parasitic resistances as lateral.   Exact calculation of parasitic resistance is very difficult, so one generally makes a number of approximations in order to obtain an estimate.

Collector Resistance

If the transistor is biased in the forward active region for small-signal operation, the parasitic collector resistance is less important than in the case where the transistor saturates.   For the latter case the collector resistance contributes a voltage drop which may be larger than the saturation voltage of the intrinsic device.

The surface layout of a single-base-stripe single-collector-stripe transistor is shown in Fig. 7-4.   It is assumed that the minimum window size and minimum registration clearance are *m*, and that the emitter window is *km* in length.   The vertical current flow in the transistor is confined to the region directly under the emitter, and the collector resistance is that resistance in the path between the edge of the emitter and the collector contact.   Lateral diffusion is neglected. The path length for collector current flow is *2m*, and the average path width is

$$\frac{(k + 2)m + (k + 4)m}{2} = m(k + 3)$$

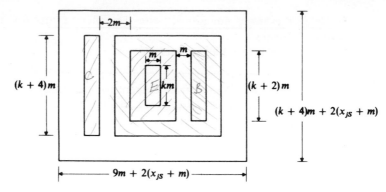

FIGURE 7-4
Surface geometry used for calculating parasitic collector and base resistances.

Therefore the number of squares is $2m/m(k + 3)$ and the collector resistance is

$$r_c = \frac{2\rho'_{sc}}{k + 3}$$

where ρ'_{sc} is the sheet resistance of the epitaxial material.

Base Resistance

Base resistance between the base contact and the edge of the emitter can be approximated in similar fashion. The path length for current flow in this case is m, while the width is $(k + 2)m$. The base resistance is

$$r_b = \frac{\rho'_{sb}}{k + 2}$$

where ρ'_{sb} is the sheet resistance of the base diffusion.

Neither r_b nor r_c includes the resistance of the contact itself, which depends upon the impurity concentration at the surface of the silicon. It will usually be smaller than r_b and r_c.

Capacitances

For the calculation of capacitances we use the bottom and sidewall capacitances employed in the calculation of diffused resistor parasitics. The emitter junction capacitance is

$$C_{jE} = C_{be}\, 3m^2(k + 2) + C_{se}(5 + k)2mx_{jE}$$

and the collector capacitance is

$$C_{jC} = C_{bc}\left[7m^2(k + 4) + \frac{\pi}{2}x_{jc}\, m(22 + 2k) \right]$$

If we now assume that the epitaxial thickness is x_{js}, the spacing between base diffusion window and isolation window must be $x_{js} + m$ and the substrate capacitance is

$$C_{js} = C_{ss}\frac{\pi}{2} x_{js}[m(2k + 34) + 8x_{js}]$$

$$+ C_{bs}[9m^2(k + 4) + 4(x_{js} + m)^2 + 2m(k + 13)(x_{js} + m)]$$

EXAMPLE 7B *Parasitics for a ½-Mil-Geometry Device.*   Consider a minimum size device in which $m = 0.5$ mil and $k = 1$.   Let

$$\rho'_{sh} = 200 \ \Omega/\text{sq}$$

$$\rho'_{sc} = 1250 \ \Omega/\text{sq}$$

$$x_{JE} = 2.0 \ \mu\text{m}$$

$$x_{jC} = 3.0 \ \mu\text{m}$$

$$x_{js} = 9.0 \ \mu\text{m}$$

$$C_{be} = 0.23 \ \text{pF}/\text{mil}^2$$

$$C_{se} = 0.65 \ \text{pF}/\text{mil}^2$$

$$C_{bc} - 0.10 \ \text{pF}/\text{mil}^2$$

$$C_{ss} = C_{bs} = 0.065 \ \text{pF}/\text{mil}^2$$

Using these values, we obtain

$$r_c = 625 \ \Omega$$

$$r_b = 67 \ \Omega$$

$$C_{jE} = 0.82 \ \text{pF}$$

$$C_{jc} = 1.10 \ \text{pF}$$

$$C_{js} = 2.45 \ \text{pF}$$

Such a large value for r_c would probably be unacceptable, even for small-signal applications.   To reduce r_c we could add a buried layer, or use a double collector stripe, or increase k.   All three measures would also result in larger substrate capacitances, the first because of an increase of C_{bs} and the other two because of an increase of the total area.   ////

Exercise 7-2 Estimate the parasitic element values for a double-base-stripe double-collector-stripe transistor with $m = 0.5$ mil and $k = 1$.

7-4   LARGE-SIGNAL AND SMALL-SIGNAL MODELS FOR THE INTEGRATED TRANSISTOR

Large-Signal Models

When all parasitic elements have been calculated, large- and small-signal models can be postulated to represent the transistor. In large-signal behavior, the four-layer nonlinear model for the intrinsic transistor can first be modified to include the intrinsic frequency effects as follows:

$$
\begin{bmatrix} I_1(s) \\ I_2(s) \\ I_3(s) \end{bmatrix} = \begin{bmatrix} 1 & -\alpha_I/(1 + \tau_{12}s) & 0 \\ -\alpha_N/(1 + \tau_{21}s) & 1 & -\alpha_{SI}/(1 + \tau_{23}s) \\ 0 & -\alpha_S/(1 + \tau_{32}s) & 1 \end{bmatrix} \begin{bmatrix} \lambda_E(s) \\ \lambda_C(s) \\ \lambda_S(s) \end{bmatrix}
$$

where $\lambda(s)$ is the Laplace transform of $I_s(e^{q\phi/mkT} - 1)$, $I(s)$ is the Laplace transform of $i(t)$, and s is the complex-frequency variable. Parasitic resistances and capacitances can be added external to the intrinsic model.

Two observations are in order regarding the calculation of large-signal behavior:

1 The frequency behavior of the intrinsic device is usually negligible in comparison with parasitic effects.

2 The equations of the four-layer model are nonlinear in the variables ϕ and i, and the parasitic capacitances are also nonlinear. The equations which describe the device and its parasitics are too complex to be solved by hand even for the simple four-layer model described above. A computer solution is generally required; for computer solution other forms of model may be more convenient than the four-layer model.

One method of analyzing large-signal behavior is to ignore the intrinsic frequency behavior, and use piecewise-linear techniques to handle the parasitics. As we shall see, in dealing with logic circuits this method permits one to make hand calculations which give reasonable estimates of transient behavior.

Small-Signal Model[5,6]

The well-known hybrid-π model is perhaps the most convenient means for representing the small-signal behavior of the device; it is shown in Fig. 7-5. In many instances, the model can be simplified to a unilateral version; this simplification and the application of the model are discussed in Chap. 9. The parameters of the model and their dependence on operating point are summarized as follows:

$r_{bb'}$ = base resistance, determined primarily by geometry and average base-region conductivity

$r_{b'e} = (\beta + 1)r'_e$

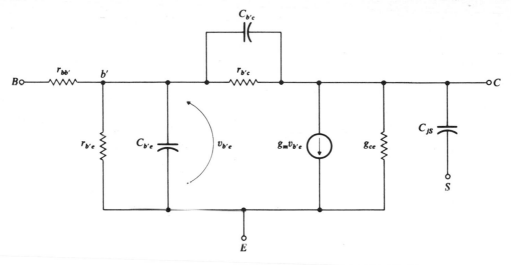

FIGURE 7-5
The hybrid-π model for small-signal behavior.

where $r_e' = \dfrac{mkT}{qI_E} = \dfrac{\alpha_N\, mkT}{qI_C}$

$\beta = \dfrac{\alpha_N}{1 - \alpha_N}$

$I_C,\, I_E =$ operating bias currents

Also,

$$C_{b'c} \approx C_{JC} = \text{collector junction capacitance}$$

$$C_{b'e} = \frac{1}{r_e'\,\omega_T} - C_{b'c} + C_{jE}$$

where $\omega_T =$ frequency of unity common-emitter current gain

$C_{js} =$ substrate junction capacitance

$g_m = \dfrac{\alpha_N}{r_e'}$

$g_{b'c} = \dfrac{1}{r_{b'c}} \approx \left(\dfrac{\eta}{\beta}\right) g_m$

$g_{ce} = \dfrac{1}{r_{ce}} = \eta g_m$

The factor η arises from base-width modulation and is given by

$$\eta \approx \frac{mkT}{q}\,\frac{1}{w}\,\frac{dw}{dV_{CB}}$$

Table 7-1 PARAMETER VALUES FOR TYPICAL COMMERCIAL DEVICES

	Small n-p-n	Large n-p-n	Lateral p-n-p	Substrate p-n-p	Double-collector lateral p-n-p		Double-emitter substrate p-n-p	
					A	B	A	B
I_s, 10^{-15} A	1.26	0.395	3.15	17.6	0.9	2.25	0.79	0.0063
$\beta_{0(\text{max})}$	290	520	95	130	0.4	1.9	94	21
$I_{C\beta_0(\text{max})}$, mA	0.3	0.3	0.11	0.11	0.03	0.07	0.015	0.015
B/β_0	0.72	0.77	0.79	0.90	0.95	0.78	0.85	0.9
B_I	2.5	6.1	3.8	4.8	1.4	1.5	1.5	1.0
r_b' 0.05 mV, Ω	670	185	500	80	1000	1600	1100	650
r_e', Ω	300	15	150	156	80	120	170	100
η, 10^{-4}	1.45	0.97	4.7	4.47	3.1	3.1	3.26	1.55
τ_t, ns	1.15	0.76	27.4	26.5	27.4	27.4	26.5	26.5
τ_n, ns	405	243	2540	2430	55	220	9550	2120
C_{je0}, pF	0.65	2.8	0.1	4.05	0.1	0.1	1.1	1.9
C_{jc0}, pF	0.36	1.55	1.05	2.8	0.3	0.9	2.4	2.4
C_{s0}, pF	3.2	7.8	5.1	—	4.8	4.8	—	—

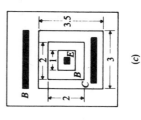

(c)

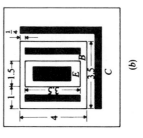

(b)

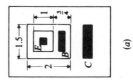

(a)

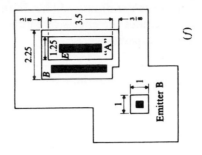

(f)

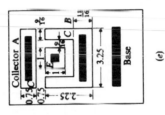

(e)

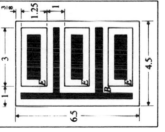

(d)

a Small n-p-n.
b Large n-p-n.
c Lateral p-n-p.
d Large substrate p-n-p.
e Dual-collector lateral p-n-p.
f Dual-emitter substrate p-n-p.   B is dc current gain, and the subscript I indicates inverse operation.
Note: All dimensions in mils.
SOURCE: Wooley et al.[24]

where w is the undepleted base width and V_{CB} is the collector-base bias voltage. Base-width modulation effects are discussed later in this chapter.

In the model, $C_{b'e}$ includes the effects of emitter depletion-layer capacitance as well as emitter diffusion capacitance.

Exercise 7-3 Calculate the hybrid-π parameters r_e', g_m, $C_{b'e}$, $C_{b'c}$ for the transistor of Example 7B if the bias current is 1 mA, and $\omega_T = 10^9$ rad.

Parameter values for typical commercial devices of various geometries are given in Table 7-1.

7-5 RELATION OF PHYSICAL PROPERTIES TO ELECTRICAL CHARACTERISTICS

In the analysis of the behavior of the integrated transistor, we will find it convenient to focus attention on various components of current which flow in the device. We confine our analysis to the steady-state case, and since in most circuits the substrate is biased so that there is no parasitic *p-n-p* action, we also ignore the substrate. Each junction current is assumed to be made up of hole and electron components I_p and I_n; these components are shown in Fig. 7-6, in which junction current polarities have been chosen consistent with the convention adopted in Sec. 7-2.

We are interested in calculating the current gain α_N where

$$\alpha_N \triangleq \frac{I_C}{I_E} \qquad \text{for } \phi_C \text{ reverse-biased}$$

The current gain can be written in terms of the various components as

$$\alpha_N = \frac{I_C}{I_E} = \frac{I_{nE}}{I_{nE} + I_{pE}} \frac{-I_{nC}}{I_{nE}} \frac{-(I_{nC} + I_{pC})}{-I_{nC}} \tag{7-10}$$

Each of the terms on the right side of (7-10) has a meaningful interpretation. The first term is the ratio of the emitter majority carrier current to the total emitter current. Since the emitter majority carriers (electrons) become minority carriers when injected into the base, the first term is a measure of the efficiency with which minority carriers are injected into the base region. It is called the *emitter efficiency* or *injection efficiency* γ.

The second term is the ratio of minority carriers leaving the base to minority carriers entering the base from the emitter. Since this is a measure of how effectively minority carriers are transported across the base, it is called the *base transport factor T'*.

Finally, the third term is the ratio of total collector current to electron current entering the base from the collector, and is called the collector efficiency.

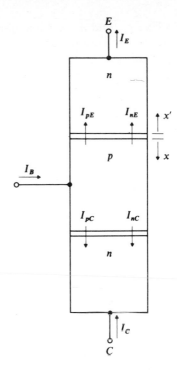

FIGURE 7-6
Current components for the transistor.

For a transistor biased in the forward active region, the hole current in the collector is negligibly small, and the collector efficiency is approximately 1. We can therefore write

$$\alpha_N \approx \gamma T' \qquad (7\text{-}11)$$

The current components in the device are controlled by the minority carrier distributions in the various regions:

In the emitter: $J_p = q\mu_p pE(x') - qD_p \dfrac{dp}{dx'}$

In the base: $J_n = q\mu_n nE(x) - qD_n \dfrac{dn}{dx}$

where μ_p, μ_n = hole and electron mobilities,
$\quad D_p$, D_n = hole and electron diffusivities,
$\qquad E(x)$ = electric field

The minority carrier distributions depend in turn upon the boundary conditions applied to the regions. These boundary conditions appear at junctions, where the minority carrier densities at the edges of the depletion regions are given by the *law of the junction*:

$$\begin{aligned} p &= p_{n0}\, e^{q\phi/mkT} \\ n &= n_{p0}\, e^{q\phi/mkT} \end{aligned} \qquad (7\text{-}12)$$

where p_{no} and n_{po} are equilibrium densities.   Other boundary conditions can also occur; for example, at a nonrectifying contact

$$n = n_{po} \quad \text{in } p \text{ material}$$
$$p = p_{no} \quad \text{in } n \text{ material}$$

(7-13)

The use of boundary conditions such as (7-12) and (7-13) generally enables one to solve for the minority carrier distributions and hence the various minority current components.

Electric Field in the Base Region

We consider first the case for thermal equilibrium.   For this case $J_p = 0$ and $J_n = 0$.   Setting $J_p = 0$ and solving for $E(x)$, we obtain

$$E(x) = \frac{kT}{q} \frac{1}{p} \frac{dp}{dx} \qquad (7\text{-}14)$$

where we have made use of the Einstein relation $D/\mu = kT/q$.   Now for highly extrinsic p-type material, $p \approx N(x)$ where

$$N(x) = N_A(x) - [N_D(x) + N_{BC}]$$

Thus we find that in equilibrium the field depends only on the impurity profile in the base:

$$E(x) = \frac{kT}{q} \frac{1}{N} \frac{dN}{dx} \qquad (7\text{-}15)$$

For nonequilibrium, if the *excess* majority carrier density is small compared with the majority carrier density, we can use the approximation

$$E(x) \approx \frac{kT}{q} \frac{1}{N} \frac{dN}{dx} \qquad (7\text{-}16)$$

That is, we assume the field to be built-in, and independent of the current flow. This simplifies the analysis considerably.

Base Transport Factor[7]

To calculate the base transport factor, we first solve for the minority carrier distribution in the base region.   In the steady state, the nonequilibrium electron current density is

$$J_n = q\mu_n nE + qD_n \frac{dn}{dx}$$

Substituting (7-16) for E, we have

$$J_n = q\mu_n n \frac{kT}{q} \frac{1}{N} \frac{dN}{dx} + qD_n \frac{dn}{dx} \qquad (7\text{-}17)$$

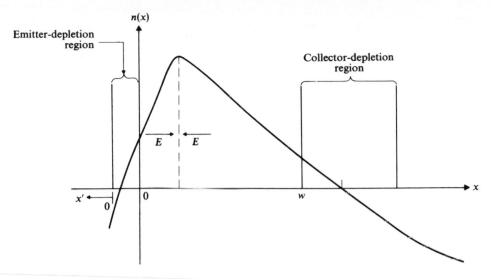

FIGURE 7-7
Net impurity profile and depletion regions.

We wish to find $n(x)$; rearranging (7-17), we obtain

$$\frac{dn}{dx} + n \frac{1}{N} \frac{dN}{dx} = \frac{J_n}{q D_n} \qquad (7\text{-}18)$$

Equation (7-18) is a first-order linear equation; it can be solved by use of an integrating factor $e^{\int u\, dx}$. The solution is

$$n e^{\int u\, dx} = \int \frac{J_n}{q D_n} e^{\int u\, dx}\, dx + C$$

where $u = (1/N)(dN/dx) = d(\ln N)/dx$. Substituting for u, we obtain

$$n(x) = \frac{1}{N} \int \frac{J_n}{q D_n} N\, dx + \frac{C}{N}$$

We now make a very important approximation: we assume that the change of transport current in the base region due to recombination is negligible. Then

$$J_n(x) \approx \text{constant} \triangleq J_n$$

We also assume D_n to be constant.

We can now write $n(x)$ in terms of the minority carrier current

$$n(x) \approx \frac{J_n}{q D_n} \frac{1}{N} \int N\, dx + \frac{C}{N} \qquad (7\text{-}19)$$

If it is assumed that $x = 0$ is the base edge of the emitter depletion region, and that the base edge of the collector depletion region is at $x = w$, as shown in Fig. 7-7, we can employ the law of the junction to find $n(w)$. For forward-active-region

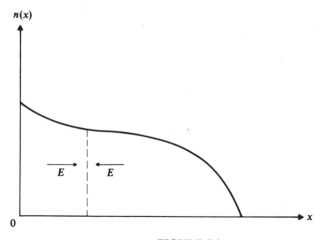

FIGURE 7-8
Electron distribution in the base.

operations, $\phi_C \ll -(mkT/q)$ and $n(w) \approx 0$. Using this in (7-19) to evaluate C, and rearranging the result, we find

$$n(x) = \frac{-J_n}{qD_n} \frac{1}{N} \int_x^w N \, dx \qquad (7\text{-}20)$$

If the emitter is forward-biased,

$$n(0) = n_{po}(0)e^{q\phi_E/mkT} = \frac{n_i^2 \, e^{q\phi_E/mkT}}{N(0)} \qquad (7\text{-}21)$$

Inserting (7-21) in (7-20), we can now find J_n as a function of ϕ_E:

$$J_n = -qD_n N(0)n_{po}(0) \frac{e^{q\phi_E/mkT}}{\int_0^w N \, dx} \qquad (7\text{-}22)$$

It is important to note that for a given ϕ_E, J_n depends only on the impurity concentration at $x = 0$, and the number of impurities in the base given by $\int_0^w N \, dx$. This latter quantity is called the _Gummel number N_G_.

Qualitatively, we see from Fig. 7-7 that the electric field is a _retarding_ field for minority carriers near the emitter side of the base, and an _aiding_ field over the rest of the base region. The distribution of minority carriers in the base is sketched in Fig. 7-8.

The total recombination current in the base is

$$J_r = \frac{q}{\tau_n} \int_0^w [n(x) - n_{po}(x)] \, dx$$

if the lifetime τ_n is constant. We have assumed that this current is small compared with J_n. If we regard J_n as the electron current collected by the collector

$$J_{nC} = J_n$$

then the current that would have to be injected at the emitter is

$$J_{nE} \approx -J_n + J_r$$

(Note that for forward-active-region operation, J_n will be negative.)   The base transport factor is then

$$T' = \frac{-J_{nC}}{J_{nE}} = \frac{-J_n}{-J_n + J_r}$$

or

$$T' = \frac{1}{1 + q\{\int_0^w [n(x) - n_{po}(x)]\, dx\}/ -J_n \tau_n}$$

from which

$$T' \approx 1 - \frac{q}{-J_n \tau_n} \cdot \int_0^w n(x)\, dx \qquad (7\text{-}23)$$

Substituting $n(x)$ in (7-23), we obtain

$$T' \approx 1 - \frac{1}{L_n^2} \int_0^w \left[\frac{1}{N(x'')} \int_{x''}^w N(x)\, dx \right] dx'' \qquad (7\text{-}24)$$

where $L_n^2 = D_n \tau_n$.   For the homogeneous base transistor, (7-24) becomes

$$T' \approx 1 - \frac{w^2}{2L_n^2}$$

Exercise 7-4   Calculate T' for an exponential impurity distribution in the base.

The Emitter Efficiency

It will be recalled that although the emitter region is n-type, it was originally the epitaxial material, which was first compensated by acceptor impurities to form the base and then compensated again by donor impurities to form the emitter.   Furthermore, the impurity concentration after the final compensation is very large. When silicon is successively compensated in this way with large impurity concentration, two degrading effects result: the carrier mobility is significantly reduced by impurity scattering effects, and the minority carrier lifetime τ_p is drastically reduced.   The result is that in the emitter region, the minority carrier diffusion length $L_p = \sqrt{D_p \tau_p}$ is of the order of only a micrometer.

With such a short diffusion length, the gradient of the minority carrier distribution is very large, and the diffusion component of hole current in the emitter will be much larger than the drift current.   This means that any electric fields in the emitter will have negligible effect on the hole current.   The conditions in the emitter are thus

$$J_p(x') \approx -q D_p \frac{dp}{dx'}$$

with

$$p(0) \triangleq P_0 = p_{n0} e^{q\phi_E/mkT}$$

$$p(L) = p_{n0}$$

If $L > L_p$, the solution for $p(x')$ is

$$p(x') = p_{n0} + (P_0 - p_{n0})e^{-x'/L} \qquad (7\text{-}25)$$

from which

$$J_{pE} \approx \frac{qD_p P_0}{L_p} \qquad (7\text{-}26)$$

The hole distribution is sketched in Fig. 7-9.

The hole current can also be written in terms of the emitter junction voltage, through the use of the law of the junction:

$$J_{pE} = q \frac{D_p}{L_p} p_{n0}(0) e^{q\phi_E/mkT}$$

$$= q \frac{D_p}{L_p} \frac{n_i^2}{-N'(0)} e^{q\phi_E/mkT} \qquad (7\text{-}27)$$

where $N'(0) \triangleq N(x' = 0)$. (The minus sign arises because $N(x)$ is negative since the net impurity is donor type.) Now the emitter efficiency γ can be calculated. Recalling that

$$\gamma = \frac{J_{nE}}{J_{nE} + J_{pE}}$$

we substitute the expressions for J_{nE} and J_{pE} to obtain

$$\gamma = \frac{1}{1 + (D_p/D_n L_p)(\int_0^w N\, dx)/-N'(0)}$$

For good transistors, the second term in the denominator will be small compared with unity, and we may approximate γ by

$$\gamma = 1 - \frac{D_p}{D_n L_p} \frac{1}{-N'(0)} \int_0^w N\, dx \qquad (7\text{-}28)$$

Since we want γ to be as close to unity as possible, we can see that the emitter region should be more heavily doped than the base in order that

$$-N'(0) \gg \int_0^w N\, dx$$

The common-base current gain is

$$\alpha_N = \gamma T'$$

while the common-emitter current gain is

$$\beta = \frac{\alpha_N}{1 - \alpha_N} = \frac{\gamma T'}{1 - \gamma T'} \qquad (7\text{-}29)$$

If we insert numbers for representative integrated transistors, we obtain

$$\beta \approx 500$$

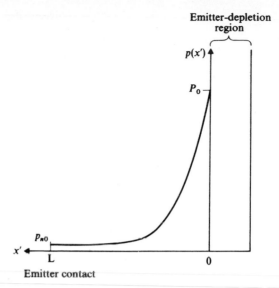

FIGURE 7-9
Hole distribution in the emitter.

This value is somewhat large, since measured values are typically of the order of 200. However, it should be noted that we have neglected surface effects, as well as three-dimensional aspects. Thus (7-29) is a reasonable indication of an upper bound for β.

EXAMPLE 7C *Estimate of the Current Gain of a Typical Transistor.* Consider a hypothetical transistor with $x_{jE} = 2.0$ μm and $x_{jC} = 3.0$ μm. Let the undepleted base width be 0.5 μm. Assume that the net impurity distribution in the unde-pleted base varies approximately linearly from 10^{16} cm^{-3} to 10^{15} cm^{-3} and that $-N'(0) \approx 10^{18}$ cm^{-3}. Let the hole diffusion length in the emitter be $L_p \approx 1$ μm and the electron diffusion length in the base be $L_n \approx 10$ μm. Then we obtain

$$\gamma \approx 1 - 0.00125$$

$$T' \approx 1 - 0.00125$$

From these, we calculate

$$\beta \approx 400 \qquad ////$$

Exercise 7-5 Calculate γ, T', and β for a homogeneous base transistor with $w = 0.5$ μm, $N_D = 10^{18}$, $N_A = 10^{16}$, $D_p/D_n \approx 0.5$, $L_p = 1$ μm, and $L_n = 10$ μm.

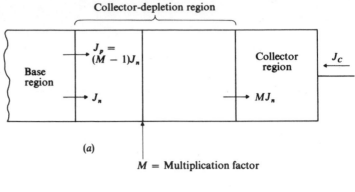

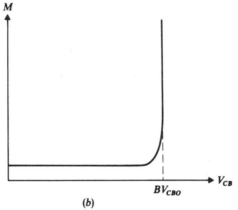

(b)

FIGURE 7-10
Avalanche breakdown. (a) Junction currents; (b) multiplication factor.

Breakdown Voltage

Breakdown in junctions occurs, as we saw in Chap. 4, when the maximum electric field in the junction exceeds a critical value E_C. In terms of physical processes, this breakdown is the result of avalanche multiplication of carriers in the depletion region. Suppose that electrons entering the collector depletion region from the base produce a current J_n as shown in Fig. 7-10. (For forward-active-region operation, J_n and J_p are negative.) This current would be the minority carrier current transported across the base. If sufficient reverse voltage is applied to the collector junction so that $E_{\max} \rightarrow E_C$, these electrons will ionize silicon atoms by impact collision, and produce a number of hole-electron pairs, say $M - 1$. Now the direction of the electric field in the depletion region is such as to force electrons into the collector region and holes back into the base region. The electrons reaching the collector thus consist of the original electrons entering the depletion region plus all the electrons created by ionization:

$$-J_C = J_n + (M - 1)J_n = MJ_n$$

while the hole current entering the base region from the collector depletion region is

$$-J_p = (M - 1)J_n$$

The *multiplication factor M* is a strong function of the electric field and hence of the applied junction voltage. It has been empirically determined that M is well approximated by[8]

$$M \approx \frac{1}{1 - (V_{CB}/BV_{CBO})^n}$$

where $n \approx 3$ for silicon, and BV_{CBO} is the collector junction breakdown voltage with zero emitter current. BV_{CBO} is the value of the breakdown voltage obtained from, for example, the Lawrence-Warner breakdown curves for the collector junction. From Fig. 7-10b it is evident that as $V_{CB} \to BV_{CBO}$, $M \to \infty$.

It is important to note that when the transistor is operated in the forward active region, the effect of M is to increase the collector current for a given transport current J_n. Thus the avalanche multiplication produces an *effective* transport factor which can be larger than unity.

Effect of Avalanche Multiplication on the *VI* Characteristics

When the transistor is connected in the common-base configuration with some bias current I_E, this current determines the level of injected electron current at the emitter edge of the base. The transport factor determines how much of this current reaches the collector, and M determines how large the resulting collector current will be. The hole current resulting from avalanche multiplication flows into the base region where it becomes majority carrier current, and ultimately base terminal current. The effective current gain is now

$$\alpha_N(M) = \alpha_N M(V_{CB}) \tag{7-30}$$

The collector characteristics appear as shown in Fig. 7-11a.

In the common-emitter configuration, base current is fixed. Now the hole current resulting from multiplication cannot flow as base current because of the external terminal constraint. Thus more electrons must flow in from the emitter to neutralize these holes. It is important to note that the hole current resulting from multiplication produces the same effect as would an equal current flowing in the base terminal without multiplication. Therefore, as far as the collector current is concerned, the effects of multiplication are multiplied by β when the transistor is in the common-emitter orientation.

This result can be quantified as follows:

$$\beta(M) = \frac{\alpha_N(M)}{1 - \alpha_N(M)} \tag{7-31}$$

This means that $\beta \to \infty$ when $\alpha_N(M) \to 1$, that is, when $M\alpha_N \to 1$. The collector current in the common-emitter orientation will thus become very large at a collector

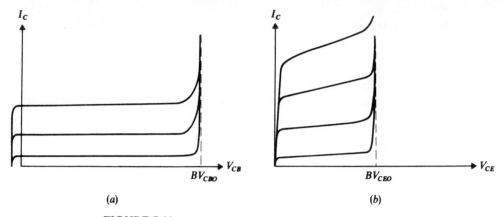

FIGURE 7-11
Effect of avalanche multiplication on transistor characteristics. (*a*) Common base;
(*b*) common emitter.

voltage much lower than BV_{CBO}. If we assume that $V_{CB} \approx V_{CE}$ when significant multiplication occurs, we can find the common-emitter breakdown voltage BV_{CEO} which is the voltage V_{CE} at which $I_C \to \infty$.

Inserting the functional behavior of M in (7-31) and setting $\alpha_N[M(BV_{CEO})] = 1$, we solve for BV_{CEO}:

$$BV_{CEO} = BV_{CBO}(1 - \alpha_N)^{1/n} \qquad (7\text{-}32)$$

The transistor characteristics in the common-emitter orientation appear as in Fig. 7-11*b*. The operating voltage for the common-emitter transistor must be maintained below BV_{CEO} if the transistor is not to be destroyed by excess power dissipation.

Equation (7-32) shows that the closer α_N approaches unity, the lower is BV_{CEO}. For devices with large β, it is necessary to have large BV_{CBO}, that is, large junction breakdown voltage, if even a reasonable BV_{CEO} is to be achieved.

EXAMPLE 7D BV_{CEO} *for a Typical Transistor.* Consider the transistor of Example 4F, for which we determined by using the Lawrence-Warner breakdown curves that $BV_{CBO} \approx 50$ V. Now suppose that the same device has $\beta = 200$. Then $1 - \alpha_N \approx 1/\beta_N = 0.005$ and

$$BV_{CEO} = BV_{CBO}(0.005)^{1/3}$$

$$= 8.6 \text{ V}$$

This device has a very small operating range in the common-emitter orientation.

$////$

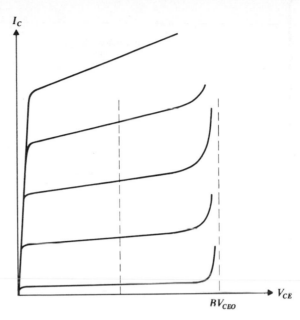

FIGURE 7-12
Base-width modulation effects.

Base-Width Modulation

If one plots the $I_C V_{CE}$ characteristics for a transistor in the common-emitter orientation, it is observed that even for voltages well below BV_{CEO}, the curves have
considerable slope, as is shown in Fig. 7-12. Moreover, at any given V_{CE}, the
slope increases as I_B (and consequently I_C) is increased. Probably the most
important single contributor to this effect is the variation of the collector depletion
region, and hence the base width, with V_{CE}. Detailed analysis of this effect for
integrated devices is difficult because it requires use of the actual impurity profiles.
However, we can obtain semiquantitative results, which are useful in determining
design guidelines, by analyzing the abrupt-junction homogeneous-base device.
    Consider qualitatively the base region of the integrated transistor. In for-
ward-active-region operation with $V_{CB} = 0$, the collector depletion region will
have its minimum width, and for a given base current I_B, the electron distribution in the base will be as shown in Fig. 7-13. Let the total number of electrons in the base be Q_B. Then the base current is

$$I_B = \frac{Q_B}{\tau_n}$$

and the collector current can be shown to be

$$I_C = \frac{Q_B}{\tau_t}$$

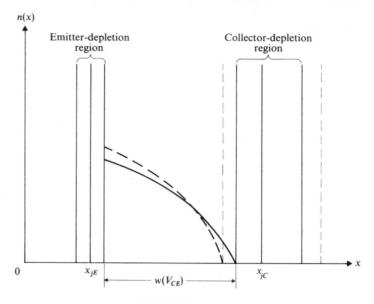

FIGURE 7-13
Electron distribution for fixed base current I_B.

where τ_t is the effective minority carrier transit time, which is a function of w. Now as V_{CB} is increased, the collector depletion region increases. Therefore the electrons must redistribute themselves over a narrower base. But since I_B does not change, neither can Q_B. Therefore the electron density must increase at the emitter edge of the base, and the slope must increase at the collector edge as shown by the broken line in Fig. 7-13. Hence the collector current increases. In terms of transit time, the transit time is reduced as w is decreased, and thus I_C increases.

To find the slope of the characteristics, we need to calculate

$$\frac{\partial I_C}{\partial V_{CE}} = \frac{\partial I_C}{\partial w}\frac{\partial w}{\partial V_{CE}} \qquad (7\text{-}33)$$

But to proceed further for the integrated transistor we must know how I_C depends on w and how w depends on V_{CE}. For nonhomogeneous-base devices, the problem is intractable. We therefore turn to the homogeneous-base device. For this case

$$\tau_t \approx \frac{w^2}{2D_n}$$

Then

$$I_C = \frac{I_B 2D_n}{w^2 \tau_n} = \beta I_B \qquad (7\text{-}34)$$

and

$$\frac{\partial I_C}{\partial V_{CE}} = -2\beta I_B\left(\frac{1}{w}\right)\frac{dw}{dV_{CE}} \qquad (7\text{-}35)$$

Solving (7-34) for w and substituting the result in (7-35), we obtain

$$\frac{\partial I_C}{\partial V_{CE}} = \frac{-\sqrt{2}}{L_n} \beta^{3/2} I_B \frac{dw}{dV_{CE}} \qquad (7\text{-}36)$$

Now we let the base width for $V_{CB} = 0$ be w_0, and the depletion width into the base be w_d. For V_{CE} much larger than the built-in voltage ϕ'_C, we can assume

1. $V_{CB} \approx V_{CE}$
2. $w_d \approx K\sqrt{V_{CE}}$

where K is a constant. Since $w = w_0 - w_d$, we have

$$\frac{dw}{dV_{CE}} = -\frac{dw_d}{dV_{CE}} = \frac{-K}{2\sqrt{V_{CE}}}$$

Inserting this in (7-36), we get

$$\frac{\partial I_C}{\partial V_{CE}} = \frac{KI_B \beta^{3/2}}{L_n \sqrt{2V_{CE}}}$$

From this we note that

1. For a given V_{CE}, the slope of the curves increases with increasing I_B.
2. At a given I_B and V_{CE}, transistors with larger β will have larger dI_C/dV_{CE}.

Integrated transistors can be expected to behave qualitatively in the same way, and dw/dV_{CE} can be crudely estimated from the Lawrence-Warner curves. We expect transistors with large β to show more pronounced base-width modulation effects than low β devices.

Inclusion of Base Width Modulation Effects in the Four-Layer Model[9,10]

For a transistor operating in the forward active region with substrate reverse-biased, the effects of base-width modulation can be included by means of a simple approximation. If the collector characteristics of the device are plotted, it is observed that each member of the family is well approximated by a straight line over a wide range of values of V_{CE}. Moreover, if the straight-line approximation for each member of the family is extrapolated, it is observed that all have the same voltage intercept $-V_A$ as shown in Fig. 7-14. This voltage is called the *Early voltage* and is found to be independent of temperature.

Now if junction saturation currents are assumed to be small compared with the terminal currents, the four-layer model equations for the collector and emitter currents in the absence of base-width modulation are

$$I_E \approx \frac{I_C}{\alpha_N} = I_C\left(1 + \frac{1}{\beta}\right)$$

$$I_C = I_{s1}(e^{qV_{BE}/mkT} - 1)\alpha_N$$

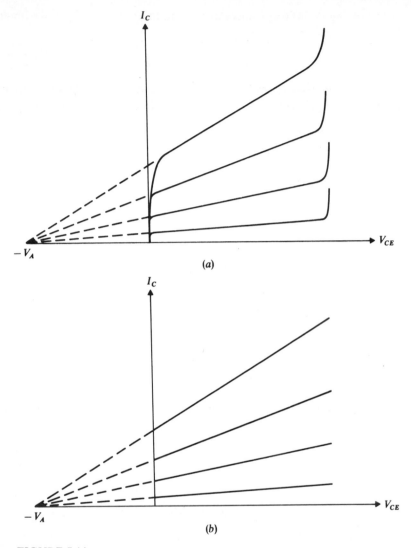

FIGURE 7-14
(*a*) Extrapolation of the collector characteristics; (*b*) idealization of the collector characteristics.

By applying simple geometric arguments to Fig. 7-14*b*, we find that with base-width modulation the equations now become

$$I_C = I_s(e^{qV_{BE}/mkT} - 1)\left(1 + \frac{V_{CE}}{V_A}\right)$$

$$I_E = I_C\left[1 + \frac{1}{\beta(1 + V_{CE}/V_A)}\right]$$

These methods provide a simple way of including base-width modulation effects in the four-layer model.

Space-Charge Layer Recombination

In calculating the emitter efficiency we assumed that the electron and hole currents at the emitter edge of the base were the same as those at the emitter edge of the depletion region. This approximation is not good at low values of forward bias because a recombination current J_{rE} occurs in the depletion region:

$$J_{rE} \propto n_i e^{qV_{BE}/2kT}$$

The electron current injected into the base, however, is

$$J_{nE} \propto n_i^2 e^{qV_{BE}/kT}$$

If we assume that hole current in the emitter is negligible, then

$$J_E = J_{nE} + J_{rE}$$

The emitter efficiency for this case is

$$\gamma = \frac{J_{nF}}{J_E} = \frac{1}{1 + J_{rE}/J_{nE}} \approx 1 - \frac{J_{rE}}{J_{nE}}$$

which can be expressed as

$$\gamma = 1 - K_1 e^{E_G/2kT} e^{-qV_{BE}/2kT} \qquad (7\text{-}37)$$

where K_1 is a constant and E_G is the gap energy for silicon. Equation (7-37) shows that γ decreases at lower bias voltage; therefore we can expect β to decrease at lower bias voltage. If one measures I_C, the result is as shown in Fig. 7-15, which is in agreement with 7-37.[11] The parameter m in the four-layer-model equations is used to account for this effect.

Qualitative Discussion of Transistor Design in Terms of Processing Variables

Several guidelines for device design can be inferred from the analyses of emitter efficiency, transport factor, breakdown, and base-width modulation.

At first it appears that large β and large BV_{CEO} are not compatible requirements. However, note that the designer has a number of variables at his disposal. He may choose N_{BC}, the junction depths, and within limits, the surface concentrations of the diffusions. To obtain large β requires $\gamma \to 1$ and $T' \to 1$.

The transport factor increases as w decreases, but base-width modulation is related to *variations* of w, and therefore to the way the *base* side of the depletion region varies. By making N_{BC} low, most of the increase of depletion width with voltage can be made to occur on the *collector* side. For this case small w can be used, and acceptable base-width modulation effects will result. Therefore, large β can be made reasonably compatible with small base-width modulation.

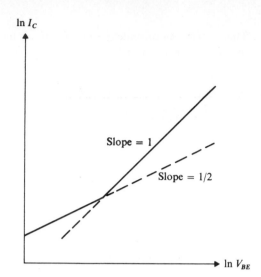

FIGURE 7-15
Effects of recombination in the emitter
depletion region.

The breakdown voltage BV_{CE0} is directly proportional to BV_{CB0}, which depends on the relative impurity concentrations in base and collector. If N_{BC} is small enough, BV_{CB0} in excess of 100 V can be achieved. For any given base diffusion, the Lawrence-Warner breakdown field curves can be used to determine what N_{BC} is necessary to achieve a given BV_{CB0}. Consequently, both large β and reasonable BV_{CE0} are compatible requirements if sufficient flexibility exists in the selection of N_{BC}.

Use of low N_{BC} increases the parasitic resistance of the collector, since this resistance varies directly with N_{BC}. This increase can be offset to some degree by increasing the epitaxial-layer thickness, but the decrease of resistance is at the expense of increased sidewall capacitance of the isolation junction, and increased lateral diffusion. The surface area for a given isolation region also increases. The collector resistance can be lowered by adding a buried-layer diffusion; note, however, that this is done at the expense of adding extra processing steps.

For good emitter efficiency, it is desirable to have a large impurity concentration on the emitter side of the emitter depletion region, and a small Gummel number for the base. But because of the diffusion boundary conditions, we cannot independently select junction depth, surface concentration, and Gummel number. If we try to achieve a low Gummel number for a given junction depth, the surface concentration will be low, and surface inversion may result. Moreover, the depletion into the base from the collector junction will be greater with low Gummel number, and hence base-width modulation will increase.

If we instead try to increase γ by increasing the impurity concentration in the emitter, the diffusion length in the emitter will decrease as a result of over-compensation of the material, and γ may actually decrease as a result. For many devices, γ is the limiting factor in determining β. An acceptable compromise is to choose the base diffusion to minimize depletion into the base from the collector, thus minimizing base-width modulation. Then the emitter diffusion can be chosen so that a reasonable γ is obtained. As a general rule, after all drive-ins have been finished, the emitter surface concentration should be at least an order of magnitude larger than the base surface concentration. The emitter surface concentration has to be of the order of 10^{19} cm^{-3} in order to obtain ohmic contacts where aluminum is used for interconnections.

7-6 HIGH-LEVEL INJECTION AND HIGH-CURRENT OPERATION

High-Level Injection

High-level injection is defined to occur when the excess majority carrier density exceeds the equilibrium majority carrier density. In the base region of the n-p-n transistor, this can be written as

$$p - p_{po} \gg p_{po} \qquad (7\text{-}38)$$

where p_{po} is the equilibrium majority carrier density. When this condition exists, the minority density is also large, and

$$p \approx n$$

In the integrated transistor, when high-level injection occurs, the carrier densities are so large that the effects of the built-in field resulting from the impurity gradient are negligible. The device then behaves approximately as a diffusion transistor, but the diffusion coefficient is approximately twice that at low-level injection.[12] This is because at high levels the drift component of current is approximately equal to the diffusion component. Thus, if one considers the total current to behave as a diffusion current, one finds that the effective diffusion coefficient doubles.

In the integrated transistor, high-level injection effects occur in the base region before they occur in the emitter, because the base impurity concentration is lower, and therefore inequality (7-38) is realized at a lower current level in the base than in the emitter region.

The law of the junction is also modified at high-level injection conditions.[13] If we assume that high-level injection occurs in the base but not in the emitter, we find that the law of the junction becomes

$$n = n_{po}(0)e^{qV_{BE}/2kT} \quad \text{in the base at } x = x_{jE} + x_{pE}$$

$$p = p_{po}(0)e^{qV_{BE}/kT} \quad \text{in the emitter at } x = x_{jE} - x_{nE}$$

where V_{BE} is the applied voltage. We can write the current densities as

$$j_{pE} = k_1(e^{qV_{BE}/kT} - 1) \qquad (7\text{-}39)$$

$$j_{nE} = k_2(e^{qV_{BE}/2kT} - 1) \qquad (7\text{-}40)$$

At high levels, a greater percentage of the total emitter current is carried by

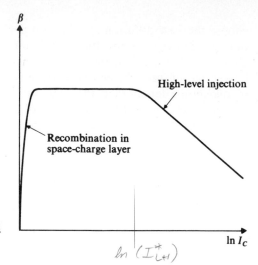

FIGURE 7-16
Decrease of β as a result of high-level injection.

holes in the emitter region than was the case for low levels, because of the change of the law of the junction. Therefore, the emitter efficiency

$$\gamma = \frac{1}{1 + (j_{pE}/j_{nE})} \qquad (7\text{-}41)$$

decreases at high levels. In fact, if we assume that the base transport factor is essentially unity, and that the common-emitter current gain is

$$\beta \triangleq \frac{\gamma}{1 - \gamma} \approx \frac{1}{1 - \gamma}$$

we can combine (7-39) to (7-41) to show that for high-level injection

$$\beta \propto \frac{1}{I_{nC}}$$

where I_{nC} is the collector current.

At high levels, the recombination rate increases in the base region, causing a further degradation of β at high current densities. A sketch of β versus collector current is shown in Fig. 7-16.

A further manifestation of high-level injection is a change of the slope of I_C versus V_{BE}, as is shown in Fig. 7-17. Again, the parameter m is used in the four-layer model to account for high-level injection.

The approximate current I_{LH}^* at which β begins to decrease because of high-level injection is

$$I_{LH}^* \approx \frac{kT}{q} \frac{WL\bar{\sigma}_b}{w} \qquad (7\text{-}42)$$

where L = emitter length (surface)
W = emitter width (surface)
$\bar{\sigma}_b$ = average conductivity of base region under emitter
w = undepleted base width

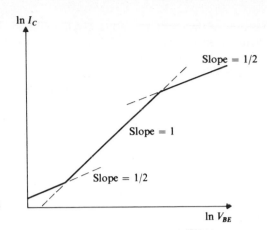

FIGURE 7-17
Effects of high-level injection on I_C vs V_{BE}.

EXAMPLE 7E *Onset of High-level Injection for a Typical Device.* Consider a device having minimum emitter window size 0.5×0.5 mil, and undepleted base width $w = 0.5$ μm. Let the average conductivity of the undepleted base region be $\bar{\sigma}_b \approx 1$ ℧/cm.

If we assume that registration clearances are also 0.5 mil, then the surface dimensions (neglecting lateral diffusion) of the emitter are 1.5×1.5 mil. The onset of high-level injection occurs at a collector current of

$$I_{LH}^* \approx 7.5 \text{ mA} \qquad ////$$

Current Crowding[14]

A sketch of the emitter and base regions of an integrated transistor is shown in Fig. 7-18. Note that the base current must flow laterally under the emitter region; this base current flows to satisfy the net majority carrier requirements in the base region. However, because of the nonzero resistivity of the base region, the lateral current flow produces a voltage drop v_b with polarity as shown. This voltage is of such a polarity as to *reduce* the forward junction voltage near the

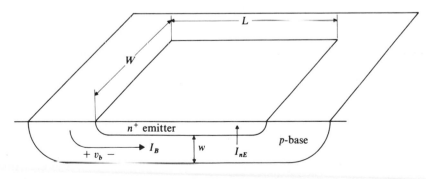

FIGURE 7-18
Sketch of the emitter and base regions.

center of the emitter. For a given applied voltage at the terminals, more injection will occur near the *edge* of the emitter than at the center. The current is therefore "crowded" toward the perimeter of the emitter, and the effective area of the emitter is reduced.

From these qualitative remarks, we anticipate that crowding effects will increase with increasing base resistivity. Since the base current increases as β decreases, we expect that a decrease of β will cause an increase of crowding effects.

If there were no current crowding effects and the base current was uniform across the base region, the lateral drop would be

$$v_b = I_B \rho_{sb}'' \frac{L}{W}$$

where ρ_{sb}'' = sheet resistance of base layer (under emitter)

$$= \frac{1}{w\bar{\sigma}_b}$$

Now if we let T' = base transport factor and γ = emitter efficiency, we can relate the base current to the emitter current by

$$I_B = I_E(1 - \gamma T')$$

and we have

$$v_b = I_E(1 - \gamma T')\rho_{sb}'' \frac{L}{W}$$

It can be shown that appreciable crowding effects occur when the lateral voltage drop is $2kT/q$. When this happens, of course, the base current is no longer uniform across the base. However, it is useful to describe crowding effects in terms of the emitter current I_{E1} which would produce a drop $2kT/q$ if there were no crowding effects:

$$I_{E1} \frac{L}{W} = \frac{2kT/q}{(1 - \gamma T')\rho_{sb}''} \triangleq I_x \qquad (7\text{-}43)$$

Note that I_x is this emitter current times the number of squares in the base region.

When appreciable crowding occurs, the effective width of the emitter is reduced, and is given approximately by

$$W_{\text{eff}} \approx W \frac{I_x}{I_E}$$

Equation (7-43) also shows that the current at which crowding is significant is

$$I_E = I_{E1} = I_x \frac{W}{L}$$

If we plot W_{eff}/W versus $(I_E/I_x)/(W/L)$, we obtain a plot similar to that shown in Fig. 7-19. Since we have defined significant crowding to occur when $(I_E/I_x)/(W/L) = 1$, we see that for this current $W_{\text{eff}} \approx 0.6W$.

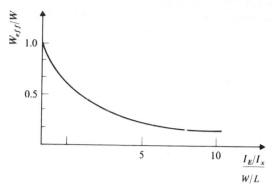

FIGURE 7-19
W_{eff}/W vs $I_E/I_x/W/L$.

Design for Maximum β[15]

Both current crowding and high-level injection cause a degradation of β. Thus in designing integrated devices, we would like to arrange for appreciable crowding effects to occur at about the same current level where high-level injection effects become significant. The maximum β would then be approximately the β we could expect to obtain at that current. We can find this current by using W_{eff} in the equation for I_{LH}^*; this yields

$$I_{LH}^* = \frac{kT}{q} L \bar{\sigma}_b \sqrt{\frac{2}{1 - \gamma T'}} \approx \frac{kT}{q} L \bar{\sigma}_b \sqrt{2\beta_{max}} \qquad (7\text{-}44)$$

When this condition is realized, detailed analysis shows that the effective base resistance is reduced by crowding effects to a value

$$R_{B(eff)} = \frac{R_B}{3}$$

where

$$R_B = \rho_{sb}'' \frac{L}{W}$$

Now if one must operate at currents above I_{LH}^*, to realize β_{max} the transistor geometry must be changed. This should be done by using the *minimum dimension* for L, and increasing W. Since crowding causes the current to flow near the perimeter of the emitter, the emitter *perimeter*, rather than the area, should be increased. Space can sometimes be conserved by using an interdigitated structure as shown in Fig. 7-20.

EXAMPLE 7F[16] *Design of a Transistor for High-current Operation.* We now consider the design of a transistor for operation at 500 mA with $\beta_{max} = 50$. Multiple emitter stripes with operation at 500 mA with $\beta_{max} - 50$. Multiple emitter stripes with interdigitated geometry are used and it is assumed that the average base resistivity is $\bar{\rho}_b = 0.2 \ \Omega \cdot cm$. Because of the interdigitated geometry, each

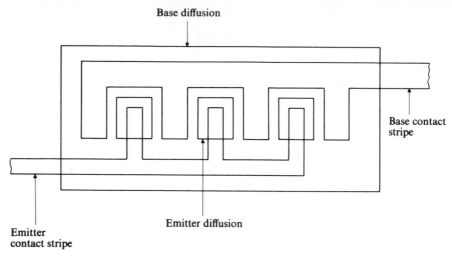

Base diffusion

Base contact
stripe

Emitter diffusion

Emitter
contact stripe

FIGURE 7-20
Interdigitated structure.

emitter stripe sees effectively a double base stripe as shown in Fig. 7-21. For each emitter stripe, we denote the emitter current by I_l. Using (7-44), we write

$$\frac{I_l}{L} = \frac{kT}{q\bar{\rho}_b}\sqrt{2\beta_{max}} = 1.3 \text{ A/cm}$$

Next we must be sure that the voltage drop along the emitter metallization stripe is negligible. For each emitter stripe, let $L = l$. Then we must have

$$I_l \rho'_{sAl} \frac{l}{1 \text{ mil}} < 0.025$$

where ρ'_{sAl} is the sheet resistance of the aluminum stripe.

For 0.5-μm thickness of aluminum, ρ'_{sAl} is approximately 0.05 Ω per square. The effective perimeter of the stripe is $2l$ since crowding forces injection to occur at the edge of the stripe. Therefore we must also satisfy the condition

$$\frac{I_l}{2l} = 1.30 \text{ A/cm}$$

Combining these, we obtain

$$l < 2.21 \times 10^{-2} \text{ cm} = 8.7 \text{ mils}$$

If we choose $l = 8$ mils, the current per stripe is then

$$I_l = 1.30 \times 2l = 53 \text{ mA}$$

Ten stripes are therefore required. The layout for the transistor might appear as shown in Fig. 7-22. ////

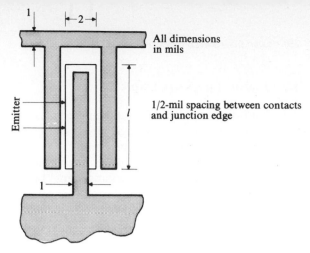

All dimensions
in mils

1/2-mil spacing between contacts
and junction edge

FIGURE 7-21
Emitter and base-stripe geometry (after Burger and Donovan[16]).

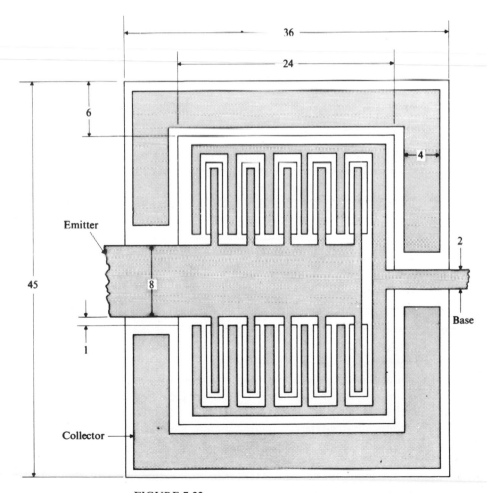

FIGURE 7-22
Topology of the high-current transistor (after Burger and Donovan[16]).

7-7 *P-N-P* TRANSISTORS

All of our discussion thus far has centered on *n-p-n* transistors. In many circuits, notably biasing circuits and logic circuits, it is desirable or even essential to have available a *p-n-p* transistor, preferably one whose characteristics complement those of the *n-p-n* device. It is difficult to achieve a complementary *p-n-p* structure without adding additional processing steps, but if the designer is willing to relax the requirement for complementary characteristics, reasonable *p-n-p* devices can be made with no extra steps. We consider first the extra processing step required for compatible *p-n-p* devices.

The Triple-diffused *P-N-P*

The triple-diffused *p-n-p* structure is perhaps the most obvious compatible structure. While there are many ways to make integrated-circuit *p-n-p* transistors using multiple buried layers, selective epitaxial layers, and complicated additional diffusion steps, the triple-diffused *p-n-p* structure is a straightforward extension of the double-diffused *n-p-n* structure. An additional *p* diffusion is used to form the *p-n-p* emitter. This *p-n-p* is doubly isolated—once by the *p*-type wafer substrate and isolation diffusions and once by the *n*-type epitaxial layer, so that all triple-diffused *p-n-p*'s in a given circuit could be placed in the same isolation region.

The primary problem in triple-diffused fabrication is the successively higher surface concentrations of the three diffusion steps. It is difficult to control the phosphorus diffusion which forms the *n-p-n* transistor emitter because of anomalous diffusion effects present with phosphorus. The second *p*-type diffusion must therefore have a very heavy surface concentration; in fact it is not always possible to get a workable structure which does not require exceeding the solid solubility of the *p* diffusant. There is also a considerable problem in specifying diffusion times and temperatures in order that the structure desired will result.

Because of all of the impurity compensation necessary for a successful triple-diffused transistor, the lifetime of holes in the *p-n-p* base will be low. The mobility will also be low both in the emitter and the base. Finally, because of the heavy concentrations involved, the emitter efficiency of the triple-diffused device is poor.

To alleviate many of the above problems, we might choose to lower the concentration of the first *p* diffusion in order to lower the concentrations required of successive diffusions. The difficulty inherent in the design compromise between impurity-level reduction and *p-n-p* design is that a lightly doped *p*-type surface will invert, producing an *n*-type conducting channel which shorts across the *p* region. In the *n-p-n* transistor the inverted surface would short collector and emitter together, while in the *p-n-p* it would short the base to the isolation region. There is always the possibility for parasitic MOS transistor action, particularly in circuits where surfaces are lightly doped or where high voltages appear on the interconnecting metallization. The channeling may be terminated by diffusing guard rings completely around all p regions. The guard rings in the triple-diffused p-n-p-n-p-n compatible structure can easily be put down by using the p^+ third diffusion.

From the foregoing discussion it should be apparent that the triple-diffused transistor structure has serious design and fabrication difficulties which are out of proportion to any performance improvements gained. Triple-diffused transistors also suffer from increased transistor size, large parasitic capacitances, and low breakdown voltages. Thus the triple-diffused p-n-p transistor structure has limited application for integrated circuits.

Vertical *P-N-P* Transistors

We have already seen that a parasitic p-n-p transistor exists between the base, collector, and substrate of the standard n-p-n device. It is difficult to make this p-n-p a complementary transistor, and there is the additional disadvantage that the collector is the substrate, which will be connected to the most negative supply voltage. This disadvantage seriously limits the applicability of the device in circuits. However, some cases exist, such as emitter followers, where a substrate p-n-p could be employed.

The cross section of the substrate p-n-p is shown in Fig. 7-23. From inspection of this figure, several reasons can be deduced for the difficulty encountered in producing complementary substrate p-n-p transistors. First, the base width is the difference between the collector junction depth of the n-p-n and the substrate junction depth. For epitaxial-layer thicknesses of 8 μm, the base width will be approximately 6 μm. When compared with the typical base width of 0.5 μm for n-p-n devices, this 6 μm seems intolerably large, since the transport factor depends on the square of the base width. However, the problem is somewhat mitigated by the fact that the epitaxial material has not been compensated, and therefore has reasonably good lifetime and diffusion length. Diffusion lengths of the order of 30 μm can be easily achieved; because of this, values of 50 for β are attainable. While this is still much lower than typical values for n-p-n devices, it is still large enough to provide reasonable performance.

The base width of the substrate p-n-p can, of course, be reduced by using thinner epitaxial material. However, this compromises the behavior of the n-p-n because the collector material of the n-p-n may be completely depleted by even low reverse collector voltages, resulting in punch-through of the collector. At the same time, the series collector resistance of the n-p-n will also be increased.

The second degrading factor of the p-n-p performance arises because of the relatively high resistivity of the collector material. We have already seen that a typical epitaxial layer (8 μm thick, $N_{BC} = 5 \times 10^{15}$ cm^{-3}) has a sheet resistance of 1250 Ω per square. The region under the p diffusion, being only 6 μm thick, has a sheet resistance of 1660 Ω per square. Thus the lateral base resistance of the device will be considerably larger than that of the n-p-n.

A third degrading factor of p-n-p performance results from its increased base resistance. Current crowding will set in at much lower current levels than is the

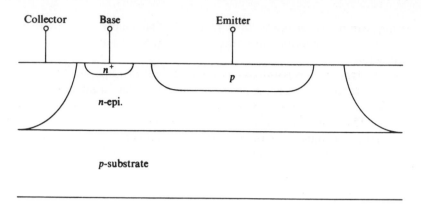

FIGURE 7-23
Cross section of the substrate *p-n-p*.

case for the *n-p-n* device. This means that for *n-p-n* and *p-n-p* devices of comparable size, the decrease of β with I_C will begin to occur at much lower values of I_C
for the *p-n-p* than for the *n-p-n*.

The emitter efficiency of the *p-n-p* is likely to be lower than the *n-p-n* because
the impurity level of the *p* diffusion has been chosen to optimize the behavior of
the *n-p-n*, and it will usually be lower than that required for good injection efficiency
of the *p-n-p*. Increase of the surface concentration of the *p* diffusion in order to
improve the injection efficiency of the *p-n-p* will result in decrease of the injection
efficiency of the *n-p-n*.

Finally, the parasitic collector capacitance of the *p-n-p* will be much larger
than that of the *n-p-n* because the sidewalls of the isolation diffusion cause the
collector junction of the *p-n-p* to have large area. The combination of large
collector capacitance and large base resistance contributes to poor high-frequency
behavior of the substrate *p-n-p*.

EXAMPLE 7G   *Onset of Current Crowding in a Typical Substrate P-N-P.*   Consider a substrate *p-n-p* with 8-μm epitaxial thickness, $N_{BC} = 5 \times 10^{15}$ cm^{-3}, *p*
diffusion junction depth of 2 μm, and emitter 1.5 × 3.0 mils. If the registration
clearance is 0.5 mil, the n^+ diffusion for the base contact will be 0.5 mil from the
edge of the *p* diffusion. The lateral base resistance which contributes to current
crowding will be that directly under the emitter. This region is 1.5 × 3.0 mils, or
0.5 square. If we assume that single-stripe geometry is used, appreciable crowding occurs at a collector current of

$$I_C \approx \frac{(W/L)(2kT/q)}{(1 - \gamma T')\rho'_{sb}}$$

Suppose that the low current β is 50; that is,

$$\beta = \frac{\gamma T'}{1 - \gamma T'} \approx \frac{1}{1 - \gamma T'} = 50$$

We have previously calculated $\rho'_{sb} = 1660\ \Omega$ per square. Thus we find

$$I_C = 6.26\ \text{mA}$$

If this transistor is required to operate at a bias current of tens of milliamperes, it will be necessary to increase the W/L ratio by an order of magnitude. This will increase the size of the device, with a resulting increase of capacitance.

This same device has a series base resistance which does not contribute to current crowding; this is the resistance between the n^+ contact diffusion and the edge of the emitter. It has an L/W ratio of $0.5/3.0 = 0.167$, and a resistance

$$R_B = 0.167 \times 1250 = 208\ \Omega \qquad ////$$

Exercise 7-6 Suppose the transistor of Example 7G has a minority carrier diffusion length $I_{·p} = 30\ \mu\text{m}$, and that the net impurity concentration on the p side of the emitter depletion region is 5×10^{16}. Assume that the base region is uniformly doped, and estimate the transport factor, emitter efficiency, and β of the device.

Exercise 7-7 Compute the current level at which the onset of high-level injection occurs for the device of Exercise 7-6.

The Lateral P-N-P[17]

The restriction imposed on circuit use by the substrate collector of the substrate p-n-p can be removed by making a lateral p-n-p, whose cross section and surface geometry are shown in Fig. 7-24. While rectangular surface geometry is shown in Fig. 7-24, circular geometry can also be used. In this device, we rely on the minority carriers injected from the sidewalls of the emitter to reach the collector and provide the desired transistor action. Since the effective base width of the device depends upon the distance between emitter and collector sidewalls, it is desirable to have emitter and collector as close together as possible. An important aspect of the fabrication of the device makes possible a narrow base width: both emitter and collector are fabricated during the same diffusion cycle. This means that no registration of masks between emitter and collector diffusions is required. Hence the tolerance on spacing between emitter and collector is determined basically by the tolerance on control of the extent of lateral diffusion, that is, control of the junction depth.

Note that parasitic substrate p-n-p transistors occur at emitter and collector regions of the lateral p-n-p. Consider forward-active-region operation. For this case the collector parasitic transistor has both junctions reverse-biased. The emitter parasitic transistor is in the foward active region, however, and it reduces the effective gain of the lateral device. In fact, since only the carriers emitted from the sidewalls of the emitter contribute to lateral p-n-p action, we can see that two guidelines should be observed in laying out the lateral p-n-p:

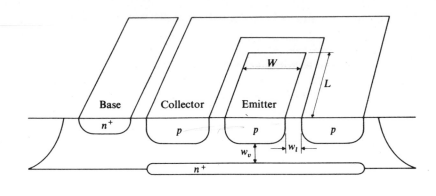

(a)

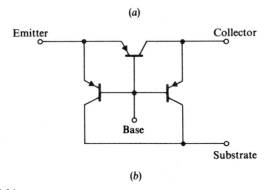

(b)

FIGURE 7-24
(a) The lateral *p-n-p* transistor; (b) parasitic transistors associated with the lateral device.

1 The emitter should be surrounded by the collector so that as much as possible of the injection from the sidewalls is collected by the collector.
2 The minimum line width should be used for the emitter diffusion window, in order to get the largest possible ratio of sidewall to bottom area, thus minimizing as much as possible the effect of the parasitic *p-n-p*.

First-order Analysis of the Lateral *P-N-P*

From the foregoing discussion, it is clear that because of the dependence of performance on sidewall emission, three-dimensional behavior will be important in the lateral *p-n-p*. In order to obtain a tractable first-order estimate of performance, we need to make a number of simplifying approximations. These are as follows:

1 The injection efficiency is unity.
2 As far as laterally injected carriers are concerned, we arbitrarily assume

that only those carriers injected between the surface and a depth $x_{jc}/2$ have a chance to reach the collector. The rest miss the collector and either recombine or are collected by the substrate. We assume the effective sidewall emitter area A_E for laterally injected carriers to be

$$A_E \approx \frac{x_{jc}}{2} (2L + 2W) = x_{jc}(L + W)$$

where L is the length of the emitter window and W is its width. The effective parasitic emitter area A_p is assumed to be that of the surface area of the emitter:

$$A_p \approx L(W + 2x_{jc})$$

3 The effective base width w_l for laterally injected carriers is approximately the base width at the surface. The base width w_v for other carriers is

$$w_v \approx x_{js} - x_{jc}$$

where x_{js} is the substrate junction depth.

4 The minority carrier diffusion length in the epitaxial material is L_p.

5 Depletion regions are neglected.

Fortunately, the lateral p-n-p is a homogeneous-base device; this greatly simplifies the analysis. The transport factor for the lateral p-n-p is

$$T'_l = \frac{1}{1 + w_l^2/2L_p^2} \approx 1 - \frac{w_l^2}{2L_p^2}$$

while for the parasitic p-n-p it is

$$T''_v = \frac{1}{1 + w_v^2/2L_p^2} = \frac{1}{1 + (x_{js} - x_{jc})^2/2L_p^2}$$

If the β of the lateral p-n-p is defined as the ratio of collector current to base current with both collector and substrate reverse-biased, β can be calculated as follows. The collector current is given by

$$I_C = \frac{T'_l I_E A_E}{A_E + A_p} \qquad (7\text{-}45)$$

The base current is made up of the recombination current of laterally and vertically injected carriers. In most cases $x_{js} - x_{jc} \gg w_l$, and vertically injected carriers will dominate in the recombination current. If this is the case,

$$I_B \approx \frac{(1 - T''_v)I_E A_p}{A_E + A_p} \qquad (7\text{-}46)$$

Taking the ratio of (7-46) and (7-45), we obtain

$$\beta = \frac{T'_l(A_E/A_p)}{1 - T''_v}$$

$$\approx \frac{A_E}{A_p}\left(1 - \frac{w_l^2}{2L_p^2}\right)\left(1 + \frac{w_v^2}{2L_p^2}\right)\frac{2L_p^2}{w_v^2}$$

For most practical cases, $w_l \ll L_p$, and β is then

$$\beta \approx \frac{A_E}{A_p}\left(1 + \frac{2L_p^2}{w_v^2}\right) \qquad (7\text{-}47)$$

Equation (7-47) shows that as far as surface geometry is concerned, β depends on A_E/A_p, as would be intuitively expected. Using our approximations for A_E and A_p, we find that

$$\frac{A_E}{A_p} = \frac{1 + W/L}{2 + W/x_{jC}}$$

This ratio shows that if $L \gg W$, there is no benefit gained from further increasing L; rather, the surface area is increased with no significant improvement of β. In designing a lateral p-n-p, it is clear that one should use the smallest possible value for W, and then choose $L \approx 10W$.

EXAMPLE 7H   *Current Gain of a Typical Lateral P-N-P.*   Consider a typical base diffusion with $x_{jc} = 2.0$ μm and assume that $x_{js} = 8.0$ μm and $L_p = 12$ μm. Assume further that the minimum line width is $W = 6.0$ μm, that $w_l = 3.0$ μm, and that $L = 60$ μm.   For these values the area ratio is

$$\frac{A_E}{A_p} = \frac{1.1}{2 + 3} = 0.22$$

and β is

$$\beta = (0.22)\left(1 + 2\,\frac{12^2}{6^2}\right) \approx 2$$

Practical lateral p-n-p transistors have β in the range 1 to 10 so our estimate, although crude, yields at least order-of-magnitude results.

For purposes of evaluating the degradation of performance caused by the vertical parasitic effects in the lateral p-n-p, it is instructive to calculate the β of a one-dimensional transistor with the same base-region material and 3.0-μm base width, but with no vertical parasitic effects.   Such a device would have

$$\beta \approx \frac{2L_p^2}{w_l^2} = 32$$

This indicates that parasitic effects predominate in the lateral p-n-p.   Even if W could be reduced to 1 μm and w_v were reduced to 4.0 μm, the resulting β would only increase to

$$\beta = (0.40)\left(1 + 2\,\frac{12^2}{4^2}\right) = 7.5$$

Although our figures may be in error by as much as a factor of 4, they at least indicate that some means other than reduction of surface geometry must be employed to improve β.   Significant improvements can be made by using a buried layer, and by using slightly lower N_{BC}.   The buried layer prevents minority carriers in the base from being collected by the substrate, and reducing N_{BC} increases the base-region lifetime. By employing these measures it is possible to fabricate lateral *p-n-p* transistors with β of the order of 50. ////

The lateral *p-n-p* also suffers from serious base-width modulation effects. This is because the collector region is much more heavily doped than the base region, with the result that most of the depletion at the collector junction occurs on the base side of the junction. Moreover, because N_{BC} is relatively low, the depletion at a given voltage will be a considerable fraction of the base width. Thus the punch-through voltage of the lateral *p-n-p* will be low, and the $\partial I_C / \partial V_C$ will be large.

EXAMPLE 7I *Punch-through Voltage of a Lateral P-N-P.* The calculation of the punch-through voltage can be accomplished by using the Lawrence-Warner curves, since the *p* diffusion is a gaussian profile in a constant background concentration. Consider the device of Example 7H, and assume $N_{sA} = 5 \times 10^{18}$ cm^{-3}. We have seen in Chap. 4 that the built-in voltage of the collector junction for such a case is about 0.6 V, so we assume that voltage. We also assume that the transistor is operating in the forward active region; with the emitter forward-biased, the emitter depletion region is negligible.

We need to calculate what junction voltage causes $x_m - x_1 = 3.0$ μm. From the Lawrence-Warner curves, we find that for $V_T = 0.6$, $x_m - x_1 = 0.77$ μm. This is the case for an applied voltage of zero. By successive trials we find that $x_m - x_1 = 3.0$ μm at $V_T \approx 8.6$ V. This device would therefore punch through at $V_{CB} = 8.0$ V. ////

Frequency Effects

The lateral *p-n-p* transistor is inherently a low-cutoff-frequency device. This is, of course, caused by the two factors which also contribute to its low current gain: large effective base width compared with vertical *n-p-n* devices, and large ratio of emitter bottom area to emitter sidewall area. The large effective base width contributes a large transit time for laterally injected carriers which are able to reach the collector. The epitaxial material under the bottom of the emitter contributes a large volume in which minority carriers which do not contribute to collector current must be stored.

We can make a crude estimate of the cutoff frequency of the lateral *p-n-p* on a first-order basis by again making the approximation that the device consists of two one-dimensional structures, a lateral *p-n-p* and a vertical *p-n-p*, with the emitters electrically connected. The cutoff frequency of the device is approximately the ratio of collector current to total minority carrier charge stored in the base region. If only the lateral transistor were present, its cutoff frequency would be

$$\omega_{Tl} \approx \frac{2D_p}{w_l^2}$$

where D_p is the hole diffusivity.

With the vertical transistor also present, much more charge is stored under the emitter than between emitter and collector; in fact, the latter charge is usually negligible in comparison with the former. The cutoff frequency, including the effects of the vertical structure, is given by

$$\omega_T = \frac{I_C}{Q_p} \approx \frac{2D_p}{w_l^2} \frac{A_E}{A_E + A_p} \frac{1 + (A_p/A_E)(w_l/w_v)}{1 + (A_p/A_E)(w_v/w_l)} \tag{7-48}$$

where Q_p is the total stored minority carrier charge in the base.

EXAMPLE 7J   *Calculation of ω_T for a Lateral P-N-P.*   Consider the device which was described in Example 7H, for which we calculated $\beta = 2$. If we assume $D_p = 12 \times 10^{-4}$ m^2/s, we find that the lateral structure by itself would have

$$\omega_T \approx \frac{2D_p}{w_l^2} = 2.66 \times 10^8 \text{ rad}$$

The device has an A_p/A_E ratio of 4.55, $w_l = 3.0$ μm, and $w_v = 6.0$ μm. Using these values in (7-48), we obtain

$$\omega_T \approx \frac{2D_p}{w_l^2} 0.058 = 0.155 \times 10^8 \text{ rad}$$

or

$$f_T = 2.46 \times 10^6 \text{ Hz}$$

For purposes of comparison we consider an experimental device which was fabricated with circular surface geometry.[18] It had emitter window diameter 50 μm, junction depth $x_{jc} = 4.2$ μm, substrate junction depth $x_{js} = 11.4$ μm, lateral base width at the surface $w_l = 6.4$ μm, and diffusion length $L_p = 28$ μm. A plot of measured β as a function of frequency is shown in Fig. 7-25, from which it is observed that the device has a low frequency $\beta = 6.2$, and a cutoff frequency $f_\beta = 0.9 \times 10^5$ Hz. Then $\omega_T \approx 2\pi\beta f_\beta = 3.5 \times 10^6$, and $f_T = 5.6 \times 10^5$ Hz.   ////

Exercise 7-8   Use Eq. (7-48) to calculate ω_T for the experimental transistor. In calculating the area ratio, take into account the circular surface geometry.

Even though our analysis of the lateral *p-n-p* involves many approximations, it at least can give order-of-magnitude estimates for β and ω_T. The results indicate that at best the lateral *p-n-p* can be expected to have f_T of several MHz. Therefore the device is useful in applications only where low-gain-bandwidth product can be tolerated. If the device is part of, for example, an operational amplifier circuit, the lateral *p-n-p* must be used with a gain near unity or excessive phase shift will result even at relatively low frequencies.

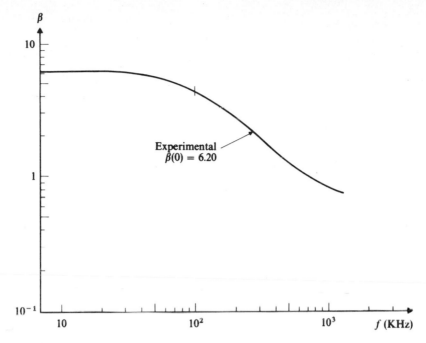

FIGURE 7-25
Measured β as a function of frequency for an experimental lateral p-n-p (after Fossum[18]).

The Composite P-N-P[19]

In some applications a p-n-p transistor is required in which only low-frequency behavior is important, but a larger current gain is necessary than can be achieved with the lateral p n p. If the lateral p-n-p is combined with an n-p-n transistor, as shown in Fig. 7 26, two results are obtained for the composite device:

1 All the bias polarities for the composite device are the same as for a p-n-p transistor.
2 The current gain of the composite device is approximately the product of the current gains of the two transistors.

As can be seen from Fig. 7-26, for biasing purposes the current polarities at the terminals of the composite are the same as would be the case for a p-n-p transistor. Note that for both transistors of the composite to operate in the forward active region, it is necessary that the collector of the composite be at least V_D more negative than the base. It is easily shown that the current gain β_C of the composite transistor is

$$\beta_C = \beta_p(1 + \beta_n)$$

where β_p and β_n are the current gains of the p-n-p and n-p-n, respectively.

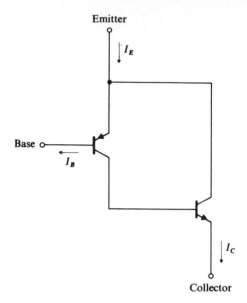

FIGURE 7-26
The composite *p-n-p* transistor.

The composite connection affords a means for improving the gain of the lateral *p-n-p* at the expense of the increase of area required by the *n-p-n*.   In addition to the area disadvantage, the frequency performance of the composite transistor is poor, being basically the same as for the *p-n-p*.   Also, the collector output conductance $g_{ce} \triangleq \partial I_C/\partial V_{CE}$ of the composite is rather large, since

$$g_{ce} \approx g_{cp}(1 + \beta_p) + g_{cn}$$

where g_{cp} and g_{cn} are the collector conductances of the *p-n-p* and *n-p-n*, respectively. As we have already seen, g_{cp} may be large for the *p-n-p* because of the excessive base-width modulation attendant in the lateral structure.

Exercise 7-9   Show that the current gain of the composite transistor is $\beta_C = \beta_p(1 + \beta_n)$.

Exercise 7-10   Calculate the cutoff frequency of β_C, and show that the frequency behavior of the composite transistor is dominated by that of the lateral *p-n-p*.

The Field-aided Lateral *P-N-P*[20]

The principal problems with the lateral *p-n-p*, as we have seen, arise from the injection occurring along the bottom of the emitter.   If this effect could be reduced, the current gain could be increased.   In fact, what is needed is enhanced current crowding at the emitter, in order to force the injection to occur at the sidewalls.

This can be accomplished, at the expense of increased area, by adding an additional base contact and applying a bias between the two bases. The bias produces a lateral field in the epitaxial material which causes the emitter sidewall toward the collector to be more forward-biased than the rest of the emitter. In fact, if the bias is large enough, the sidewall can be made to be forward-biased and the rest of the emitter reverse-biased. Enhancements of β of an order of magnitude are possible. The cross section and surface geometry are shown in Fig. 7-27.

The analysis of the field-aided lateral p-n-p clearly must include three-dimensional effects. Computer-aided modeling techniques have been applied to devices with circular surface geometry; the resulting calculated and measured β as a function of applied base voltage are shown in Fig. 7-28.[21]

The Multiple-collector Lateral P-N-P

The lateral p-n-p structure affords a means of obtaining current division determined by the surface geometry. If the artwork is arranged so that the collector is divided into several segments, these collector segments are approximately independent of each other as long as all have the same reverse bias relative to the base of the p-n-p. The collector current in each segment is now proportional to the area of the segment. Thus for the two-collector structure of Fig. 7-29a, for example, we see that

$$I_{C1} = \frac{\beta I_B A_{C1}}{A_C}$$

$$I_{C2} = \frac{\beta I_B A_{C2}}{A_C}$$

where A_{C1} and A_{C2} are the respective areas of the collectors, $A_C = A_{C1} + A_{C2}$, and β is the current gain with both collectors connected together: $\beta = (I_{C1} + I_{C2})/I_B$. Note that the ratio I_{C2}/I_{C1} is A_{C2}/A_{C1}; since these areas depend on surface geometry they can be rather precisely controlled. The precision is reduced somewhat if the biases are different, since different depletion widths at each collector will change the ratio slightly.

A particular application for multiple-collector devices involves the use of one collector to provide negative feedback in order to stabilize the current gain of the p-n-p. This is shown in Fig. 7-29b. The current gain $\beta_F \triangleq I_{C1}/I_B$ is now given by

$$\beta_F = \frac{(A_{C1}/A_C)\beta}{1 + (A_{C2}/A_C)\beta}$$

Correspondingly, the area ratio required for a given β_F and β is

$$\frac{A_{C1}}{A_{C2}} = \frac{\beta_F \beta + \beta_F}{\beta - \beta_F}$$

If reasonably large values of β can be obtained by careful processing, then β_F can be made less dependent upon processing than β.

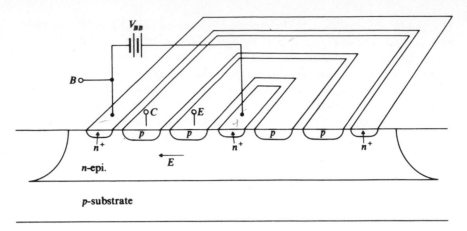

FIGURE 7-27
Field-aided lateral *p-n-p*.

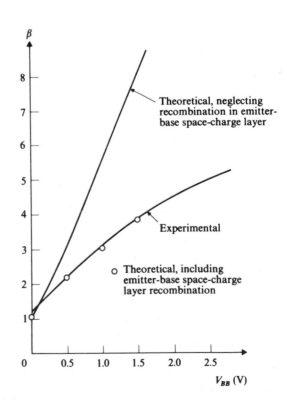

FIGURE 7-28
β vs base bias for the field-aided lateral *p-n-p* (after Fossum[21]).

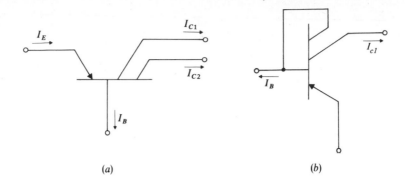

FIGURE 7-29
(a) Multiple-collector lateral p-n-p; (b) multiple-collector p-n-p using negative feedback to stabilize β.

EXAMPLE 7K *Design of a Lateral P-N-P with Stable β_F.* Suppose we require a p-n-p transistor with β_F which may be low, but must be relatively independent of processing variations. In particular, suppose that $\beta_F = 2$ is sufficient. Assume, for example, that processing variations yield values of 15 to 60 for β. We therefore design the layout so that the lateral p-n-p has two collector segments with $A_{C1}/A_{C2} = 2 \times \frac{10}{13} = 2.46$. Then we obtain

$$\beta_F = 2.0 \qquad \text{for } \beta = 15$$
$$\beta_F = 2.21 \qquad \text{for } \beta = 30$$
$$\beta_F = 2.36 \qquad \text{for } \beta_N = 60$$

The variation of a factor of 4 in β has produced only 18 percent variation of β_F. It should be emphasized, however, that a lateral p-n-p having a β_N as large as 15 will certainly exhibit large base-width modulation effects. If the two collector segments are operated at significantly different bias voltages, the *effective base* widths for each collector segment will be different, resulting in a change of the division ratio of collector current between the two segments. The value of β_F then becomes somewhat dependent upon the collector bias voltages. ////

Use of the Four-layer Model to Represent the Lateral *P-N-P*

In our analysis of the lateral p-n-p, we treated the problem in piecewise-one-dimensional fashion, by considering the device to be basically a one-dimensional vertical transistor combined with a one-dimensional horizontal transistor. Since we considered only the forward active region of operation, the regions of the device under the collector and between collector sidewalls and isolation region were not important. For inverse-active or saturated operation, these regions must be taken into account. This can be done by the same piecewise-one-dimensional method we have already employed. Since no n^+ emitter diffusion is present, the

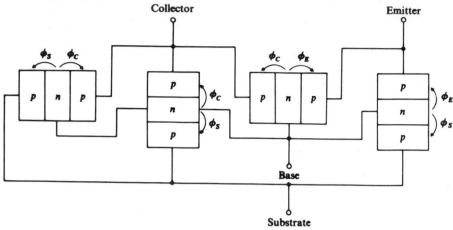

FIGURE 7-30
Representation of the lateral *p-n-p* by a composite of three-layer models.

device is really a three-layer structure, and the model is as shown in Fig. 7-30. Four one-dimensional models are involved; however, note that since only four terminals are present, common junction voltages exist among the various models. In the forward active region both ϕ_C and ϕ_S are negative; the two models on the left have both junctions reverse-biased and do not contribute to device behavior. In the inverse active region, similar remarks apply to the two models on the right.

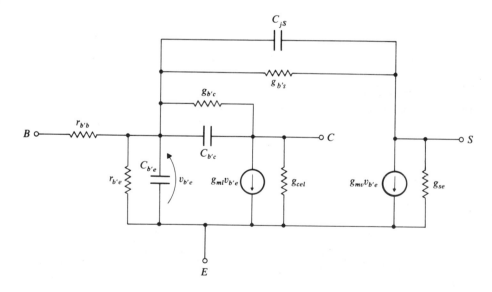

FIGURE 7-31
Hybrid-π model for the lateral *p-n-p*.

Only for saturated operation is it necessary to consider all of the models; even then the substrate will generally be reverse-biased, and this will simplify the analysis.

Small-signal Behavior of the Lateral P-N-P

For small-signal operation the lateral p-n-p is operated in the forward active region, and a modified hybrid-π model can be used to represent the device. Since both lateral and vertical transistors are in the forward active region, the hybrid-π is a four-terminal model, as shown in Fig. 7-31. Since the substrate is connected to a power supply voltage, C_{js} and $g_{b's}$ will generally appear in parallel with $C_{b'e}$ and $r_{b'e}$ for common-emitter operation, and $r_{b'e} \ll 1/g_{b's}$.

7-8 DIODES[22]

It is obvious that diodes can be obtained by using any of the various junctions which are formed as part of the standard fabrication process. However, except for the diode made by the emitter n^+ diffusion into an isolation diffusion, all diodes will not be two-terminal devices, but will have at least three terminals. Since we are only using two of the terminals, we have some options regarding how the remaining terminals are connected. If we assume that the substrate is always connected to the most negative power supply voltage, this removes one option. With the remaining options, there are six possible ways in which diodes can be obtained, excluding the n^+ isolation diode which is not a useful option. These six configurations are shown in Fig. 7-32; type f is made by omitting the emitter diffusion. All have slightly different electrical characteristics. The forward characteristics are shown in Fig. 7-33, from which it is evident that type a has the lowest voltage and type d the highest voltage at moderate forward currents.

The diodes also have two parasitic capacitances which must be considered: the capacitance which appears between the two terminals of interest, and the capacitance between one terminal and the substrate. These capacitances are shown in Fig. 7-34.

Storage time and breakdown voltage are also of interest. In Fig. 7-35 are sketched the relative charge storage modes for each configuration.

(a)   (b)   (c)   (d)   (e)   (f)

FIGURE 7-32
The six basic diode connections (after Meyer et al.[22]).

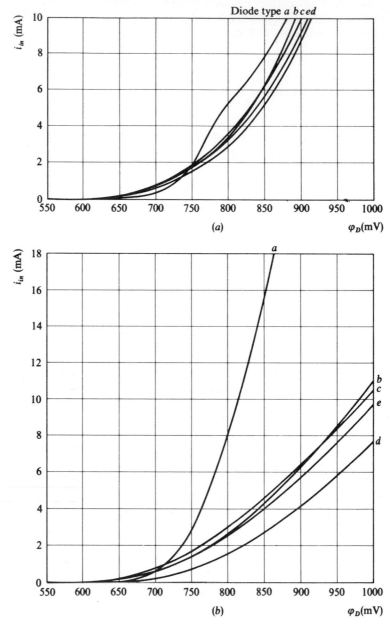

FIGURE 7-33
Forward characteristics of integrated-circuit diodes. (a) 1.2 $\Omega \cdot$ cm collector, 200 Ω per square base; (b) 0.1 $\Omega \cdot$ cm collector, 200 Ω per square base (after Meyer et al.[22]).

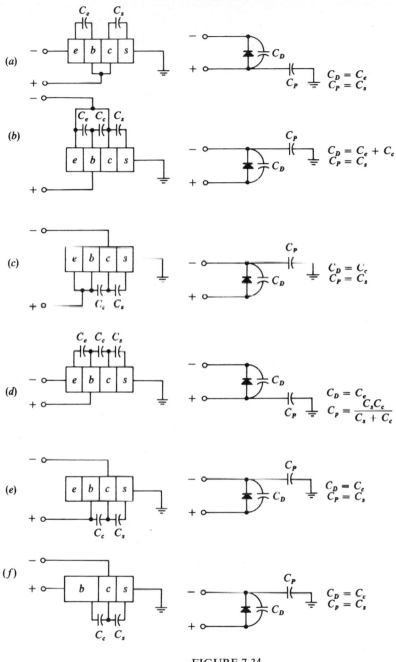

FIGURE 7-34
Diode capacitances (after Meyer, et al.[22]).

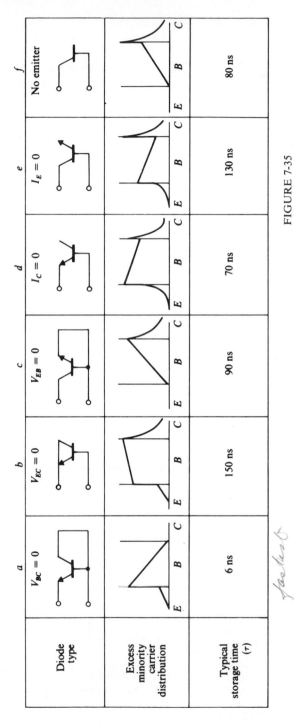

FIGURE 7-35
Storage times for the six diode connections.

The choice of diode configuration clearly depends upon the circuit application, which in turn determines the relative importance of forward voltage, capacitance, storage time, and breakdown voltage.   As we shall see in a later chapter, some applications require large forward voltage and large storage time, while others require low forward voltage and small storage time.

7-9   SCHOTTKY-BARRIER DIODES[23]

We have seen that regions of the epitaxial material that are to be contacted must receive an n^+ diffusion in order that the contact be ohmic rather than rectifying. This is, of course, essential for contacts to collector regions of n-p-n transistors. However, a metal-semiconductor contact which is rectifying rather than ohmic can be used to advantage as a diode.   Such a device is called a Schottky-barrier diode (SBD).

Schottky-barrier diodes are fabricated in their simplest form by opening windows in SiO_2 over n-type epitaxial material and sputtering platinum onto the silicon surface.   The substrate is heated to 600°C in a vacuum chamber, and the platinum and silicon react to form Pt_5Si_2.   Aluminum is then deposited to contact the Pt_5Si_2 as shown in Fig. 7-36a.   Diodes fabricated in this way do not have ideal VI characteristics; the reverse breakdown is of the order of a few volts, and the forward characteristic is not sharp.   This is because the p^+-n junction formed between Pt_5Si_2 and the silicon is very shallow and consequently the radius of curvature at the edges is very small, causing very large electric fields to exist at the edges.   The high field concentration at the edges causes breakdown to occur at low voltages, and also causes larger leakage currents than would be observed in a planar junction.

The problems associated with these high field regions can be virtually eliminated by adding a guard ring, as shown in Fig. 7-36b.   Here the p^+ diffused guard ring removes the regions of small radius of curvature, and the VI characteristic follows the ideal $I = I_s e^{qV/kT}$ for eight decades of current.   The breakdown voltage is now determined by the lower of the metal-semiconductor planar breakdown or p^+-n junction breakdown with its larger radius of curvature.   The metal-semiconductor junction behaves almost like an ideal p^+-n junction, and has typical breakdown voltages of the order of 40 V.   The breakdown voltage of the guard-ring junction depends upon the junction depth and the impurity concentrations in n and p^+ regions, and will usually be larger than 40 V.

The guard-ring diffusion can be eliminated by applying an MOS overlay around the Pt_5Si_2 as shown in Fig. 7-36c, and applying a negative bias between silicon and overlay.   This produces a depletion region in the silicon at the edges of the Pt_5Si_2 and reduces the edge-field effects which caused the low breakdown voltage.   However, in order to achieve the desired effects from the MOS overlay at reasonable bias voltages, the oxide must be thin, as was the case for the MOS transistor.

Schottky-barrier diodes have low forward voltage and very low storage

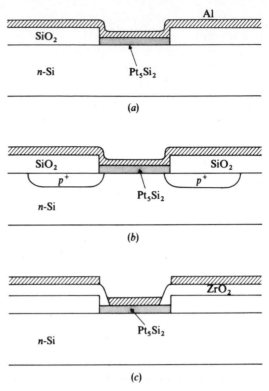

FIGURE 7-36
Schottky-barrier diodes.   (*a*) Basic structure; (*b*) guard-ring diffusion; (*c*) MOS overlay.

time—typically less than 0.1×10^{-9} s.   They are not obtained, however, without additional processing steps, so their need in a given circuit application must be weighed against the cost of the extra processing steps.

PROBLEMS

7-1   The transistor of Example 7B is used in an amplifier circuit in which the bias current is $I_C = 2.0$ mA and the load resistance is 500 Ω.   If the transistor has a cutoff frequency $f_T = 500$ MHz, calculate the voltage gain and estimate the cutoff frequency.

7-2   For the circuit of Fig. P7-2, use the four-layer model for each transistor and calculate the current I_B required to produce a given saturation voltage $V_{CE(sat)}$.   The substrate is grounded.   Insert numerical values of Sec. 7-2 for the four-layer-model parameters and find I_B for $I_C = 3.5$ mA.

7-3   Consider a transistor in which the impurity concentration in the emitter is kept low enough so that the diffusion length for minority carriers is much larger than the junction depth x_{jE}.   Derive an expression for the emitter efficiency γ.   Compare this with the emitter efficiency of a homogeneous-base device.

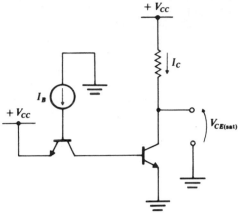

FIGURE P7-2

7-4 For the substrate *p-n-p* of Example 7G, calculate the emitter current at which the onset of high-level injection occurs.

7-5 Derive an expression for β_N of the lateral *p-n-p* including the effects of γ.

7-6 A lateral *p-n-p* is made with the dimensions given in Example 7H, but the minimum line width used is 1 μm. Assume that the emitter and collector junctions are step junctions, and calculate the collector conductance arising from base-width modulation at a collector voltage of $V_{CE} = -5$ V and a current $I_C = 2$ mA.

7-7 For a transistor with the dimensions given in Example 7H, find what w_i would have to be used if avalanche breakdown and punch-through are to occur at the same voltage. Calculate the β which would result at zero volts if this w_i is used.

7-8 A lateral *p-n-p* is made with circular surface geometry. Derive expressions for the normal current gain β and the inverse current gain β_I.

7-9 Use the four-layer model to derive expressions for the forward voltage of type *a* and type *d* diodes (Fig. 7-32). Assume a forward current of 1 mA, and insert the parameter values of Sec. 7-2 to obtain numerical results.

REFERENCES

1 ENGL, W. L., and O. MANCK: Two-dimensional Analysis of Bipolar Transistor Transients, IEEE International Electron Devices Meeting, Washington, D.C., Dec. 5, 1973.

2 EBERS, J. J., and J. L. MOLL: Large-signal Behavior of Junction Transistors, *Proc. IRE*, vol. 42, pp. 1761–1772, 1954.

3 NARUD, J. A., W. C. SEELBACH, and C. S. MEYER: Microminiaturized Logic Circuits: Their Characterization, Analysis and Impact upon Computer Design, IEEE International Convention, New York, March 1963.

4 MEYER, C. S., D. K. LYNN, and D. J. HAMILTON: "Analysis and Design of Integrated Circuits," chap. 4, McGraw-Hill Book Company, New York, 1968.

5 MILLMAN, J., and C. HALKIAS: "Integrated Electronics," chap. 11, McGraw-Hill Book Company, New York, 1972.

6   SEARLE, C. L., A. R. BOOTHROYD, E. J. ANGELO, JR., P. E. GRAY, and D. O. PEDERSON: "Elementary Circuit Properties of Transistors," vol. 3, "SEEC Notes," chap. 3, John Wiley & Sons, Inc., New York, 1964.

7   LINDMAYER, J., and C. Y. WRIGLEY: "Fundamentals of Semiconductor Devices," chap. 4, D. Van Nostrand, Inc., Princeton, N.J., 1965.

8   BURGER, R. M., and R. P. DONOVAN: "Fundamentals of Silicon Integrated Device Technology," vol. 2, pp. 134–136, Prentice-Hall, Inc., Englewood Cliffs, N.J., 1968.

9   LINDHOLM, F. A., and D. J. HAMILTON: Incorporation of the Early Effect in the Ebers-Moll Model, *Proc. IEEE*, vol. 59, pp. 1377–1378, 1971.

10 LOGAN, J., F. A. LINDHOLM, P. ROHR, and D. J. HAMILTON: Modeling the "Early Effect" in Bipolar Transistors Using an Empirical but Effective Parameter, *Proc. IEEE*, vol. 60, pp. 335–336, 1972.

11 FOSSUM, J. G.: A Theory Predicting the Forward Current-voltage Characteristics of P-N Junctions, *Tech. Mem.* SC-TM-720820, Sandia Laboratories, February 1973.

12 WEBSTER, W. M.: On the Variation of Junction Transistor Current-amplification Factor with Emitter Current, *Proc. IRE*, vol. 42, pp. 914–920, 1954.

13 GIBBONS, J. F.: "Semiconductor Electronics," p. 212, McGraw-Hill Book Company, New York, 1960.

14 BURGER and DONOVAN, op. cit., pp. 103–117.

15 Ibid.

16 Ibid., pp. 249–255.

17 LIN, H. C.: "Integrated Electronics," pp. 274–283, Holden-Day, Inc., Publisher, San Francisco, 1967.

18 FOSSUM, J. G.: "Systematic Computer-aided Three-dimensional Modeling of Integrated Bipolar Devices," Ph.D. dissertation, p. 90, The University of Arizona, Tucson, Ariz., 1971.

19 LIN: op. cit., pp. 283–290.

20 LONG, E. L., and T. M. FREDERIKSEN: High-gain 15-W Monolithic Power Amplifier with Internal Fault Protection, *IEEE J. Solid-State Circuits*, vol. SC-6, pp. 35-44, 1971.

21 FOSSUM: op. cit., p. 122.

22 MEYER, LYNN, and HAMILTON: op. cit., pp. 248–258.

23 LEPSELTER, M. P., and S. M. SZE: Silicon Schottky Barrier Diode with Near-ideal *I-V* Characteristic, *Bell System Tech. J.*, vol. 47, pp. 195–208, 1968.

24 WOOLEY, B. A., S.-Y. J. WONG, and D. O. PEDERSON, A Computer-Aided Evaluation of the 741 Amplifier, *IEEE J. Solid-State Circuits*, vol. SC-6, pp. 357–365, 1971.

8

THERMAL EFFECTS IN INTEGRATED CIRCUITS

Two aspects of thermal effects in integrated circuits are important: the first is the behavior of integrated devices as the chip temperature changes; the second is the manner in which the chip temperature varies. Device behavior is determined by the temperature dependence of certain important bulk properties of silicon. Chip temperature is determined by the ambient temperature, the thermal properties of the chip and its mounting, and the total power dissipated on the chip.

In this chapter we consider first the temperature dependence of such bulk properties as mobility, intrinsic carrier density, and energy gap, and from this we deduce the effects of temperature on device terminal behavior. This information is used to establish electrothermal models (ETM) for the devices. Such models provide a convenient means of analyzing the behavior of a circuit as chip temperature varies. We then consider the effects of power dissipation, ambient temperature variations, and header mountings on chip temperature. Finally, through the design of a temperature-stabilized substrate (TSS) system, we illustrate how device electrical-thermal interactions can be used to control chip temperature.

8-1 THERMAL BEHAVIOR OF SILICON BULK PROPERTIES[1]

Carrier Mobility

As mobile carriers move through the crystal lattice, they are scattered by collisions with lattice atoms and with impurity atoms. This scattering effect determines the value of the carrier mobility. As temperature increases, the collision cross section of the atoms increases, resulting in a decrease of the mobility. One would expect that for low impurity levels the scattering would be primarily the result of collisions between carriers and lattice atoms. This is indeed the case: for impurity levels less than about 10^{16} cm^{-3}, lattice scattering dominates; for higher impurity levels, both lattice and impurity scattering take place. Since both effects must be included for most of the impurity levels of interest in integrated circuits, the analysis becomes quite involved. It is therefore customary to approximate the mobility by

$$\mu = C'T^{-\eta} \qquad (8\text{-}1)$$

where T = temperature, °K

C' = constant

η = parameter whose value is determined experimentally

Experimental data for mobility are given in Figs. 8-1 and 8-2.

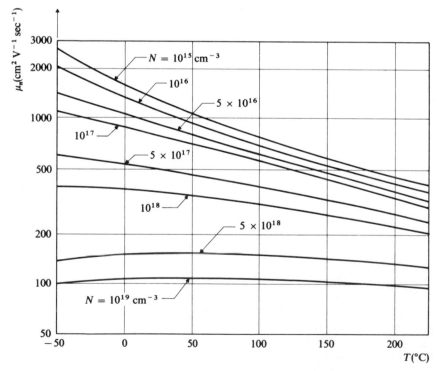

FIGURE 8-1
Electron drift mobility in silicon.

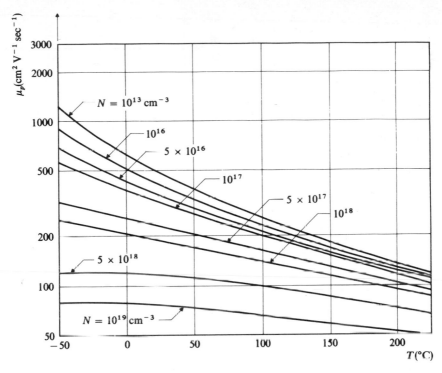

FIGURE 8-2
Hole drift mobility in silicon.

If (8-1) were accurate, the slope of the curves in these figures would be constant; however, we note from the curves that if (8-1) is to be used, η must be considered to be a function of both impurity level and temperature. The best procedure is to use these figures to determine the approximate value of η to use in (8-1).

Once the mobility is determined, carrier diffusivity D can be determined from the Einstein relation

$$D = \frac{kT}{q}\mu \qquad (8\text{-}2)$$

From (8-2) it can be seen that the variation of D with temperature is negligible compared with that of μ.

Conductivity

The conductivity of a semiconductor depends in general on the mobility and density of both types of carriers. However, for all cases of interest in integrated circuits, the material used is sufficiently extrinsic that majority carrier behavior dominates. We may therefore write

$$\sigma = q\mu_p p \qquad \text{for } p\text{-type}$$
$$\sigma = q\mu_n n \qquad \text{for } n\text{-type}$$

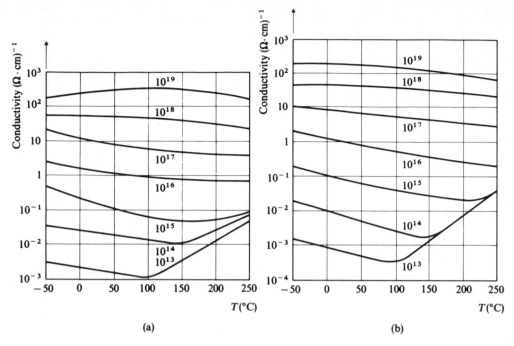

FIGURE 8-3
Conductivity versus temperature. (a) For n-type silicon; (b) for p-type silicon.

Moreover, the majority carrier density is approximately equal to the impurity density. If the temperature is sufficiently large that all impurities are ionized, then the majority carrier density is independent of temperature, and the temperature dependence of the conductivity becomes that of the mobility. Curves of conductivity versus temperature for various impurity concentrations are given in Fig. 8-3.

Intrinsic Carrier Concentration and Gap Energy

The performance of junctions, and hence diodes and transistors, depends on both the intrinsic carrier density n_i and the gap energy E_g. It can be shown that

$$n_i^2 = K_i T^3 e^{-E_g/kT} \text{ cm}^{-6} \qquad (8\text{-}3)$$

where K_i is a constant. The gap energy is also temperature-dependent, and is given by

$$E_g = 1.205 - 0.28 \times 10^{-3}T$$

Thus the intrinsic carrier concentration is

$$n_i^2 = K_i T^3 \exp\left(\frac{0.28 \times 10^{-3}}{k} - \frac{1.205}{kT}\right) \qquad (8\text{-}4)$$

Lifetime

For temperatures of interest in integrated circuits, the bulk lifetime can be approximated by

$$t = t_1 e^{-E_r/kT} \qquad (8\text{-}5)$$

where t_1 is a constant and E_r is the recombination level energy difference. Typical values for E_r range from 0.03 to 0.12 eV for non-gold-doped material. A representative value is 0.1.

The functional dependences we have given for various quantities involve many approximations. Fortunately, with the exception of diffused resistors, device performance is not critically dependent upon the values of such parameters as η and E_r. The functional forms we have given are useful in evaluating the way in which temperature affects device performance. Where values of η are required, they can be obtained from the curves of Figs. 8-1 and 8-2.

8-2 ELECTROTHERMAL MODELS FOR INTEGRATED DEVICES[2]

Resistors

Consider a resistor made in uniformly doped material, such as the epitaxial layer. For a resistor of length L with rectangular cross section of area A, the resistance is

$$R = \frac{L}{\sigma A} = \frac{L}{q\mu_n nA} = \frac{LT^{\eta}}{qnAC'} \qquad (8\text{-}6)$$

Now suppose the temperature varies from T_0 to T_1. Since η is a function of temperature, it will also vary. The fractional variation of R, defined by

$$\Delta_R \triangleq \frac{R(T_1) - R(T_0)}{R(T_0)}$$

is found from (8-6) to be

$$\Delta_R = \frac{(T_1)^{\eta 1}}{(T_0)^{\eta 0}} - 1 \qquad (8\text{-}7)$$

where $\eta 1$ and $\eta 0$ are evaluated at T_1 and T_0.

For cases where variations of the order of 100 degrees are encountered, it is necessary to use Fig. 8-1 or 8-2 to find $\eta 1$ and $\eta 0$, and then calculate Δ_R. If only a small variation dT about some operating point T_0 is required, we can proceed as follows. Let $R(T)$ be expressed in terms of a Taylor's series about the operating temperature T_0:

$$R(T) = R(T_0) + \frac{\partial R(T_0)}{\partial T} dT + \frac{\partial^2 R(T_0)}{\partial T^2} dT^2 + \cdots$$

which can be written as

$$R(T) = R(T_0)\left[1 + \frac{1}{R(T_0)} \frac{\partial R(T_0)}{\partial T} dT + \frac{\eta - 1}{2T_0} \frac{1}{R(T_0)} \frac{\partial R(T_0)}{\partial T} dT^2 + \cdots \right] \qquad (8\text{-}8)$$

Now we define δ_R, the temperature coefficient of resistance at T_0, to be

$$\delta_R \triangleq \frac{1}{R(T_0)} \frac{\partial R(T_0)}{\partial T}$$

Then (8-8) becomes

$$R(T) = R(T_0) \left[1 + \delta_R \, dT + \frac{(\eta - 1) \, \delta_R \, dT^2}{2T_0} + \cdots \right] \quad (8\text{-}9)$$

For sufficiently small dT, we can make the approximation

$$R(T) \approx R(T_0)(1 + \delta_R \, dT) \quad (8\text{-}10)$$

Typical values of δ_R for several resistor types are given in Table 8-1.

EXAMPLE 8A   *Useful Range of Eq. (8-10).*   Consider a resistor with $\eta \approx 2.0$. Suppose we are willing to accept (8-10) as long as the third term of (8-9) is no larger than 0.1 times the second term.   Then $(\eta - 1) \, dT/2T_0 < 0.1$, from which we find

$$dT < \frac{0.2T_0}{\eta - 1}$$

If $T_0 = 300°K$ (27°C), we have

$$dT < 60°C$$

For temperature variations larger than this, it would be necessary to use (8-7).

////

Note that δ_R is a function of T_0; Figs. 8-4 and 8-5 give measured values of $\delta_R(T_0)/\delta_R(300°K)$ for several types of resistors.   In these figures, the normalized *temperature coefficients* of resistance are plotted, not the resistance itself.   The temperature coefficients are positive; and in Fig. 8-4 two curves are shown for emitter resistors, each for different values of sheet resistance.

Table 8-1   TEMPERATURE COEFFICIENT FOR SEVERAL TYPES OF RESISTORS

Region and type of material	Sheet resistance, Ω/sq, or concentration, cm^{-3}	δ_R at 300°K, ppm
Emitter, n	5	1500
	15	2500
Base, p	100	1000
	200	2500
Collector, n	10^{15}	8000
	10^{16}	4000
	10^{17}	3500
Base pinch, p	...	8000

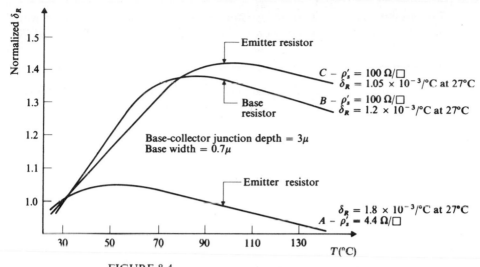

FIGURE 8-4
Normalized temperature sensitivity $\delta_R/\delta_{R(27°C)}$ for base and emitter resistors.

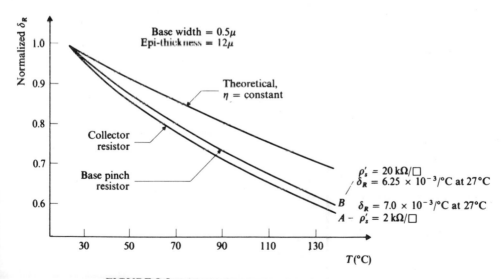

FIGURE 8-5
Normalized temperature sensitivity $\delta_R/\delta_{R(27°C)}$ for base-pinch and collector resistors.

Exercise 8-1 Assume that all resistors have $\eta = 2$. A base resistor with $\rho'_s = 200\ \Omega$ per square is fabricated with $R = 20\ k\Omega$ at 27°C. Find R at 70 and 125°C.

We now consider the effect of temperature variations on the circuit performance of resistors. Let small-signal quantities v and i be defined by

$$v \triangleq V(T_0 + dT) - V(T_0)$$
$$i \triangleq I(T_0 + dT) - I(T_0)$$

At a temperature $T = T_0 + dT$,

$$V(T) = I(T)R(T)$$

If we make the approximation

$$R(T) \approx R(T_0)[1 + \delta_R\, dT]$$

we can write

$$V(T_0) + v = R(T_0)(1 + \delta_R\, dT)[I(T_0) + i] \qquad (8\text{-}11)$$

Neglecting $R(T_0)\, \delta_R\, dT\, i$ in comparison with other terms, and making use of the fact that $V(T_0) = R(T_0)I(T_0)$, we find

$$v = R(T_0)i + R(T_0)I(T_0)\, \delta_R\, dT \qquad (8\text{-}12)$$

The second term of (8-12) is a temperature-dependent voltage whose value also depends upon the bias current and the resistance.

An equivalent circuit such as that shown in Fig. 8-6 can be used to represent the resistor for small-signal variations resulting from temperature variation dT. Such an equivalent circuit is called an *electrothermal model* (ETM).

EXAMPLE 8B *Use of the ETM.* Two resistors are connected as shown in Fig. 8-7a; the resistors have values R_1 and R_2 at T_0. R_1 is a base resistor and R_2 a collector resistor; they have different temperature coefficients δ_{R_1} and δ_{R_2}. We wish to determine the voltage variation v as a function of temperature variation dT.

First we note that the bias currents at T_0 are $I_1 = IR_2/(R_1 + R_2)$ and $I_2 = IR_1/(R_1 + R_2)$. The ETMs for the resistor can now be used as shown in Fig. 8-7b. Analyzing this circuit we obtain

$$v = \frac{IR_1 R_2}{R_1 + R_2}\left[\frac{R_2\, \delta_{R_1} + R_1\, \delta_{R_2}}{R_1 + R_2}\right] dT$$

For $I = 1\ mA$, $R_1 = 2\ k\Omega$, $R_2 = 1\ k\Omega$, $\delta_{R_1} = 2.5 \times 10^{-3}/°C$, $\delta_{R_2} = 8 \times 10^{-3}/°C$, we find:

$$v = 4.1\ mV/°C \times dT \qquad ////$$

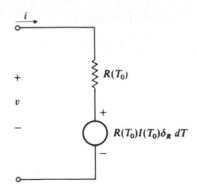

FIGURE 8-6
Electrothermal model for a resistor.

Exercise 8-2 A base resistor ($\delta_R = 2.5 \times 10^{-3}/°C$) and a collector resistor ($\delta_R = 8.0 \times 10^{-3}/°C$), each having resistance of 1 kΩ, are connected in series. A voltage of 2 V is connected to the series combination. Find the current variation i resulting from a temperature variation dT.

Electrothermal models can be derived for other integrated devices as well as resistors. They provide a useful means of evaluating the temperature effects in a complex circuit; this can be done by replacing each device with its ETM and analyzing the resulting network for the voltage or current variation of interest. Since the circuit with ETMs embedded in it is an electrical network, conventional analysis techniques can be applied to it. For very complex networks, computer-aided analysis programs can be used.

The basic techniques which were used to derive the ETM for resistors can also be applied to other devices. We describe the results without detailed derivation.

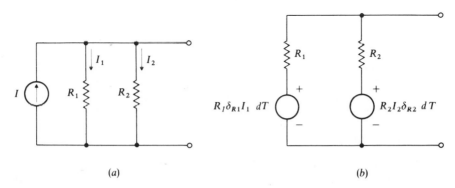

(a)   (b)

FIGURE 8-7
(a) Two dissimilar resistors and bias current; (b) use of ETMs to find voltage variation v with temperature variation dT.

Junction Capacitors

For nonabrupt junctions, the capacitance can be reasonably well approximated by

$$C \simeq C_0 \left(1 + \frac{V}{\phi_c'}\right)^{-r} \qquad (8\text{-}13)$$

where V = reverse bias
ϕ_c' = built-in voltage
r = constant, depending on impurity profile at junction
C_0 = zero-bias capacitance
Typically $\frac{1}{2} > r > \frac{1}{6}$. Capacitance variation with temperature results from variation of ϕ_c' with temperature; this variation is approximately $d\phi_c'/dT = -2\ \text{mV}/°\text{C}$. The temperature coefficient λ_C for the capacitance can be obtained from (8-13); it is

$$\lambda_C \triangleq \frac{1}{C}\frac{dC}{dT}$$

$$= \frac{rV}{\phi_c'}(V + \phi_c')^{-1}\frac{d\phi_c'}{dT} \qquad (8\text{-}14)$$

The ETM for the junction capacitor is shown in Fig. 8-8; note that (8-14) shows that to minimize capacitance variations, large reverse bias is desirable.

The Diode

For the intrinsic step junction diode, in which there are no bulk resistance effects, it is possible to derive an expression for the temperature coefficient of the forward voltage when the diode is biased with a current source. Integrated junctions are much more difficult to analyze, but fortunately experimental results indicate that for first-order temperature behavior they can be treated as step junctions. The temperature coefficient γ_D for diode voltage V_D can be shown to be

$$\frac{dV_D}{dT} \approx \gamma_D \triangleq -2\ \text{mV}/°\text{C}$$

approximately independent of temperature. Fig. 8-9a shows the ETM for the diode with bias current I. In practical integrated circuits, bulk resistance is always present; it can be included as shown in Fig. 8-9b by adding the ETM of the bulk resistor r_b. The effects of r_b will differ slightly for the different diode connections.

Breakdown Voltage

The temperature coefficient γ_Z of breakdown voltage depends upon the value of the breakdown voltage, as is shown in Fig. 8-10, but γ_Z is relatively independent of temperature. For typical emitter junctions with breakdown voltage of approximately 7 V,

$$\gamma_{Ze} \approx 2.3\ \text{mV}/°\text{C}$$

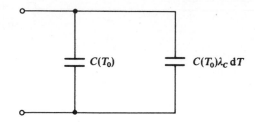

FIGURE 8-8
ETM for the junction capacitor

while for collector junctions with breakdown voltage of 40 V

$$\gamma_{Zc} \approx 35 \text{ mV/°C}$$

The ETM for the intrinsic junction in breakdown is given in Fig. 8-11. For practical diodes, bulk resistance effects can also be included by adding the ETM for the bulk resistance.

The Intrinsic N-P-N Transistor

We consider here only the intrinsic transistor in the forward active region with reverse-biased substrate junction, and we assume the effects of the substrate to be negligible. From the four-layer model of Chap. 7 we can obtain the equations which describe the device under these conditions:

$$I_1 = I_E = I_{s1}(e^{q\phi_E/kT} - 1) - \alpha_I I_{s2} \qquad (8\text{-}15)$$
$$I_2 = -I_C = -\alpha_N I_{s1}(e^{q\phi_E/kT} - 1) - I_{s2} \qquad (8\text{-}16)$$

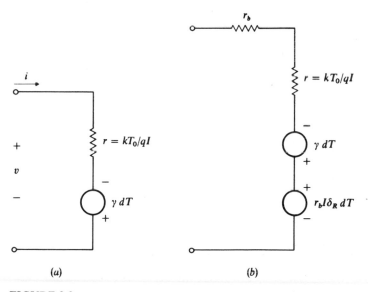

(a)　　　　　　　　　　　　(b)

FIGURE 8-9
(a) ETM for the intrinsic diode; (b) ETM including bulk resistance r_b.

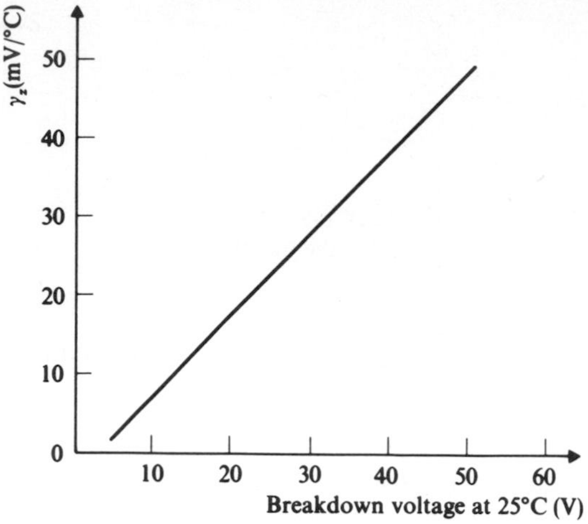

γ_z(mV/°C)

FIGURE 8-10
Temperature coefficient γ_z as a function of breakdown voltage.

For the emitter junction, it can be shown that the ETM is approximately the same as that for a diode. The small-signal collector current can be shown to be

$$i_c = \frac{1}{\beta(T_0)} \frac{\partial \beta(T_0)}{\partial T} \beta(T_0) I_B(T_0) \, dT + \beta(T_0) i_B + \frac{1}{I_{s2}(T_0)} \frac{\partial I_{s2}(T)_0}{\partial T} I_{s2}(T_0) \, dT \qquad (8\text{-}17)$$

where $\beta = \alpha_N/(1 - \alpha_N)$ and $i_B = [\partial I_B(T_0)/\partial T] \, dT$.

Let temperature coefficients ε_s and b be defined by

$$\varepsilon_s \triangleq \frac{1}{I_{s2}(T_0)} \frac{\partial I_{s2}(T_0)}{\partial T}$$

and

$$b \triangleq \frac{1}{\beta(T_0)} \frac{\partial \beta(T_0)}{\partial T}$$

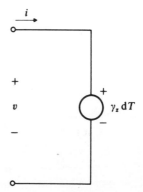

FIGURE 8-11
ETM for the intrinsic junction in breakdown.

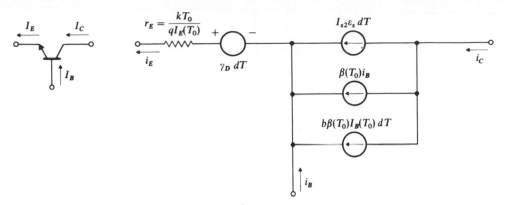

FIGURE 8-12
ETM for the intrinsic transistor.

Then (8-17) can be written as

$$i_c = b\beta(T_0)I_B(T_0)\,dT + \beta(T_0)i_B + \varepsilon_s I_{s2}(T_0)\,dT \qquad (8\text{-}18)$$

The ETM for the transistor is then as shown in Fig. 8-12. In order to insert explicit values for the temperature coefficients b and ε_s, it is necessary to know on what physical processes these depend.

It can be shown that the saturation current temperature coefficient ε_s is

$$\varepsilon_s \approx \frac{4 - \eta}{T} + \frac{1.205}{kT^2}$$

where η applies to the base region; thus ε_s is of the order of $0.1/°C$ for silicon devices. Nevertheless, I_{s2} is usually sufficiently small that $\varepsilon_s I_{s2}\,dT$ can be neglected except for large values of dT.

The value for b depends upon whether injection efficiency or base transport factor is dominant in determining the value of $\beta(T_0)$. If the transport factor, that is, base recombination, dominates, then

$$b \approx \frac{-(\eta - 1)}{T} + \frac{E_r}{kT^2} \qquad (8\text{-}19)$$

For this case $b \approx 8 \times 10^{-3}/°C$ at $T_0 = 300°K$. If injection efficiency dominates, then

$$b \approx \frac{-(\eta - 1)}{2T} + \frac{E_r}{2kT^2} \qquad (8\text{-}20)$$

which for typical devices is $b \approx 4 \times 10^{-3}/°C$ at $T_0 = 300°K$.

If it is necessary to characterize the collector current in terms of α_N, it can be shown that the temperature coefficient

$$a \triangleq \frac{1}{\alpha_N(T_0)} \frac{\partial \alpha_N(T_0)}{\partial T}$$

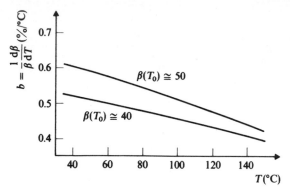

FIGURE 8-13
Measured values of b as a function of T for two experimental transistors.

is related to b by

$$a = \frac{b}{\beta(T_0)}$$

Thus we see that for reasonably large $\beta(T_0)$, a will often be negligible in comparison with other temperature coefficients.

Measured values for b as a function of T_0 are shown in Fig. 8-13 for two experimental transistors, each biased at a collector current $I_C = 1$ mA. For any given device, the dependence of b can be calculated from (8-19) or (8-20).

Extrinsic effects such as bulk resistances can easily be included in the ETM by adding the ETMs for these effects in appropriate places.

Lateral *P-N-P* Transistors

As one would intuitively expect, the form of the ETM for the lateral *p-n-p* is the same as for the *n-p-n* transistor. However, for the lateral *p-n-p*, the base transport factor usually dominates the determination of β because of the wide base, and therefore

$$b \approx \frac{E_r}{2kT^2} - \frac{(\eta - 1)}{2T}$$

For the intrinsic transistor, we would expect $b \approx 4 \times 10^{-3}/°C$. For the extrinsic transistor it should be noted that:

1 Because β is very low, the base current will be much larger than for the *n-p-n*, so base resistance will be more important for the *p-n-p*.
2 Because the base of the *p-n-p* is the relatively lightly doped epitaxial layer, the base resistance itself will be larger than for the *n-p-n*.

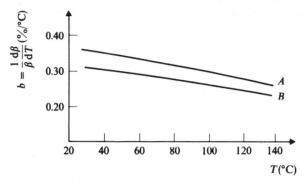

FIGURE 8-14
Measured values of b for two experimental lateral devices.

These two facts indicate that it will usually be necessary to add the ETM for the base resistance to the ETM of the intrinsic p-n-p. Moreover, we can see intuitively that because δ_R and γ_D have opposite signs, the presence of resistance should cause b for the p-n-p to be less than that for the n-p-n. That this is the case is indicated by the measured values shown in Fig. 8-14 for two experimental devices operated at bias current $I_C \approx 1$ mA. Curve A is for a device with $\beta = 0.46$ and base width 0.80 mil, while curve B is for a device with $\beta = 2.4$ and base width 0.6 mil.

8-3 APPLICATION OF ELECTROTHERMAL MODELS

EXAMPLE 8C *Electrothermal Analysis of a Bias Circuit.* A widely used integrated bias circuit, to be discussed in detail in Chap. 9, is shown in Fig. 8-15a. We wish to calculate the variation of I with temperature. To do this we define

$$i \triangle I(T_0 + dT) - I(T_0)$$

and we replace the devices in Fig. 8-15a with their ETMs; the result is shown in Fig. 8-15b. Note that Q_1 is a diode-connected transistor; its emitter junction voltage will have a temperature coefficient γ_D. We assume that the emitter junction voltage of Q_2 has temperature coefficient γ_T.

It can be shown that

$$I_R(T_0) = 1 \text{ mA} \approx I \approx I_{E1} \approx I_{E2}$$

Then $r_E = 26\ \Omega$ at $T_0 = 300°$K. We also assume that $\beta = 50$, $\delta_{\dot{R}} = 2 \times 10^{-3}/°C$, and $b = 6 \times 10^{-3}/°C$. For these values, we calculate $I_B(T_0) = 0.02$ mA and $\beta(T_0)I_B(T_0)b = 6\ \mu A/°C$. The problem of finding i is now merely that of analyzing the circuit of Fig. 8-15b using network analysis techniques.

If the two transistors are perfectly matched, as they should be if they have the same surface geometry and the same temperature, then $\gamma_D = \gamma_T$ and we obtain

$$i = \frac{6 \times 10^{-6}}{°C} + \frac{\beta}{2+\beta} \left(\frac{-6 \times 10^{-6}}{°C} - \frac{6.8 \times 10^{-3}/°C}{R} \right) dT \qquad (8\text{-}21)$$

from which we find

$$\frac{i}{I} = \frac{-1.7 \times 10^{-3}}{°C} dT \qquad \qquad ////$$

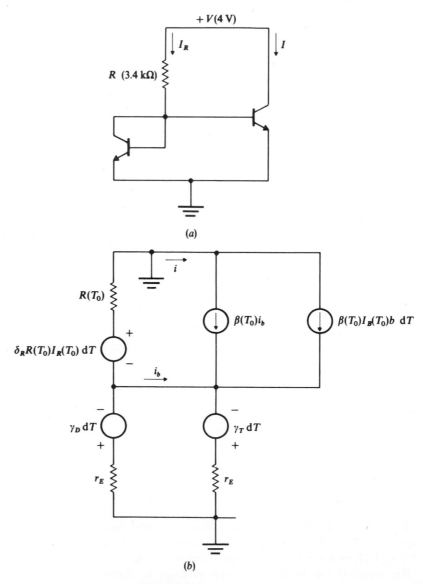

(a)

(b)

FIGURE 8-15
(a) Integrated bias circuit; (b) ETM for the bias circuit.

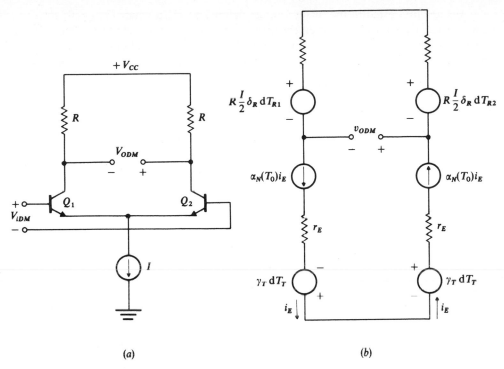

FIGURE 8-16
(*a*) Differential amplifier; (*b*) ETM for the differential amplifier with $V_{iDM} = 0 = v_{iDM}$.

EXAMPLE 8D *Effects of Temperature Differences on the Offset Voltage of a Differential Amplifier.* Differential amplifiers such as that shown in Fig. 8-16 are frequently employed as the input stage of operational amplifiers. Because they are usually followed by gain stages having voltage gains of the order of 40 to 60 dB, no offset voltages in the differential amplifier can be tolerated. This means that the input transistors must be very well matched, and care must be taken to ensure that they both experience the same temperature. We wish to evaluate the effects of small temperature differences on the performance of the differential amplifier. Such temperature differences can result from having devices which are dissipating large amounts of power on the same chip as the differential amplifier.

We assume that a bias current I is used for the differential amplifier, as shown. Let the temperature at Q_1 be $T_0 + dT_T$ and that at Q_2 be $T_0 - dT_T$. Here T_0 is the average temperature, which will be larger than the ambient temperature because power is being dissipated on the chip. Since the resistors take up more surface area than the transistors, it may not be possible to locate them in close proximity to the transistors; therefore the temperatures of the two resistors may not be equal, and they may each differ from the temperatures of the transistors. Therefore we let their temperatures be $T_0 + dT_{R1}$ and $T_0 + dT_{R2}$.

We are interested in the offset voltage defined as V_{ODM} when $V_{iDM} = 0$; thus we wish to analyze the circuit when the transistor bases are connected together. If transistors are perfectly matched, and if resistors are perfectly matched, there is no offset voltage if all components experience temperature T_0. Moreover, the emitter currents for this case are both $I/2$. The voltage we wish to calculate is v_{ODM} resulting from dT_T, dT_{R1}, and dT_{R2}.

The circuit is most conveniently analyzed by focusing attention on i_E, the variation of emitter current. The ETM circuit is shown in Fig. 8-16b. Here we have obtained the dependent sources in terms of α_N rather than β, since emitter current is the independent variable. We have neglected the dependent source involving I_{s2}, since it is assumed to be small in comparison with $\alpha_N i_E$; the term involving $\alpha_N(I/2)a\,dT$ has also been neglected because of the small value of a. For the circuit of Fig. 8-16b, we can write

$$v_{ODM} = R\frac{I}{2}\,\delta_R(dT_{R1} - dT_{R2}) + 2R\alpha_N(T_0)i_E \qquad (8\text{-}22)$$

and

$$i_E = \frac{2\gamma_T\,dT_T}{2r_E}$$

$$= \frac{2\gamma_T\,dT_T\,qI}{4kT_0}$$

$$= \frac{qI\gamma_T\,dT_T}{2kT_0} \qquad (8\text{-}23)$$

Combining (8-22) and (8-23), we find

$$v_{ODM} = R\frac{I}{2}\left[\delta_R(dT_{R1} - dT_{R2}) + \frac{2\alpha_N q\gamma_T\,dT_T}{kT_0}\right] \qquad (8\text{-}24)$$

For representative values we choose $R = 10$ kΩ, $V_{cc} = 10$, $I = 1$ mA, $T_0 = 300°$K, $\alpha_N = 0.995$, and we assume that

$$\delta_R = 2 \times 10^{-3}/°C \qquad \gamma_T = 2 \times 10^{-3}\ V/°C$$

$$dT_T = 0.25°C \qquad dT_{R1} = 2.0°C \qquad dT_{R2} = 1.0°C$$

Note that this corresponds to a temperature difference of only $0.5°$C between transistors and $1.0°$C between resistors. Inserting these values in (8-24), we obtain

$$v_{ODM} = 203\ \text{mV}$$

Of this, 10 mV is contributed by the resistor variations and 193 mV by the transistor variations.

This example illustrates the importance of designing the layout of a differential amplifier so that the transistor emitters are located on isotherms, as temperature differences of the order of a tenth of a degree can produce intolerable offset voltages.   It also indicates that placement of resistors is much less critical than that of transistors.   ////

Exercise 8-3   Suppose that all devices are matched and all are at the same temperature.   What voltage V_{iDM} would have to be applied to produce $V_{ODM} = 203\,\text{mV}$?

8-4 THE ISOTHERMAL CHIP

In Example 8D we saw a case for which small differences of the temperature of devices caused large variations of circuit performance.   In most integrated circuits, however, temperature differences are of less importance than the actual temperature of the devices.   Such is the case when large power dissipation on the chip and large ambient temperature variations occur.   For example, circuits which must operate over the military temperature range face large variations of ambient temperature. If the temperature variations are sufficiently large, the component value variations that result can alter the circuit power dissipation which, in turn, can change the chip temperature.   In poorly designed circuits, this can cause "thermal runaway" and catastrophic failure.

How power dissipation and ambient temperature affect chip temperature is perhaps best seen by means of an analog model in which

Current → power
Voltage → temperature
Conductance → thermal conductance

If we assume that power dissipation is uniformly distributed over the surface of the chip, heat flow and temperature will be one-dimensional.   We can then represent the thermal situation with the electrical analog shown in Fig. 8-17.   Here G_C represents the thermal conductance of the silicon chip, G_B the thermal conductance of the die bond, G_H the thermal conductance from the header surface to the outside of the header, and G_S the thermal conductance from header to ambient environment. Note that G_S will be a strong function of the means by which heat is carried away from the header to the environment.   If a heat sink is used, G_S will be larger than for the header alone.

From the model of Fig. 8-17, we can see that

$$I_C = P\left(\frac{1}{G_C} + \frac{1}{G_B} + \frac{1}{G_H} + \frac{1}{G_S}\right) + T_a$$

Typical values for several headers are shown in Table 8-2.[3]

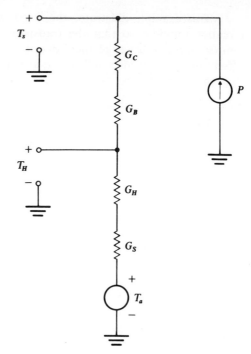

FIGURE 8-17
Electrical analog for the isothermal chip
and header.

Table 8-2 THERMAL RESISTANCE FOR SEVERAL
INTEGRATED-CIRCUIT PACKAGES*

Test condition	TO-99 TO-100	TO-86 TO-91 flat pack	Dual in-line plastic package
Still air, socket mounting	185	187	150
Still air, soldered to printed circuit board	157	165	145
Forced air, socket mounting	130	152	135
Heat sink on case, socket mounting, still air	133		

$$* \ \frac{1}{G_C} + \frac{1}{G_B} + \frac{1}{G_H} + \frac{1}{G_S}, \ °C/W.$$

EXAMPLE 8E   *Variation of a Resistance with Temperature and Power.*   Consider a circuit which has a power dissipation of 1 W.   We wish to calculate the maximum change of resistance of an integrated resistor as the ambient temperature changes from -25 to $+125°C$ and the power changes from 1 to 0.5 W.   The circuit is gold-bonded to a TO-99 header in free air.

From the model of Fig. 8-17, we see that the maximum change of chip temperature is

$$\Delta T_{s(\text{max})} = \Delta T_a + \Delta P \left(\frac{1}{G_C} + \frac{1}{G_B} + \frac{1}{G_H} + \frac{1}{G_S} \right)$$

Inserting the above values, we obtain

$$\Delta T_{s(\text{max})} = 150°C + 99°C = 249°C$$

If the resistor had a constant temperature coefficient $\Delta_R = 2 \times 10^{-3}/°C$, the fractional change of resistance would be

$$\frac{\Delta R}{R} = \Delta_R \, \Delta T - 0.5$$

or a change of 50 percent.   ////

8-5   THE TEMPERATURE-STABILIZED SUBSTRATE

It is possible to stabilize the chip temperature if one is willing to expend some additional power for temperature control.   To accomplish this, an insulator is inserted between chip and header as shown in Fig. 8-18a.   A sensor, such as a diode, is used to detect temperature variation and convert it to electrical power. This power is dissipated on the chip as control power.   As ambient temperature increases, the control power decreases.   Such a negative-feedback system results in a temperature-stabilized substrate (TSS).[4-6]

Let P_H be the power dissipated by other circuits on the chip, and P_C the control power dissipated on the chip by the sensor-heater circuit.   We assume P_H and P_C to be uniformly distributed over the surface of the chip.   If the thermal resistance R_i of the insulator is much larger than that of the silicon, the temperature change from top to bottom of the chip will be negligible.   The thermal resistance of the header can also be ignored.   Then the analog model for the system becomes that shown in Fig. 8-18b.   From this model, we see that

$$T_s = (P_H + P_C)R_i + T_a \qquad (8\text{-}25)$$

where T_s is the chip temperature

and

$$P_C = K(T_{\text{ref}} - T_s) \qquad (8\text{-}26)$$

where K is the transfer gain of the sensor-heater in watts per degree.

Substituting (8-26) in (8-25), we find

$$T_s = \frac{R_i(P_H + KT_{\text{ref}}) + T_a}{1 + KR_i} \qquad (8\text{-}27)$$

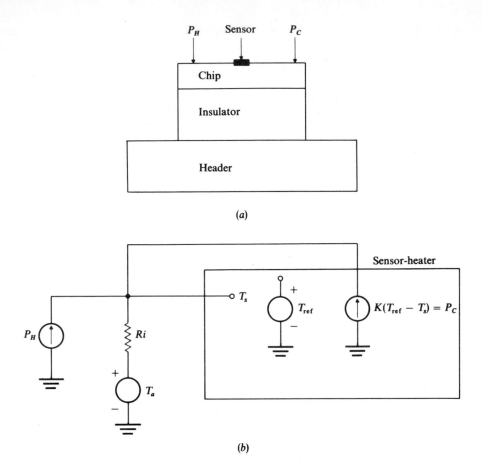

FIGURE 8-18
(*a*) Cross section of the temperature-stabilized system; (*b*) model for the system.

Now we can calculate the variation of T_s with T_a:

$$\frac{\partial T_s}{\partial T_a} = \frac{1}{1 + KR_i} \qquad (8\text{-}28)$$

If the sensor heater has sufficient gain that

$$KR_i \gg 1$$

we see that the temperature T_s is essentially independent of T_a. If both power and ambient temperature vary, the change of substrate temperature is

$$dT_s = \frac{\partial T_s}{\partial P_H} dP_H + \frac{\partial T_s}{\partial T_a} dT_a = \frac{1}{1 + KR_i} (R_i \, dP_H + dT_a)$$

Thus we see that if K is sufficiently large, dT_s can be made negligibly small.

Design of a TSS

The stabilization of the temperature is accomplished at the expense of control power P_C. In designing a TSS we wish to minimize P_C. Qualitatively, we can see that when T_a is maximum, P_C should be zero, and P_H is then sufficient to maintain the temperature at $T_s = T_{ref}$. However, when T_a is minimum, P_C must be added so that the temperature T_s can still be maintained at $T_s = T_{ref}$. Clearly the maximum P_C is required when T_a is minimum. The design procedure is therefore as follows:

1 Choose T_{ref} as *large* as the chip will tolerate (about 200°C). Note that T_{ref} must be larger than $T_{a(max)}$.
2 Choose R_i so that when $T_a = T_{a(max)}$, $T_s = T_{ref}$ in order to make $P_C = 0$ for this condition. Thus

$$T_{ref} = T_{a(max)} + P_H R_i \qquad (8\text{-}29)$$

3 For $T_a = T_{a(min)}$, T_s is still required to be T_{ref}. Therefore

$$T_s = T_{ref}$$

$$= T_{a(min)} + P_H R_i + P_C R_i$$

$$= T_{a(min)} + [T_{ref} - T_{a(max)}] + \frac{P_H}{P_C} [T_{ref} - T_{a(max)}]$$

Then the required control power P_C can be found:

$$P_C = P_H \frac{T_{a(max)} - T_{a(min)}}{T_{ref} - T_{a(max)}} \qquad (8\text{-}30)$$

Note that (8-30) shows that to minimize P_C, T_{ref} should be as large as permissible.

4 Design a sensor-heater to have gain K sufficiently large to produce the required stability $\partial T_s/\partial T_a$. From (8-28) we have

$$K = \frac{1}{R_i} \left(\frac{1}{\partial T_s/\partial T_a} - 1 \right) \qquad (8\text{-}31)$$

Substituting (8-31) in (8-30), we obtain

$$K = \frac{P_H}{T_{ref} - T_{a(max)}} \left(\frac{1}{\partial T_s/\partial T_a} - 1 \right) \qquad (8\text{-}32)$$

The system design of the TSS is now complete.

Design of a Reference Temperature and Sensor

From the diagram of Fig. 8-18b, the system appears to have a serious drawback in that it requires a reference temperature source T_{ref}. However, this disadvantage can be easily overcome by combining the functions of reference temperature and sensor on the chip. The basic circuit is shown in Fig. 8-19, in which the equivalent

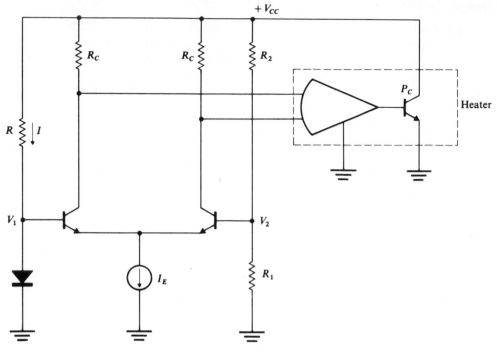

FIGURE 8-19
Circuit for the sensor-heater reference-temperature circuit.

reference temperature is provided by the voltage divider R_1 and R_2, and the diode acts as the sensor.

Since both resistors are on the chip, they have the same temperature, and V_2 is therefore independent of temperature. V_1, however, decreases with temperature. By choosing the proper V_2, we can make $V_1 = V_2$ at the temperature we require for T_s. At this temperature the output of the differential amplifier will be zero, and this will cause the heater power to be zero. The design of the circuit proceeds as follows:

1 Choose a value for I, say 1 mA. Then

$$V_1 = V_D(I, T_{a(\min)}) \triangleq V_D$$

and

$$R = \frac{V_{cc} - V_D}{I}$$

2 Calculate ΔV_1 for $T_s = T_{ref}$, where T_{ref} is the desired reference temperature. Let $\Delta T = T_{ref} - T_{a(\min)}$. For a diode, $\Delta V_1 \approx \gamma_D \Delta T$. Then at $T_s = T_{ref}$, let

$$V_{1s} = V_1[T_{a(\min)}] + \Delta V_1$$

3   At $T_s = T_{ref}$, we want $V_2 = V_{1s}$; that is,

$$V_{1s} = V_2 = \frac{V_{CC}}{1 + R_2/R_1}$$

from which

$$\frac{R_2}{R_1} = \frac{V_{CC}}{V_1 + \Delta V_1} - 1$$

4   Choose a value for R_1; calculate R_2.
5   Choose R_C and I_E for the desired voltage gain in the differential amplifier.

EXAMPLE 8F   *First-order Design of a TSS for a Reference Temperature of* 200°.
Suppose we have a system with $V_{CC} = 10.7$, and we assume $V_D = 0.7$ at 0°C, and
$\gamma_D = -2 \times 10^{-3}/°C$.   Suppose also that the circuits on the chip have a total dissipation of 40 mW and that we wish to stabilize the chip temperature to $\pm 0.5°C$
when the ambient temperature varies from 0 to 100°C.
First we calculate $P_C = P_H(100 - 0)/(200 - 100) = P_H = 40$ mW.   Then for
this P_C, the thermal resistance of the insulator must be $R_i = 100/40$ mW = 2.5°C/
mW.   Glass, ceramic, or other insulator can be selected with the proper thickness
to produce this R_i.
The stability we require indicates that $\partial T_s/\partial T_a = \Delta T_s/\Delta T_a = 1 \cdot 100 = 10^{-2}$.
Using (8-32) we find

$$K = \frac{40 \text{ mW}}{100°} \; 99 \approx 40 \text{ mW/°C}$$

Now we choose $I = 1$ mA for the sensor circuit; then $R = 10$ kΩ.   We calculate

$$V_{1s} = V_D + \gamma_D \, \Delta T = 0.7 + (-2 \times 10^{-3})200 = 0.3 \text{ V}$$

Then $V_2 = 0.3$ and

$$\frac{R_2}{R_1} = \frac{10.7}{0.3} - 1 = 35$$

Arbitrarily selecting $R_1 = 1000$ Ω, we find

$$R_2 = 35 \text{ k}\Omega$$

If we now select $R_C = 18$ kΩ, and choose $I_E = 1$ mA so that the transistors do
not saturate, the voltage gain of the differential amplifier is $18 \times 10^3/52 = 33$.   This
means that with a diode temperature coefficient of γ_D mV/°C, the sensor gain at
the output of the differential amplifier is $33\gamma_D$ mV/°C.   For an overall gain $K = 40$ mW/°C, the heater must have a transfer function 0.61 mW/mV.
For the ambient temperature range specified in this design, the required control

power is equal to the circuit power.   Our first-order design has neglected the power dissipated in the sensor circuit; this should be taken into account since it may be comparable to the power in the circuit being controlled.   ////

8-6 LATERAL TEMPERATURE VARIATIONS

We have thus far assumed that the chip surface is isothermal; this assumption is valid only if the power dissipation is uniform over the surface of the chip.   It is unlikely that the power dissipation will be uniform in most circuits.   For chips which are mounted directly to the header by means of gold solder or similar die bonds, moderate power dissipations of the order of 100 mW, even though localized, will not produce much temperature gradient at the surface of the chip.   However, we have already seen that temperature variations of a fraction of a degree are sufficient to influence certain special circuits such as differential amplifiers.   The detailed analysis of steady-state temperature distribution in the chip involves the solution of Laplace's equation in three dimensions.   Because of the boundary conditions involved, the solution is likely to be intractable for most cases, and computer-aided analysis is necessary.[7]

For chips mounted on insulators as in the case of the TSS or of electrothermal filters, temperature variations of the order of 3 degrees can occur across the surface of the chip.   For electrothermal filters, layout of the circuit is critical, and a two-dimensional analysis in the frequency domain is required; again a computer-aided analysis is necessary.   Even for temperature-stabilized systems, the effect of these relatively small temperature gradients at the chip surface makes the placement of heaters and sensors critical, and a computer-aided analysis is necessary if precise control of temperature at a particular location on the chip is required.

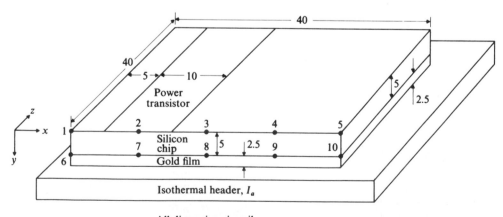

All dimensions in mils

FIGURE 8-20
Hypothetical chip and mounting used to illustrate lateral temperature variations.

Table 8-3 CALCULATED TEMPER-
ATURES $T_n - T_a$ PER
WATT DISSIPATION OF
THE TRANSISTOR IN
THE HYPOTHETICAL
CHIP OF FIG. 8-20

n	$T_n - T_a$ °C/W	
1	1.19	
2	5.18	
3	0.60	silicon chip surface
4	0.072	
5	0.017	
6	0.19	
7	0.64	
8	0.10	silicon-gold interface
9	0.012	
10	0.003	

Some idea of the problems encountered with variations of temperature across the chip can be obtained from the hypothetical chip shown in Fig. 8-20. This chip is 40 × 40 mils, 5 mils thick, using a 2.5-mil gold die bond to a header which is assumed to be a perfect heat sink at temperature T_a. A single power transistor is located with its centerline 10 mils from one edge of the chip. Because the transistor is uniformly distributed in the y direction, the temperature distribution in the y direction will be uniform. Temperatures at the numbered points have been computed using a simple finite-difference solution of Laplace's equation in two dimensions; the results are given in Table 8-3.

In high-power circuits involving localized power dissipations of several watts, conventional die bonds to ordinary headers are not sufficient to reduce temperature gradients to an acceptable level. For example, in an amplifier with 15-W output capability, temperature variations of approximately 100°C across the chip surface are observed. For such high-power circuits, special headers of stainless steel with copper overlay must be used, and considerable care must be taken in the layout of the circuit in order to minimize the effects of temperature gradients on the circuit. In this case the designer attempts to use geometrical symmetry in order to place critical devices along isotherms.

PROBLEMS

8-1 A diode is operated with bias current of 1 mA. Use ETMs to determine what bulk resistance it must have if the temperature coefficient of the total voltage is zero.

8-2 A Zener diode is needed with breakdown voltage of 20 V, and temperature coefficient as near zero as possible. It is to be obtained by putting forward-biased diodes in series with a reverse-biased diode. How many forward-biased diodes should be used, and what breakdown voltage should be specified for the reverse-biased diode?

8-3 In Chap. 7, it was shown that feedback could be used in a lateral *p-n-p* transistor with two collector segments in order to stabilize the β. Suppose the *p-n-p* has $\beta = 15$, and the collector segments are chosen as in Chap. 7 to produce $\beta_F = 2$. Assume that only intrinsic effects need be considered, and use ETMs to calculate the temperature coefficient of β_F.

8-4 The TSS design of Example 8F ignored the power dissipation of the sensor circuit. Reformulate the general design equations to take into account the sensor power, and recalculate the values for Example 8F.

8-5 Design a heater circuit to be connected to the output of the differential sensor amplifier of Example 8F, with the required transfer function of 0.61 mW/mV.

REFERENCES

1 FLETCHER, A. R.: "Performance of Integrated Electrothermal Circuits," Ph.D. dissertation, chap. 4, The University of Arizona, Tucson, Ariz., 1970.

2 Ibid.

3 DANIELS, G. R.: Heat Transfer and Integrated Circuits, *Electro-Technology*, pp. 22–30, January, 1969.

4 PROSSER, T. F.: An Integrated Temperature Sensor-Controller, *IEEE J. Solid-State Circuits*, vol. SC-1, pp. 8–13, 1966.

5 GREENHOUSE, H. M., and R. L. MCGILL: Design of Temperature-controlled Substrates for Hybrid Microcircuits, *Bendix Tech. J.*, pp. 18–27, Winter, 1972–73.

6 GRAY, P. R.: "Electrothermal Integrated Circuits," Ph.D. dissertation, chap. 6, The University of Arizona, Tucson, Ariz., 1969.

7 ———: A 15-watt Monolithic Amplifier, *IEEE J. Solid-State Circuits*, vol. SC-7, pp. 870–880, 1972.

9

BASIC LINEAR INTEGRATED CIRCUITS

Through common usage, the term *linear* has come to denote the class of integrated circuits which are not digital or switching circuits. In reality the term is misleading because many of the circuits in the linear class depend heavily upon certain device nonlinearities for their operation. An excellent example of this is the general biasing problem, in which the nonlinearities of junctions are combined with the matched-component property of integrated circuits to produce bias circuits that require no capacitors and that are stable with temperature. What is really meant by the term "linear" is that class of circuits in which the signal variables are treated as continuous as contrasted with those circuits, such as digital circuits, in which quantized variables are a functional requirement.

The circuit designer accustomed to discrete-component circuit design, upon viewing the circuit diagram of an integrated operational amplifier for the first time, is undoubtedly staggered by the large number of transistors employed and by the complexity of the circuit. As we have previously pointed out, transistors are no more difficult to fabricate than resistors, and the former generally require less area. Thus the integrated-circuit designer recognizes no economic preference for resistors; rather, the opposite is the case. The complexity of the circuit arises from the sophisticated circuit techniques which have been developed to take advantage of the matched-component and approximate isothermal properties of the integrated circuit in order to provide simultaneously stable biasing and large gain without the use of capacitors or large resistors.

Because of the complexity of linear circuits it is important that the integrated-circuit designer be well acquainted with certain techniques which are widely used for biasing and gain in such circuits as operational amplifiers. In this chapter we examine these techniques in detail.

9-1 TRANSISTOR MODELS

In Chap. 7, we made use of the four-layer nonlinear model for the transistor for dc calculations, and showed how the frequency dependence of the device could also be included in the model. For forward-active-region operation with reverse-biased substrate, the model can be simplified considerably. If we assume the effects of the substrate to be negligible, the equations for the *n-p-n* transistor for direct current become

$$\begin{bmatrix} I_E \\ -I_C \end{bmatrix} \approx \begin{bmatrix} 1 & -\alpha_I \\ -\alpha_N & 1 \end{bmatrix} \begin{bmatrix} I_{s1}(e^{q\phi_E/mkT} - 1) \\ -I_{s2} \end{bmatrix} \qquad (9\text{-}1)$$

where the currents are defined as shown in Fig. 9-1a. Small-signal low-frequency behavior can be obtained by taking $\partial I_E/\partial \phi_E$, etc., and frequency behavior can be obtained by inserting parasitic capacitances and the frequency behavior of α_N. For design purposes, however, it is more convenient to have a progression of models of varying complexity, each tailored to specific needs.

A Model for Biasing

In many cases the dc behavior is determined primarily by the external circuit and the salient features of the device, and only secondarily by detailed device behavior. For example, (9-1) shows that the collector current is basically independent of collector voltage. We have also seen that the emitter voltage is a weak function of the emitter current, and for even an order-of-magnitude variation of emitter current, the emitter junction voltage is approximately V_D. A model which incorporates these two properties is shown in Fig. 9-1b. Even in cases where the circuit performance may depend upon detailed device behavior, it is usually worthwhile to make use of the model of Fig. 9-1 first in order to establish approximate operating conditions. This procedure will help the designer determine what approximations can reasonably be used in detailed analysis. Once these approximations are made, the analysis using (9-1) generally is simplified.

Small-Signal Models

Probably the worst way to proceed in analyzing the behavior of a small-signal circuit is to use the most complex model available for the device. Not only does such a procedure lead to lengthy computation, but by its complexity it also obscures

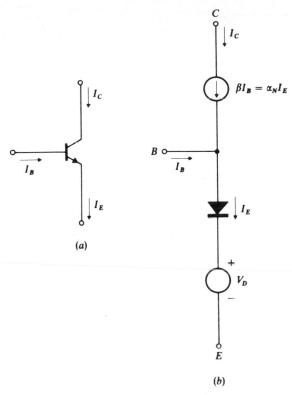

FIGURE 9-1
(a) Definition of current polarities; (b) first-order active-region model applicable for most bias situations.

the really important aspects of the circuit behavior. Instead, one should begin with a highly idealized and simple model for the device. One can then focus on important circuit behavior and obtain some idea as to reasonable approximations; one can usually determine also what features of detailed device behavior will next be important. The complexity of the model should then be increased to include only those features which seem to be of first-order importance. Consistency tests can often be made at this stage to determine whether further complexity is required. The procedure of using a progression of models of increasing complexity is especially important in design. Here the designer must evolve a circuit configuration to meet certain performance specifications. He must therefore evaluate the configuration in the early stages of design to determine if it must be modified in order to meet these specifications. If a particular configuration will not meet the specifications when analyzed with a highly idealized device model, it certainly cannot meet them with a more complex and less ideal model. Before wasting considerable analysis time and effort on a circuit, the designer must be reasonably certain that it has a chance of meeting the required specifications.

A zero-order small-signal model for the transistor is shown in Fig. 9-2*a*; this model incorporates only the most outstanding properties of the device. It can be obtained from the model of Fig. 9-1*b* by taking small variations of the currents involved. Note that it is the model for what could be defined as the "ideal" transistor.

In some cases, the small-signal behavior of the emitter junction is important in the circuit being analyzed; this behavior can be incorporated on a low-frequency basis as shown in Fig. 9-2*b*. The resistors in these models represent the dynamic behavior of the emitter junction, and are dependent upon the bias currents:

$$g_m = \frac{qI_C}{mkT}$$

$$r_{b'e} = \frac{\beta}{g_m} = (\beta + 1)r'_e$$

$$r'_e = \frac{mkT}{qI_E} = \frac{\alpha_N mkT}{qI_C}$$

These two models are equivalent, as can be seen by comparing their terminal behavior and making use of the fact that $I_C/I_E \approx \alpha_N$. The second model of Fig. 9-2*b* will be recognized as a simplified form of the hybrid-π model; it is more useful when i_b is the independent variable.

If it becomes necessary to include base resistance and base-width modulation effects, the model of Fig. 9-2*c* is used. In this model the conductances $g_{b'c}$ and g_{ce} represent the effects of base-width modulation, and are dependent upon the bias, the impurity profile in the base, and the resistivity of the epitaxial layer. The base resistance $r_{bb'}$ depends upon the geometry and the base-layer resistivity but is at least first-order-independent of bias. As guidelines, the following approximations[1] can be employed for $g_{b'c}$ and g_{ce}:

$$g_{ce} = \eta g_m$$

$$\eta \approx \frac{kT}{q} \frac{1}{w} \frac{dw}{dV_{CB}}$$

$$g_{b'c} \approx \frac{\eta}{\beta} g_m$$

where β and η are evaluated at the particular bias voltage used. The factor η is clearly a measure of the variation of the base width with collector-base voltage; it is typically of the order of 10^{-4}. Both $g_{b'c}$ and g_{ce} vary with bias current in the same way as does g_m. Since β is of the order of 10^2, $g_{b'c}$ will in many cases be negligible in comparison with g_{ce}. If β is limited by emitter efficiency, as is the case in most integrated transistors, $g_{b'c}$ will be much smaller than $\eta g_m/\beta$.

Frequency effects may also be incorporated in a progression of models. The intrinsic cutoff frequency of the device is represented in Fig. 9-2*d* by the diffusion capacitance given by

$$C_b \approx \frac{1}{r'_e \omega_\alpha}$$

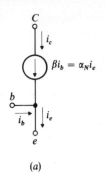

(a)

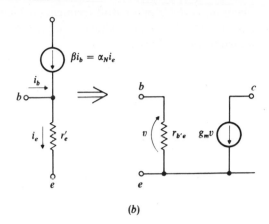

(b)

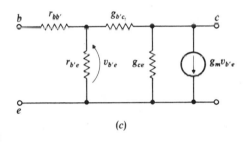

(c)

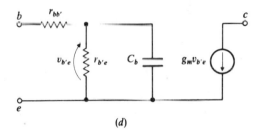

(d)

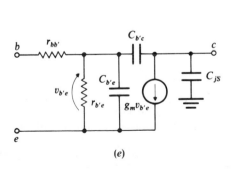

(e)

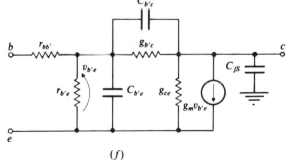

(f)

FIGURE 9-2
(a) Zero-order model; (b) first-order model; (c) complete low-frequency model; (d) first-order model including intrinsic cutoff frequency; (e) model including parasitic capacitances; (f) model including parasitic capacitances and base-width modulation effects.

where ω_α is the alpha cutoff frequency which is approximately ω_T, the frequency at which $\beta = 1$. When parasitic capacitances are included, the model becomes as shown in Fig. 9-2e where

$$C_{b'e} = C_b + C_{jE}$$
$$C_{b'c} \approx C_{jC}$$

The substrate capacitance C_{jS} is shown grounded, since it is assumed that the substrate is connected to a supply voltage.

Base-width modulation effects may be included as shown in the complete model of Fig. 10-2f.

Exercise 9-1 Find the Thévenin equivalent at the collector-emitter terminals for the model of Fig. 9-2c if the small-signal base current is zero. What does this become if $g_{b'c} \ll 1/r_{b'e}$?

Exercise 9-2 A transistor has negligible $r_{bb'}$. If a load resistance $R_L = 1/g_{b'c}$ is connected between collector and emitter, what is the small-signal low-frequency voltage gain $A_v = v_{ce}/v_{be}$ for the model of Fig. 9-2c? How does this differ from the result obtained with the model of Fig. 9-2b?

Miller-Effect Multiplication

Under certain circumstances $g_{b'c}$ and $C_{b'c}$ may be magnified, and thus no longer be negligible. This is of particular importance in the design of amplifiers with large input impedance, and in the design of amplifiers with large bandwidth.

Consider the circuit of Fig. 9-3, which is a simple common-emitter-amplifier circuit. If we analyze the circuit using the model of Fig. 9-2b, which includes only the effect of the emitter junction, we find

$$v_0 = -(g_m r_{b'e})i_b R_L = -\beta i_b R_L$$

$$A_v \triangleq \frac{v_0}{v_{b'e}} = -g_m R_L$$

$$G_{in} \triangleq \frac{i_b}{v_{b'e}} = \frac{1}{r_{b'e}}$$

To focus attention on the effect of $g_{b'c}$ on input conductance, we use the model of Fig. 9-2c and neglect $r_{bb'}$ and g_{ce}; the resulting small-signal equivalent circuit is shown in Fig. 9-3b. We note that $g_m \gg 1/r_{b'e}$, and also $1/r_{b'e} \gg g_{b'c}$; if it is assumed that $G_L \gg g_{b'c}$, we find

$$v_0 = \frac{-\beta i_b R_L}{1 + (g_{b'c} r_{b'e})(1 + g_m R_L)}$$

$$A_v \approx -g_m R_L$$

$$G_{in} = g_{b'c}(1 + g_m R_L) + \frac{1}{r_{b'e}}$$

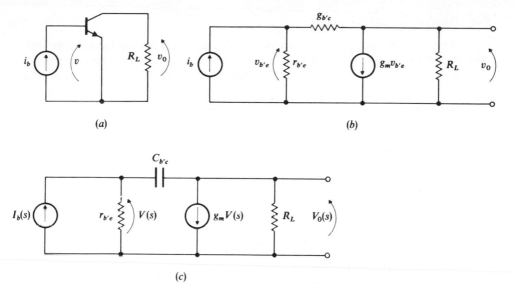

(a)   (b)

(c)

FIGURE 9-3

Miller-effect multiplication. (a) Small-signal common-emitter amplifier circuit; (b) equivalent circuit for evaluating effects of $g_{b'c}$ on input conductance; (c) equivalent circuit for evaluating effects of $C_{b'c}$ on bandwidth.

Note that wherever $g_{b'c}$ appears in these expressions it is multiplied by the factor

$$1 + g_m R_L = 1 - A_v$$

Now let us focus attention on the effect of $C_{b'e}$. For this purpose we use the equivalent circuit of Fig. 9-3c. Analysis of this circuit yields

$$V_0(s) \approx \frac{-\beta I_b(s) R_L}{1 + s(C_{b'c} r_{b'e})(1 + g_m R_L)}$$

$$A_v(0) \approx -g_m R_L$$

$$Y_{in}(s) \approx \frac{1}{r_{b'e}} + sC_{b'c}(1 + g_m R_L)$$

As we expected, the effect of $C_{b'c}$ is greatly magnified, since it appears everywhere multiplied by the factor $1 + g_m R_L$. This causes an increase of Y_{in}, and a reduction of bandwidth of V_0. Of course, we have not considered the effects of $C_{b'e}$ here, but we can see intuitively that if large $A_v(0)$ is used, the effects of $C_{b'c}$ on Y_{in} and the bandwidth of V_0 may be comparable to the effects of $C_{b'e}$.

9-2 BIASING CIRCUITS

In order to set the stage for the discussion of integrated biasing circuits, we first consider the problem of obtaining a bias current source. In discrete-component circuits this is done by using negative feedback to stabilize the emitter current of a transistor, as shown in Fig. 9-4. Here resistors R_1 and R_2 provide a voltage divider which fixes the voltage across R_E, thus determining the emitter current of Q. The collector current I_C is thus stabilized against temperature variation and variation of transistor parameters. If the circuit is to be employed as an amplifier, a capacitor is added as shown, a load resistor R_L is inserted in the collector circuit, and the input signal is inserted through a capacitor at the base.

This circuit is not acceptable for integration for two reasons:

1 It requires two capacitors, one of which has a capacitance of several μF.
2 It uses three resistors. Moreover, to obtain maximum gain, R_1 and R_2 will have large values and therefore require large area.

A more appropriate circuit employs shunt-shunt feedback as shown in Fig. 9-5. This method uses no capacitors and only two resistors, but has the disadvantage that it does not exhibit current-source behavior. However, the negative feedback employed ensures that the operating point will be stable with variations of temperature and transistor parameters.

The Widlar Circuit[2]

A basic biasing circuit, many variations of which are widely used in integrated circuits, was originated by R. J. Widlar and is shown in Fig. 9-6. This circuit capitalizes on the fact that Q_1 and Q_2 can be made to be identical and to have the same temperature. It also makes use of large negative feedback to stabilize the operating point of Q_1.

To analyze the circuit, we assume active-region operation, and make use of the model of Fig. 9-1, together with knowledge that the transistors are identical.

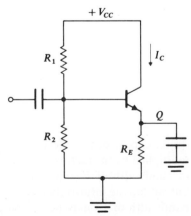

FIGURE 9-4
Discrete-component biasing circuit.

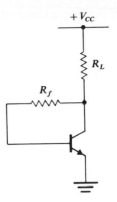

FIGURE 9-5
Feedback biasing circuit.

Since the transistors are identical and have the same base-emitter voltage, they must have approximately equal collector currents. The bias current I is

$$I = \frac{V_{CC} - V_D}{R}$$

Applying Kirchhoff's current law to the base node, we find

$$I = I_C + 2I_B$$

But from the model of Fig. 9-1, we also have

$$I_C = \beta I_B$$

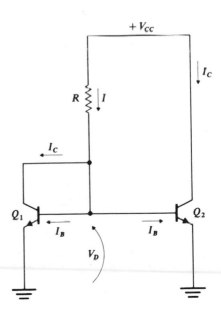

FIGURE 9-6
Widlar's basic bias circuit.

Combining these relations, we obtain

$$I_C = \frac{I\beta}{\beta + 2} = \frac{\beta(V_{CC} - V_D)}{R(\beta + 2)} \qquad (9\text{-}2)$$

If β is of the order of 50 or larger,

$$I_C \approx I$$

This circuit, which is sometimes called a *current mirror*, provides a current source whose current is relatively independent of transistor parameters. The circuit uses only one resistor.

Exercise 9-3 Suppose Q_1 has twice the emitter area of Q_2. Find I_C.

Exercise 9-4 Our analysis used the model of Fig. 9-1, which neglected I_{s2}. For a more exact analysis we could use the model of Eq. (9-1). Note that since the collector voltages of Q_1 and Q_2 are not identical, the collector currents will be slightly different. Use Eq. (9-1) to find I_C of Q_2.

Exercise 9-5 A resistor $R_L < R$ is inserted in series with the collector of Q_2. Find I_C of Q_2.

Exercise 9-6 Suppose Q_2 is replaced by six identical transistors connected in parallel. Find I_C of each transistor.

Small-Value Current Sources

Suppose we have a requirement in an integrated circuit to produce a small-value current source; say 10 μA. If a supply voltage of 10 V is available, the resistance required in the circuit of Fig. 9-6 would be

$$R = \frac{10 - 0.7}{10 \times 10^{-6}} = 930 \text{ k}\Omega$$

Such a large-value resistance would occupy far too much area on the chip. However, we can modify the circuit to overcome this disadvantage.

Examining (9-1) we see that if I_{s2} is negligible in comparison with the collector current, we can write

$$I_C = \alpha_N I_{s1}(e^{q\phi_E/mkT} - 1) \approx \alpha_N I_{s1} e^{q\phi_E/mkT} \qquad (9\text{-}3)$$

This indicates that the collector current is a strong function of the emitter voltage. If we can slightly unbalance the emitter voltages of the two transistors, we can cause a large difference between collector currents to occur. This is done by inserting a resistance in the emitter circuit of Q_2 as shown in Fig. 9-7. To obtain a first-order analysis of the circuit we assume $\alpha_N \to 1$ ($\beta \to \infty$). Then $I_{C1} = I \approx (V_{CC} - V_D)/R$. From (9-3) we see that

$$\frac{I_{C2}}{I_{C1}} = e^{(q/mkT)(\phi_{E2} - \phi_{E1})} \qquad (9\text{-}4)$$

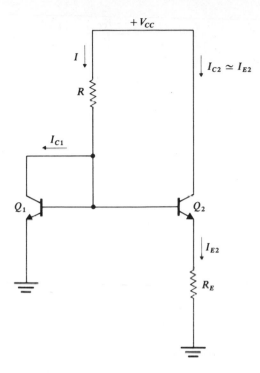

FIGURE 9-7
The small-value current source.

But the resistor R_E causes

$$\phi_{E2} - \phi_{E1} = -I_{E2} R_E \approx -I_{C2} R_E$$

Thus we find

$$R_E = \frac{mkT}{qI_{C2}} \ln \frac{I_{C1}}{I_{C2}} \qquad (9\text{-}5)$$

EXAMPLE 9A *Design of a 10-μA Current Source.* If we require a 10-μA current source, and a supply voltage of 10 V is available, we might choose 1 mA as a value of I which will lead to a reasonable value for R. Then $I_{C1} = 1$ mA, $R = 9.3$ kΩ, and if $m = 1$ we calculate

$$R_E = \frac{0.026}{10^{-5}} \ln \frac{10^{-3}}{10^{-5}} = 12 \text{ k}\Omega$$

Although the circuit uses two resistors, note that the total resistance (which would determine the area) is 21.3 kΩ, while the single-resistor circuit required 930 kΩ.

////

Exercise 9-7 Do not assume $\alpha_N \rightarrow 1$, and calculate R_E for the circuit of Fig. 9-7.

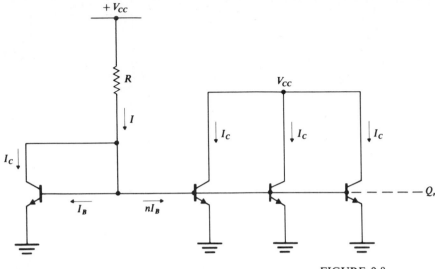

FIGURE 9-8
Multiple current sources.

Multiple-Current Sources

It often occurs that several stages in an amplifier require a certain ratio between the bias currents.

Since direct coupling is used to avoid the necessity for capacitors, it is also required that these ratios be stable with temperature in order to provide bias stability for the entire amplifier. The circuit of Fig. 9-6 can be expanded to perform this function, as shown in Fig. 9-8. Since all transistors have the same base voltage, they must all have identical collector currents I_C. By inspection of Fig. 9-8, we see that

$$I_C = \frac{\beta I}{\beta + n + 1} \qquad (9\text{-}6)$$

If several different values of current are required, they can be obtained by inserting resistors in the emitter circuits of the appropriate transistors.

Exercise 9-8 Currents of 1, 2, and 3 mA are required in an integrated circuit. If $V_{CC} = 10.7$, show how these currents can be obtained with four transistors, and calculate the value of all resistors.

We have thus far only discussed the use of *n-p-n* transistors for current sources. It is obvious that if currents of the opposite polarity are required, *p-n-p* transistors can be used. However, it should be noted from (9-6) that if β is not

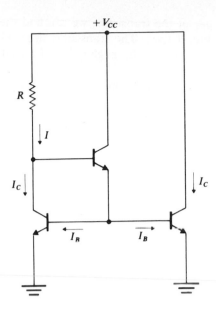

FIGURE 9-9
Improved feedback to better balance I_C
and I.

large, or if n increases, the current becomes more influenced by variations of β. This will be particularly a problem with lateral p-n-p transistors, which inherently have low β. It can also be a problem with n-p-n transistors where precise balance of two currents is required.

By adding another transistor in the feedback path of the circuit of Fig. 9-6, we can make the current less dependent on β. This is shown in Fig. 9-9. For this circuit

$$I = I_C + \frac{2I_B}{\beta + 1}$$

Since $I_C = \beta I_B$, we find

$$I_C = \frac{I\beta(\beta + 1)}{\beta(\beta + 1) + 2}$$

Even for moderate values of β, I_C and I will be well balanced.

Exercise 9-9   The feedback technique of Fig. 9-9 is applied to the circuit of Fig. 9-8.   Calculate the ratio I_C/I.

Output Conductance of Current Sources

When a current source is required in a circuit it is often also necessary that the incremental conductance dI_C/dV of the current source be very small.   Since the output conductance of the current source arises from the collector conductance

$g_{b'c}$ of the transistor, we can find the output conductance by using the circuit of Fig. 9-10a. The transistors are replaced by the model of Fig. 9-2c, and $r_{bb'}$ is assumed to be negligible. To analyze the circuit we must first determine $r_{b'e}$ and g_m.

Since it is known that $I_C \approx I_E \approx I$ for both transistors, we can write

$$g_m = \frac{qI}{mkT}$$

and

$$r_{b'e} = \frac{\beta}{g_m} = \frac{\beta mkT}{qI}$$

If we assume $\beta \gg 1$, and $g_m \gg g_{b'c}$, we find

$$v \approx dV \frac{g_{b'c}}{g_{m,}}$$

and

$$dI_C = dV(g_{b'c} + g_{ce}) + g_m v$$

Combining these results we obtain

$$\frac{dI_C}{dV} \approx 2g_{b'c} + g_{ce}$$

If the output conductance is too large for the application at hand, as it may be for values of I of several milliamperes, one can increase I and add a resistor R_E in the emitter circuit of Q_2. This provides negative feedback which decreases the output conductance. The values of I and R_E can be chosen so that the required I_C is obtained.

An alternative is to provide the necessary feedback in a different manner by using an additional transistor in the circuit as shown in Fig. 9-11.[3] Note the similarities and differences between this circuit and that of Fig. 9-9.

Exercise 9-10   Find the dc current I_C for the circuit of Fig. 9-11.

The V_D Multiplier Circuit[4]

We have centered our discussion of biasing primarily on current sources. Some circuits, such as class AB output stages of operational amplifiers, also require stabilized bias voltage sources. Perhaps the most convenient method of obtaining such a voltage source is to use a forward-biased diode, or diode-connected transistor. We have seen that the forward voltage is approximately 0.7 V, and that it changes only slightly with current. The model of Fig. 9-2c can easily be used to show that the small-signal conductance of the diode-connected transistor is

$$g \approx g_m = \frac{qI}{mkT}$$

which can be made large enough for most purposes by using moderate values of I. If voltages larger than V_D are required, several diodes can be connected in series.

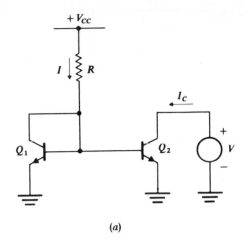

(a)

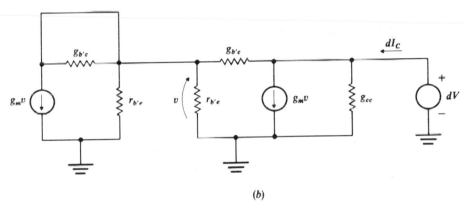

(b)

FIGURE 9-10
(a) Circuit for calculating the output conductance of the current source; (b) small-signal equivalent circuit.

Unfortunately this forces the designer to accept voltages which are integral multiples of V_D.

An alternative to the use of multiple diodes is the feedback circuit of Fig. 9-12, called a V_D multiplier circuit. The basic performance of this circuit can best be understood by making use of the model of Fig. 9-1 and assuming that β is very large. If β is very large, negligible base current will flow, but the base voltage will still be V_D. If negligible base current flows, then

$$V \frac{R_2}{R_1 + R_2} = V_D$$

and we have

$$V = V_D \left[1 + \frac{R_1}{R_2} \right]$$

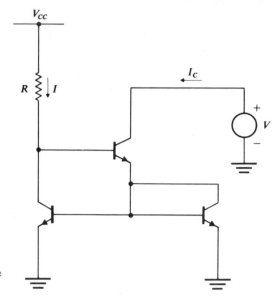

FIGURE 9-11
Circuit for reduced output conductance
of the current source.

By properly selecting R_1 and R_2 we can thus choose any multiple of V_D we wish. In practice the current source I would be replaced by either a resistor and supply voltage or a *p-n-p* current source.

Exercise 9-11 Do not assume β to be very large, and calculate V.

Exercise 9-12 Use electrothermal models to show that the temperature coefficient of V is $\gamma_D[1 + (R_1/R_2)]$.

The V_D multiplier circuit can be fabricated in integrated form in a very small space by making use of pinch resistors.[5] Both resistors are connected to the transistor base, so both can be fabricated as extensions of the base region. The transistor is laid out with elongated base region on both sides of the emitter, as shown in Fig. 9-13. Emitter diffusions are performed over both elongated regions, as well as in the normal emitter location. Metal is applied to short one end of the elongated base to the collector and to connect the other end of the base to the emitter. The elongated regions under the emitter diffusion form pinch resistors; since the voltage obtained depends on the ratio of these resistors, the desired voltage can be obtained by choosing an appropriate ratio of the lengths of the elongated regions.

The Saturated Transistor

Voltages less than V_D can be obtained by using the circuit of Fig. 9-14a in which the transistor is allowed to partially saturate. This circuit takes advantage of the

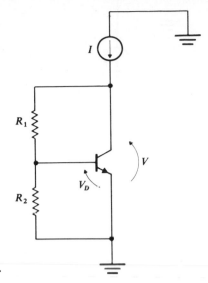

FIGURE 9-12
The V_D multiplier circuit.

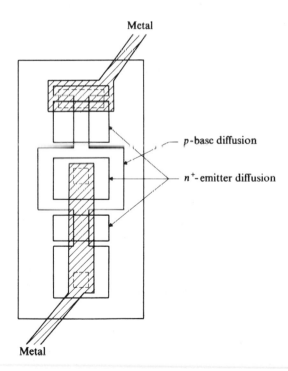

Metal

p-base diffusion

n^+-emitter diffusion

Metal

FIGURE 9-13
Layout of the V_D multiplier circuit.

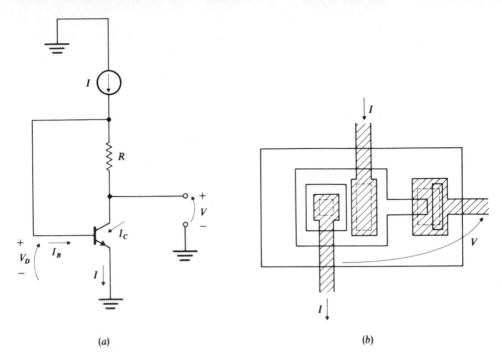

(a)   (b)

FIGURE 9-14
The partially saturated transistor.   (a) Circuit; (b) layout.

fact that until the collector-base junction is forward-biased by about 0.5 V the collector current is $I_C = \beta I_B$; that is, the transistor behaves as though it were in the forward active region.   The collector current for the transistor in Fig. 9-14 is easily
seen to be

$$I_C = \alpha_N I = \frac{\beta I}{\beta + 1} \approx I$$

while the output voltage is

$$V = V_D - I_C R \approx V_D - IR$$

This simplified analysis gives reasonable results for

$$0.2 \leqq V \leqq V_D$$

Note that if I is of the order of a milliampere, R will be only a few hundred ohms;
it can usually be fabricated by slightly elongating the base region, as shown in
Fig. 9-14b.

Level-shifting Stages

In amplifiers of several stages, the dc level of the output of each stage is not the
same as that of the input of the stage.   For n-p-n stages, the base is more positive

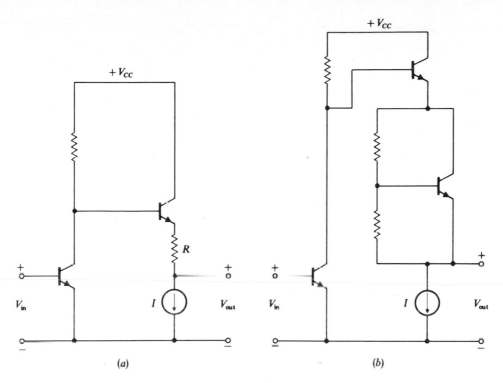

FIGURE 9-15
Level shifting (a) with resistor and current source, and (b) with V_D multiplier circuit.

than the emitter, and the collector is more positive than the base, in order to ensure operation in the forward active region. In some applications this is of no consequence, but in others, such as operational amplifiers used as voltage followers, it is necessary that the output of the amplifiers be at the same level as the input. This means that at some point in the circuit, means must be provided to shift the dc level so that input and output voltages are compatible. Depending upon the particular circuit, the amount of level shifting required varies from V_D to several volts.

In some cases, the required level shift can be obtained simply by adding several diodes in series between the output of one stage and the input of the next, and providing appropriate bias current to ensure that the diodes are forward-biased. For larger voltages, the circuit of Fig. 9-15a can be used. Here an emitter follower is used to isolate the output of the gain stage from loading effects of the next stage. Level shifting is provided by the resistor R and the bias current source I. Proper design can ensure that the output voltage depends on the ratio of resistors and is therefore independent of temperature.

This type of level shifting is simple, and has the advantage of providing current gain. Its principal disadvantage is that its output resistance is approximately R; the bandwidth will be low because of the input capacitance of the next stage. An improvement of the circuit can be made by using a V_D multiplier in place of R, as is shown in Fig. 9-15b. This produces the desired level shift at a much lower output resistance.

If a voltage of $2V_D$ or $3V_D$ is required, cascaded emitter followers can provide both current gain and the required level shift.

9-3 SIMPLE GAIN STAGES

If a transistor is to be used for gain purposes, not only must it be biased with stable collector current, but means must also be provided for injecting the input signal and for obtaining an output signal. We can arrange for the latter by including a load resistor in the collector of Q_2 in our two-transistor bias circuit, but the injection of the input signal is more complicated.

If we assume for the moment that the input is a current source, and connect it to the base as shown in Fig. 9-16a, we note that most of the input current will flow through the diode-connected transistor, and will not contribute to the gain of the circuit. In fact, it is easily shown that the small-signal base current of Q_2 is approximately i_s/β and the transimpedance of the circuit is $v_0/i_s = -R_L$. The value of R_L is, of course, limited to

$$R_L < \frac{V_{CC} - V_D}{I}$$

If the original bias circuit is modified as shown in Fig. 9-16b, this shunting effect can be virtually eliminated. Note that for dc bias conditions, both transistors have identical resistors in their base circuits, and both resistors are connected to the same point. Both Q_1 and Q_2 have the same base current and therefore the same collector current I_C. Now, however, the adverse shunting effect on the base of Q_2 is only that of R_1. Thus by choosing $R_1 \gg r_{b'e}$ we can ensure that almost all i_s appears as base current for Q_2. The transimpedance of the circuit is

$$\frac{v_0}{i_s} = \frac{-R_L \beta R_1}{R_1 + r_{b'e}}$$

EXAMPLE 9B *Design of a Gain Stage.* To design the circuit, we begin by noting that if β is very large, the base bias current of each transistor will be very small and the voltage across R_1 will be negligible. Therefore Q_1 will be operating with $V_{C1} \approx V_D$, and with $I_C \approx I$. Now for maximum dynamic range of the output, we want Q_2 to operate midway between saturation and cutoff, that is, with $V_{C2} = (V_{CC} + V_D)/2$. Since $I_C = I$, we therefore choose

$$R_L = \frac{R}{2}$$

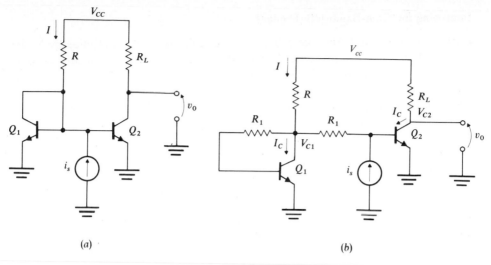

FIGURE 9-16
(a) Input signal applied to current source transistor; (b) modification of the circuit to improve the gain.

Next we calculate $r_{b'e} = \beta mkT/qI$, and select

$$R_1 \gg r_{b'e} \qquad (9\text{-}7)$$

The design is now complete. We note that if (9-7) is satisfied, the transimpedance will be given by

$$\frac{v_0}{i_s} \approx \frac{-\beta R}{2}$$

For typical numbers, we assume that $V_{CC} = 10.7$, $I = 1$ mA, and $m = 1$. Then $R \approx 10$ kΩ and $R_L = 5$ kΩ. With $I = 1$ mA we have $r_{b'e} = 2.6$ kΩ. Note that the input resistance seen by i_s is approximately $r_{b'e}$ in parallel with R_1. The transimpedance of the amplifier is -0.5×10^6. ////

Exercise 9-13   Find the input resistance of the circuit of Fig. 9-16b; show that for large β it becomes $R_1 r_{b'e}/(r_{b'e} + R_1)$. Assume $R_L \ll 1/g_{ce}$.

If the input resistance is not large enough for the application at hand, it can be increased by adding a resistance R_E of several hundred ohms in the emitter circuit of Q_2. To maintain stable bias $I_C = I$, an identical resistor must be added in the emitter circuit of Q_1. Note also that the value of R_1 must be increased so that

$$R_1 \gg r_{b'e} + (\beta + 1)R_E$$

Improving the Gain-Bandwidth Product

In the circuit of Fig. 9-16b, the collector capacitance of Q_2 is multiplied by Miller effect and therefore has the same effect on the output as would a capacitor of value $C_{jc}(1 + g_m R_L)$ connected directly across the output terminals. As was pointed out in Sec. 9-1, Miller multiplication of the capacitance severely limits the bandwidth of the circuit. The bandwidth of the circuit can be increased by reducing R_L and/or g_m, but this is done at the expense of gain. What is required is some way to obtain the current gain of Q_2 but to simultaneously use a low effective load resistance. This is done by using a cascode connection, the circuit for which is shown in Fig. 9-17a without bias circuits.

Transistor Q_1 is connected in the common-emitter configuration, but its effective load resistance is the emitter-base input resistance of Q_2, which is $1/g_{m2}$. Because of the low value of $1/g_{m2}$, the voltage gain of Q_1 is approximately unity and its Miller multiplication factor is unity. However, the current gain of Q_1 is β. The current gain of Q_2 is $\alpha_N \approx 1$, but the Miller multiplication factor is also unity. Thus the overall voltage gain is the same as would be obtained with Q_1 having a load resistance R_L, while the bandwidth resulting from $C_{b'c}$ is $1/C_{b'c} R_L$. For small values of R_L the intrinsic bandwidth $1/\omega_T$ must also be taken into account.

Exercise 9-14   Suppose Q_1 and Q_2 each have collector current I. Show that the voltage gain for the cascode circuit is $g_m R_L$.

EXAMPLE 9C   *The Fairchild μA701 Cascode Circuit.*   The Fairchild μA701 provides an illustration of the practical application of both the cascode circuit and the biasing techniques we have previously described; it is shown in Fig. 9-17b. In analyzing the circuit we consider first the bias conditions. It will be noted that Q_1 and Q_2 form a bias combination like that of Fig. 9-16b, in which the collector current of Q_2 is stabilized. Resistors R_1 and R_2 determine the bias current. Note that in Fig. 9-17b, emitter resistors R_4 and R_6 have been added to increase the input resistance of the circuit.

If we assume β to be large, we can easily determine the bias conditions, since base currents will be negligible. The collector current I_{C1} is

$$I_{C1} = \frac{10 - V_D}{12.3 \text{ k}\Omega} = 0.75 \text{ mA} = I_{C2}$$

Then $V_{E1} = V_{E2} = 0.18$ V, and $V_{B3} = 3.9$ V, making $V_{E3} = 3.2$ V. With a collector current $I_{C2} = 0.75$ mA, the current I_{C3} will also be 0.75 mA; this will produce $V_{C3} = 8.5$ V. Transistor Q_4 is simply an emitter follower to provide low output resistance, and its emitter junction voltage lowers the output voltage to $V_{E4} = 7.8$ V, producing an emitter bias current $I_{E4} = 2.6$ mA. All bias currents and voltages have now been determined.

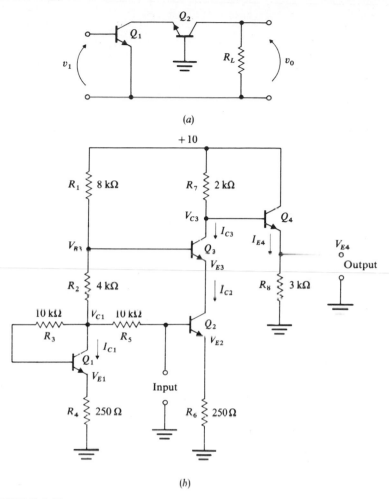

FIGURE 9-17

(a) The basic cascode circuit; (b) the Fairchild μA701 integrated cascode amplifier.

The midband gain of the circuit can now be calculated by replacing the critical transistors with their small-signal models. Note that Q_1 serves only for biasing purposes and can be ignored for small-signal calculations. We can also assume the voltage gain of Q_4 to be unity. The essential parts of the circuit are shown in Fig. 9-18. Note that we have used a T model for Q_3 because emitter current is the independent variable, and we have also used a T model for Q_2 because of the presence of R_6. For all transistors we assume $r_{bb'}$ to be negligible. From the bias conditions above, we have $I_{E2} \approx I_{E3} = 0.75$ mA, so for both Q_2 and Q_3

$$g_m = \tfrac{1}{34}$$

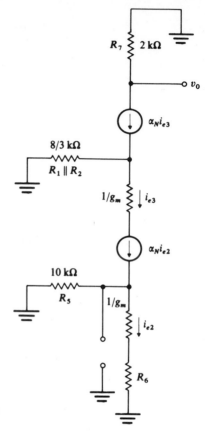

FIGURE 9-18
Small-signal equivalent circuit for the
cascode amplifier.

The small-signal voltage gain of the amplifier is easily seen to be

$$A_v \triangleq \frac{v_0}{v_1} = \frac{-\alpha_N{}^2 R_7}{R_6 + 1/g_m} \approx \frac{-2 \text{ k}\Omega}{284} = -7$$

The input resistance is

$$R_{\text{in}} = \frac{R_5(R_6 + 1/g_m)(\beta + 1)}{R_5 + (R_6 + 1/g_m)(\beta + 1)}$$

which for $\beta = 100$ is

$$R_{\text{in}} = 7.4 \text{ k}\Omega$$

This assumes $r_{b'e}$ to be negligible.   To find the output resistance we can include
the model for the emitter follower Q_4; the result is, for $\beta = 100$,

$$R_{\text{out}} = \frac{[R_7/(\beta + 1) + 1/g_{m4}]R_8}{R_7/(\beta + 1) + R_8 + 1/g_{m4}} = 52.9 \text{ }\Omega$$

We can also calculate the Miller multiplication factor for Q_2. Note that the effective load resistance for Q_2 is

$$R_{L(\text{eff})} = \frac{R_1 \parallel R_2}{\beta + 1} + \frac{1}{g_m} = 61\ \Omega$$

And the voltage gain of Q_2 is

$$A_{v2} = \frac{-R_{L(\text{eff})}}{R_6 + 1/g_m} = -0.215$$

Since the Miller multiplication factor is $1 - A_{v2}$, we have

$$1 - A_{v2} = 1.215$$

The bandwidth of Q_2 will thus be determined by the intrinsic bandwidth of the transistor rather than by parasitics.

The Common-collector Common-base Circuit

An amplifier which is quite similar to the cascode circuit is the common-collector common-base circuit shown, without bias elements, in Fig. 9-19a. As was the case for the cascode circuit, the first stage of this amplifier has approximately unity gain, and therefore no Miller multiplication effects. It produces a current gain $\beta + 1$, however, and while the second stage has only a current gain α_N, the second stage produces a voltage gain. As was the case in the cascode circuit, the second stage is a common-base stage and thus has no Miller multiplication effects.

The voltage gain of this circuit is almost identical to that of the cascode circuit, but in this case the biasing is much simpler. It can easily be arranged so that both transistors have the same bias current. This has been done in the Fairchild μA703 circuit which is shown in Fig. 9-19c. This circuit is designed for RF-IF amplifier applications and is intended primarily for use with transformer input and output. Here Q_3 is the common-collector transistor and Q_4 is the common-base stage. The other transistors are merely for biasing purposes; note that Q_2 and Q_5 form a bias current source in which the current is determined by R_2. An additional offset voltage is provided by Q_1, in order that Q_5 not be saturated. If transformer coupling is used at the input, then

$$V_{B1} = V_{B3}$$

and with the emitter offset voltages of Q_1 and Q_2

$$V_{B1} = 2V_D$$

The collector voltage of Q_5 is then

$$V_{C5} = 2V_D - V_D = V_D$$

while its base voltage is

$$V_{B5} = V_D$$

Thus we see that Q_5 is operating with zero collector-base voltage.

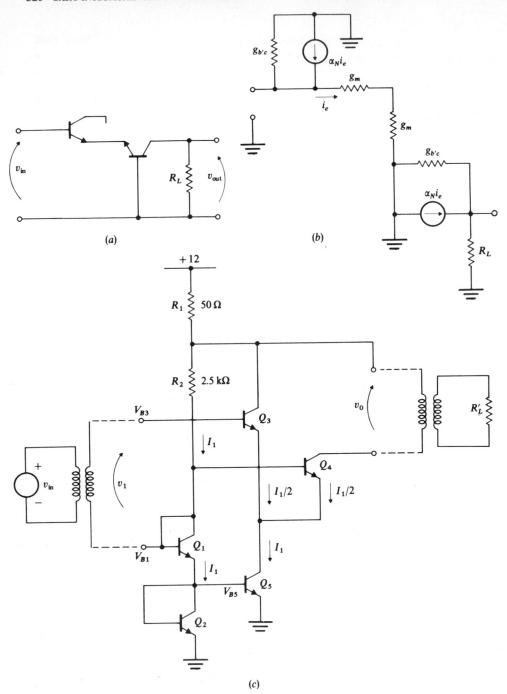

FIGURE 9-19
(a) Common-collector common-base circuit; (b) small-signal equivalent circuit; (c) the μA703 amplifier.

If Q_3 and Q_4 are matched, they will have equal emitter currents since their base-emitter voltages are equal. Thus they have equal values of g_m, and the equivalent circuit is as shown in Fig. 9-19b, in which R_L is the transformed value of R'_L.

For a 12-V supply and the values shown in Fig. 9-19c, $I_1 = 4.08$ mA and $g_m = 79 \times 10^{-3}$ ℧. The gain of the circuit is then

$$A_v = \frac{v_0}{v_1} = \frac{g_m}{2} R_1 = 40 \times 10^{-3} R_L$$

9-4 DIFFERENTIAL AMPLIFIERS[6,7]

Many applications, such as biomedical instrumentation, require the amplification of very-low-level signals in the presence of noise which is many times larger than the signals. Often this noise is produced external to the signal source and is therefore a *common-mode* signal, appearing equally at both terminals of the device at which measurements are being made. What is required therefore is a circuit which will amplify voltage differences, and reject common voltages. Such a circuit is called a *differential amplifier;* the basic differential amplifier is shown in Fig. 9-20a. Transistors Q_1 and Q_2 with resistors R_C form the differential amplifier; R_E serves to provide bias current.

First-order Analysis

For a first-order analysis, we assume:

  1. Identical transistors
  2. $r_{bb'} \approx 0$
  3. $\alpha_N \approx 1$
  4. Matched resistors R_C

If these assumptions are met, the small-signal equivalent circuit becomes that shown in Fig. 9-20b. Note that because the transistors are identical, their emitter bias currents are equal, as are their g_m's. In Fig. 9-20b we have used a T-model representation for the transistor since emitter current is the independent variable.

Because we wish to examine differential-mode and common-mode properties of the circuit, we define differential- and common-mode voltages as

$$v_{iDM} \triangleq \frac{v_1 - v_2}{2} \qquad (9\text{-}8)$$

$$v_{iCM} \triangleq \frac{v_1 + v_2}{2} \qquad (9\text{-}9)$$

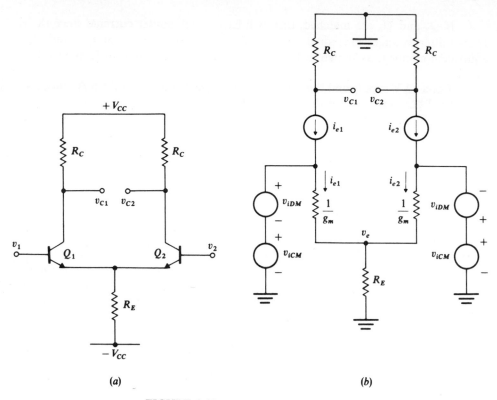

FIGURE 9-20
(a) Basic differential-amplifier circuit; (b) small-signal equivalent circuit.

We can now express v_1 and v_2 in terms of differential- and common-mode components by solving (9-8) and (9-9) for v_1 and v_2:

$$v_1 = v_{iDM} + v_{iCM} \qquad (9\text{-}10)$$

$$v_2 = v_{iCM} - v_{iDM} \qquad (9\text{-}11)$$

These are shown pictorially in Fig. 9-20b.

Consider first the differential-mode case, with $v_{iCM} = 0$. Note that with $v_{iCM} = 0$, $v_e = 0$ because of symmetry; we can now work with only half the circuit since v_e is a virtual ground and R_E has no effect on the circuit. Let v_{C1} for this case be

$$v_{C1} \triangleq v_{ODM}$$

We see by inspection that

$$v_{ODM} = -v_{iDM} g_m R_C$$

and we can define a differential-mode gain A_{DM} as

$$A_{DM} \triangleq \frac{v_{ODM}}{v_{iDM}} = -g_m R_C \qquad (9\text{-}12)$$

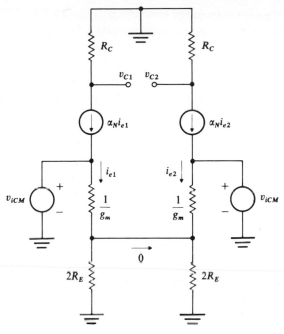

FIGURE 9-21
Equivalent circuit for common-mode calculations.

Next consider the common-mode case, for which $v_{iDM} = 0$. The equivalent circuit is redrawn in Fig. 9-21, in which we note that no current flows in the connection between the two $2R_E$ resistors. This connection may thus be broken for purposes of analysis, and once again we have only to analyze half the circuit.

Let v_{C1} for this case be

$$v_{C1} \triangleq v_{0CM}$$

and define a common-mode gain A_{CM} as

$$A_{CM} \triangleq \frac{v_{0CM}}{v_{iCM}}$$

Since

$$i_e = \frac{v_{iCM}}{2R_E + 1/g_m}$$

we obtain

$$A_{CM} = \frac{-g_m R_C}{1 + 2g_m R_E} \qquad (9\text{-}13)$$

A figure of merit for differential-amplifier performance is the discrimination ratio D, defined as

$$D \triangleq \frac{A_{DM}}{A_{CM}}$$

This is a measure of how well the circuit rejects common-mode signals and amplifies differential-mode signals; clearly a large D is desirable. Using (9-12) and (9-13), we find

$$D = 1 + 2g_m R_E \qquad (9\text{-}14)$$

As far as D is concerned, (9-14) shows that it is desirable to have R_E and g_m as large as possible. Since $g_m = qI/2mkT$, we wish to operate with as large a bias current as possible.

In analyzing the differential mode of operation, we saw that R_E does not influence the differential-mode gain; its only function is to provide bias current for the circuit. What is more desirable is to have a current source for bias; this provides the necessary current, but with very large effective R_E. Such a circuit is shown in Fig. 9-22. Here the current-source circuit of Fig. 9-6 is employed for bias current; since its incremental resistance is very high, large D is obtained. Note also that the dc collector voltage of the differential-amplifier transistors depends upon the ratio R_C/R_E rather than the absolute values of the resistors.

Exercise 9-15 The circuit of Fig. 9-22 has $V_{CC} = 10.7$, $R_E = 10\ \mathrm{k\Omega}$, $R_C = 5\ \mathrm{k\Omega}$. Calculate D. Assume $\eta = 10^{-4}$.

EXAMPLE 9D *Gain Limitations of the Differential Amplifier.* Although use of a current source makes possible $D \to \infty$, the differential-mode gain can still be limited by the common-mode signal amplitude. Consider a circuit in which the bias current is provided by a perfect current source of value I. Let the maximum value of the common-mode signal be $v_{iCM(max)} = V_{iCM}$, and assume that the common-mode voltage is much larger than the differential-mode voltage. Now if the circuit is to perform as a differential amplifier, the transistors must not saturate even when $v_{iCM} = V_{iCM}$. For a common-mode input, the current source ensures that the collector voltages v_{C1} and v_{C2} will not change, but the base voltage is increasing. Thus the resistors R_C must be chosen so that

$$V_{CC} - V_{iCM} \geq \frac{R_C I}{2} \qquad (9\text{-}15)$$

But R_C and I also determine A_{DM}:

$$A_{DM} = -g_m R_C = -\frac{qI}{2mkT} R_C \qquad (9\text{-}16)$$

If the maximum value of R_C is chosen from (9-15) and inserted in (9-16), we find the maximum A_{DM} to be

$$A_{DM(max)} = \frac{-(V_{CC} - V_{iCM})}{mkT/q} \qquad (9\text{-}17)$$

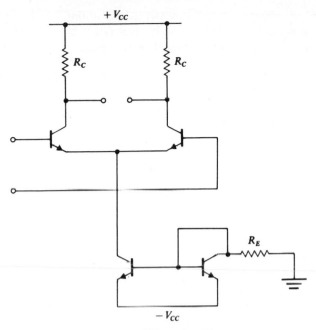

FIGURE 9-22
Current-source bias to improve D.

Most commercial differential amplifiers have supply voltages of 15 V or less. If the maximum common-mode input voltage is 10 V, the gain is limited to less than 200 for a single stage.

Differential Input Resistance

The differential input resistance R_{iDM} is defined as

$$R_{iDM} \triangleq \frac{(v_1 - v_2)}{i_b} = \frac{2v_{iDM}}{i_b}$$

We can easily calculate i_b by recalling that for differential-mode signals the emitter node of the circuit is a virtual ground.   Thus

$$i_b = \frac{v_{iDM}}{r_{b'e}} = \frac{v_{iDM}\, g_m}{\beta}$$

Since g_m depends on I, we find

$$R_{iDM} = \frac{4\beta mkT}{qI} \qquad (9\text{-}18)$$

This analysis has neglected $g_{bb'}$ and $g_{b'c}$. In most practical cases the former is negligible; however, the latter may not be. Usually $g_{b'c} \ll 1/R_C$; for this case we obtain, by using the model of Fig. 9-2c,

$$R_{iDM} \approx \frac{2}{g_m/\beta + g_{b'c}(1 - A_{DM})} \qquad (9\text{-}19)$$

Both g_m and $g_{b'c}$ vary directly with bias current; (9-19) shows that it is desirable to have I as small as possible in order to obtain large R_{iDM}.

EXAMPLE 9E *Input Resistance of a Typical Stage.* Consider a circuit with $I = 52~\mu\text{A}, R_C = 50~\text{k}\Omega, m = 1, g_{ce} = 10^{-5}, \beta = 100$. For this case $g_m = 26 \times 10^{-6}/26 \times 10^{-3} = 1 \times 10^{-3}$. Then $A_{DM} \approx -g_m R_C = -50$. If $g_{b'c}$ at this current is 1×10^{-7}, we obtain

$$R_{iDM} = \frac{2}{10^{-5} + 51 \times 10^{-7}} = \frac{2 \times 10^6}{15.1} = 133~\text{k}\Omega \qquad ////$$

Differential Amplifier with Emitter-Follower Input

Even for low bias currents, the input resistance of the differential amplifier may not be sufficiently large for some applications. The input resistance can be increased by using emitter-follower inputs as shown in Fig. 9-23a. To calculate the gain and input resistance, we first find the bias currents. The emitter current of Q_2 is $I/2$, and thus $I_{B2} = I/2(\beta + 1)$. We obtain

$$g_{m2} = \frac{qI}{2mkT}$$

$$g_{m1} = \frac{qI}{2(\beta + 1)mkT} \approx \frac{qI}{2\beta mkT}$$

The equivalent circuit for the calculation of A_{DM} is shown in Fig. 9-24a. Note that $g_{m2} = \beta g_{m1}$ and that $r_{b'e2} = \beta/g_{m2}$. Thus we see that $r_{b'e2} = 1/g_{m1}$; then

$$v_{b2} = \frac{v_{iDM}}{2}$$

and
$$A_{DM} = -\frac{g_{m2} R_C}{2}$$

The use of emitter followers in this manner has reduced A_{DM} by a factor of 2. To calculate the input resistance, we use the circuit of Fig. 9-24b, noting that $g_{b'c1}/g_{b'c2} = g_{m1}/g_{m2}$, and $g_{m2} = \beta g_{m1}$. Analyzing the circuit, we find

$$R_{iDM} \approx \frac{2}{g_{b'c1} + \beta^{-1}\left[\dfrac{1}{g_{m1}} + \dfrac{1}{g_{m1} + g_{b'c1}\beta(1 - 2A_{DM})}\right]^{-1}}$$

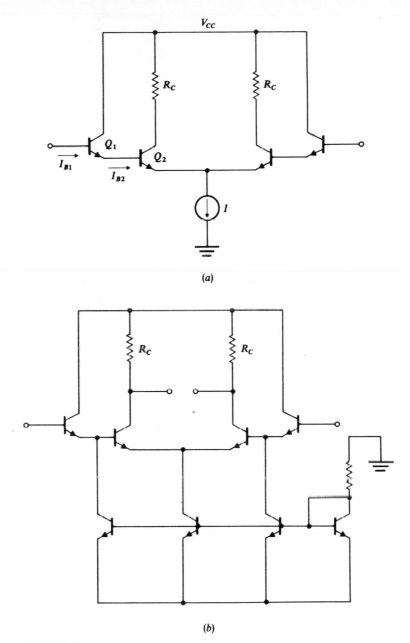

FIGURE 9-23
(a) Emitter-follower inputs; (b) separate bias sources for the emitter followers.

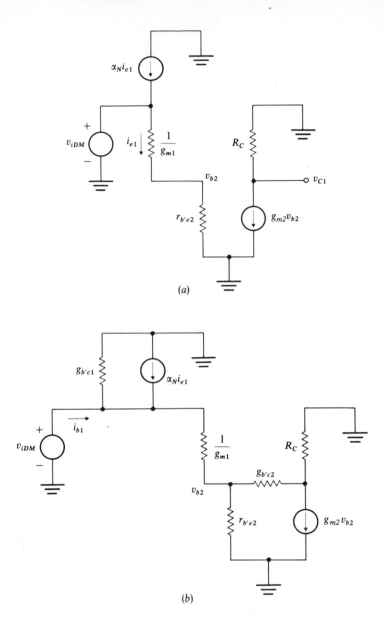

(a)

(b)

FIGURE 9-24
Equivalent circuits for (a) calculating A_{DM}; (b) calculating R_{iDM}.

The gain can be improved by increasing g_{m1}; this can be accomplished by providing separate current sources for the emitter followers, as shown in Fig. 9-23b. Now $g_{m1} \approx g_{m2}$, and

$$-A_{Dm} \approx g_{M2} R_C$$

However, $g_{b'c1} \approx g_{b'c2}$, and R_{iDM} is now

$$R_{iDM} \approx \frac{2}{g_{b'c2} + \beta^{-1}\left[\dfrac{1}{g_{m2}} + \dfrac{1}{g_{m2}/\beta + g_{b'c2}(1 - A_{DM})}\right]^{-1}}$$

Sometimes the input transistors have their collectors in common with the differential transistors, forming Darlington pairs. This should not be done, because Miller-effect multiplication of C_{jc} occurs, with the result that the input impedance of the Darlington connection is little larger than that of the two-transistor differential amplifier.

We have seen that in order to get large R_{iDM}, it is necessary to use small I in the two-transistor circuit; this small I results in small g_m. Since (9-17) shows that the maximum obtainable gain depends only on V_{CC} and V_{iCM}, we see that what must be done to obtain both maximum gain and large R_{iDM} is to use (9-18) to choose I small enough to produce the desired R_{iDM}, use (9-17) to calculate $A_{DM(max)}$, and insert I and $A_{DM(max)}$ in (9-16) to calculate the required R_C. Note that as R_{iDM} increases, so does R_C for a given $A_{DM(max)}$. Since the differential-mode output resistance is $R_{ODM} = 2R_C$ for $R_C \ll 1/g_{ce}$, we see that although the gain is not reduced by increasing R_{iDM}, both the area required and the output resistance are increased. This disadvantage can be offset by using emitter-follower output transistors.

EXAMPLE 9F *First-order Design of a Differential Amplifier with Large R_{iDM} and Low R_{ODM}.* Consider a case for which $V_{CC} = +12$ and $V_{iCM} = 2$ V. Suppose we require $R_{iDM} \geq 1$ MΩ, R_{ODM} as low as possible, and maximum gain. We assume for first-order design that $g_{b'c}$ is negligible and that $\beta = 100$. The circuit to be used is that shown in Fig. 9-25. If $g_{b'c}$ is negligible, the input resistance is

$$R_{iDM} \approx 2\beta \left(\frac{1}{g_{m1}} + \frac{\beta}{g_{m2}}\right)$$

and since

$$I_{E1} \approx \frac{I_{E2}}{\beta} = \frac{I}{2\beta}$$

we find

$$g_{m2} \approx g_{m1}\beta$$

and

$$R_{iDM} \approx \frac{4\beta^2}{g_{m2}}$$

The bias current required is thus

$$\frac{I}{2} = \frac{mkT}{q} g_{m2} = \frac{mkT}{q} \frac{4\beta^2}{R_{iDM}} \approx 1 \text{ mA}$$

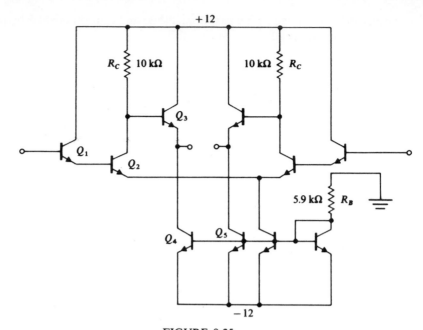

FIGURE 9-25
Differential amplifier with large R_{iDM} and low R_{oDM}.

To find the maximum gain, we note that when the input is V_{iCM}, the base voltage of Q_2 is $V_{iCM} - V_D$. In order to provide some voltage margin for differential voltages, we design the circuit so that the collector-base voltage of Q_2 is never less than V_D when the input is V_{iCM}. Then the maximum gain of the differential pair is

$$A_{DM(max)} = \frac{-q(V_{CC} - V_{iCM})}{mkT} \qquad (9\text{-}20)$$

Note, however, that the overall gain of the amplifier is half this value because no separate bias currents are used for the input emitter followers. Thus we have for the amplifier

$$-A_{DM} = \frac{-q(V_{CC} - V_{iCM})}{2mkT} \approx 192$$

The value of R_C required is obtained from (9-15) to be

$$R_C = 10 \text{ k}\Omega$$

Since we have already constructed a bias current source to supply 2 mA to the differential-amplifier transistors, we add two additional transistors Q_4 and Q_5 to provide bias current for the emitter followers. Since their bias current is not critical, the use of 2 mA for each enables us to obtain these bias currents without the use of additional resistors. The output resistance is now

$$R_{ODM} = 2\left(\frac{R_C}{\beta} + \frac{1}{g_{m3}}\right) \approx 226\ \Omega$$

To obtain the bias current $I = 2$ mA, we require

$$R_B = \frac{V_{CC} - V_D}{I} = 5.65\ k\Omega$$

It is interesting to compare this design with that of a two-transistor differential amplifier having the same gain and R_{iDM}. For negligible $g_{b'c}$, we see that

$$R_{iDM} = \frac{2\beta}{g_m}$$

and thus we require

$$\frac{I}{2} = 5.2\ \mu A$$

With this value of $I/2$, g_m is 2×10^{-4}. To obtain a gain of 192, the value of R_C required is now

$$R_C = 96 \times 10^4 = 960\ k\Omega$$

and the output resistance is

$$R_{ODM} = 1.92\ M\Omega$$

To obtain the bias current of 10.4 μA, a small-value current source would have to be designed.

It is important to note that with the two-transistor circuit and this value of gain, $g_{b'c}$ would not be negligible. If $g_{b'c} = 10^{-7}$, the Miller-effect multiplication causes $R_{iDM} = 94$ kΩ, while for the circuit of Example 9F we still obtain $R_{iDM} \approx 1$ MΩ.

Super-gain Transistors[8]

In our discussion of input resistance, we have focused attention on the small-signal differential-mode input resistance. In many applications, the input bias current is equally important. A straightforward method of reducing the offset current is to use a small emitter bias current in the differential amplifier and then add emitter-follower input transistors. The base current of the emitter followers can in this way be made very low, but at a considerable sacrifice of the gain of the differential amplifier.

Most integrated-circuit manufacturers have a so-called *super-gain* or *super-β* processing sequence in which an extra emitter diffusion is added. This is used to produce transistors whose base width is less than 0.1 μm in thermal equilibrium.

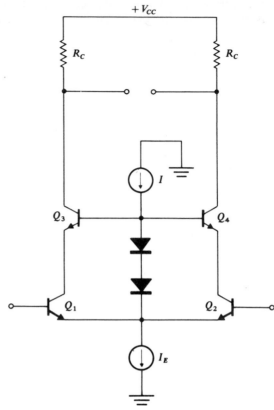

FIGURE 9-26
Cascode differential amplifier using super-gain input transistors Q_1 and Q_2.

By using this extra diffusion and exercising careful process control so that surface effects do not degrade device performance, it is possible to produce transistors with β in excess of 2000. With such a large β, even for moderate emitter bias currents the input base current of super-gain transistors in a differential amplifier is of the order of 1 nA. Thus it is in principle possible to obtain reasonable gain with low input bias current.

Special circuit techniques must be employed in the use of super-gain transistors. Because their base width is so small, punch-through occurs at collector-base voltages of the order of 1 V. Moreover, base-width modulation effects are extreme, and care must therefore be taken to avoid Miller-effect multiplication of $g_{b'c}$ lest the small-signal input resistance be unacceptably low. It is therefore necessary that the super-gain transistors be operated with approximately zero collector-base bias voltage, and also with low voltage gain. While this appears contrary to the objective of reasonable gain with low offset current, these objections can be overcome by combining super-gain transistors with ordinary transistors. A circuit which accomplishes this is the differential cascode amplifier of Fig. 9-26. Transistors Q_1 and Q_2 are super-gain transistors forming a differential amplifier, and Q_3 and Q_4 perform, for differential-mode signals, as common-base stages. Diodes D_1 and D_2 are diode-connected transistors used to provide offset voltage so that the collector-base voltage of Q_1 and Q_2 is zero. Current I_E is obtained from a small-value current-source bias circuit using n-p-n transistors, while I is obtained from a p-n-p small-value current-source circuit. If the emitter currents of Q_1 and Q_2 are held to 1 μA or less, the base bias current is less than 1 nA.

Since the load resistance for Q_1 and Q_2 is $1/g_m$, the voltage gain of Q_1 and Q_2 is approximately unity, and no severe Miller multiplication results. Reasonable gain can be obtained by making R_C sufficiently large.

9-5 ACTIVE LOADS[9]

The limitation imposed upon maximum differential-mode gain by power supply and common-mode input voltages arose because the small-signal load resistance and the collector bias resistor were in fact the same resistor. Ideally, we would like to use a large small-signal load resistance and a small bias resistor. This is, of course, what is accomplished by transformer coupling. However, transformers are impractical for many applications. What is required, therefore, is a nonlinear resistor which can be operated at a point such that its small-signal resistance dV/dI is much larger than its voltage-current ratio V/I.

The collector circuit of another transistor can be used to provide such a non-linear resistor. This is particularly convenient for a differential amplifier in which the output can be single-ended. The circuit is shown in Fig. 9-27a; here Q_4 provides the nonlinear resistance for the collector of Q_2, and Q_3 provides bias. The combination Q_3 and Q_4 is called an active load. In order that neither Q_4 nor Q_2 saturate, it is necessary that the collector currents be almost perfectly balanced; this is accomplished by using the p-n-p version of the biasing circuit of Fig. 9-6. Now if $v_{in} = 0$, and if Q_1 and Q_2 are matched, $I_{C1} = I_{C2}$. If Q_3 and Q_4 are matched, $I_{C4} \approx I_{C1}$; thus $I_{C4} \approx I_{C2}$. The match between I_{C4} and I_{C2} can be improved if the p-n-p version of the circuit of Fig. 9-9 is used.

The differential-amplifier circuit is no longer symmetrical; note that the load resistance seen by Q_1 is the diode-connected Q_3, while that seen by Q_2 is the collector circuit of Q_4. Moreover, there is a signal path from Q_1 through Q_3 and Q_4 to the output as well as from Q_1 through Q_2 to the output. Since the small-signal resistance presented by Q_3 will be very low, the circuit is similar to a common-collector common-base amplifier, with Q_1 functioning approximately as the common-collector stage and Q_2 as the common-base stage.

To analyze the circuit we assume that Q_3 and Q_4 are matched, and that Q_1 and Q_2 are matched. Then all emitter bias currents will be approximately the same and all four transistors will have $g_m \approx qI/2mkT$. We form the small-signal equivalent circuit by replacing each transistor with its small-signal model; for

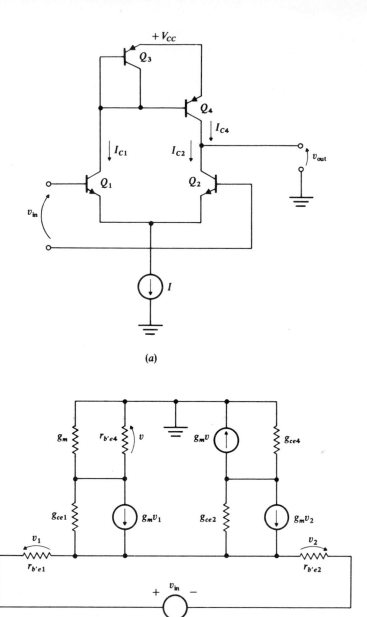

FIGURE 9-27
(a) Differential amplifier (Q_1,Q_2) with active load transistors (Q_3,Q_4); (b) approximate small-signal equivalent circuit.

simplicity the effects of $g_{b'c}$ are assumed to be negligible in comparison with g_{ce}. Note that since Q_3 is connected as a diode, its equivalent circuit is simply a conductance g_m. Since Q_1 and Q_2 are matched, $g_{ce1} = g_{ce2}$; however, g_{ce4} may differ from g_{ce1} and g_{ce2} since Q_4 is a p-n-p transistor. To simplify the analysis, we assume that β of Q_4 is large enough that $r_{b'e4} \gg 1/g_m$ and therefore $r_{b'e4}$ can be neglected. We also assume $r_{bb'}$ to be negligible for all transistors and β of Q_1 and Q_2 to be very large. The small-signal equivalent circuit is shown in Fig. 9-27b.

Analyzing the equivalent circuit, we find the voltage gain A_v to be

$$A_v \triangleq \frac{v_{out}}{v_{in}} = \frac{2g_m + g_{b'e4}}{g_{ce2}\left(2 + \frac{g_{b'e4}}{g_m}\right) + 2g_{ce4}\left(1 + \frac{g_{b'e4}}{g_m}\right)} \tag{9-21}$$

The input resistance is

$$R_i \approx r_{b'e1} + r_{b'e2} \tag{9-22}$$

and the output resistance is

$$R_0 \approx \frac{2 + \frac{2g_{b'e4}}{g_m}}{g_{ce2}\left(2 + \frac{g_{b'e4}}{g_m}\right) + 2g_{ce4}\left(1 + \frac{g_{b'c4}}{g_m}\right)} \tag{9-23}$$

If $g_{b'e4} \ll g_m$, we see that

$$A_v \approx \frac{g_m}{g_{ce2} + g_{ce4}}$$

$$R_0 \simeq A_v/g_m$$

EXAMPLE 9G *Comparison of Differential Amplifiers with Active and Passive Loads.* Consider a differential-amplifier circuit in which $V_{CC} = \pm 12$, employing transistors with $\beta = 100$, $g_{b'c} = 10^{-7}$. Let the maximum common-mode input voltage be $V_{iCM} = 2$ V and the bias current be $I = 52$ μA. If an ordinary two-transistor differential amplifier with current-source bias is used, the following results are obtained:

$$-A_{DM(max)} = \frac{10}{0.026} = 384$$

$$g_m = 10^{-3}$$

$$R_C = 384 \text{ k}\Omega$$

$$R_{iDM} \approx \frac{2}{10^{-5} + 385 \times 10^{-7}} - \frac{2}{4.85 \times 10^{-5}} = 41.2 \text{ k}\Omega$$

$$R_{ODM} \approx 768 \text{ k}\Omega$$

If a differential amplifier with active load is now used with the same bias current, and if we assume that $g_{ce4} \approx g_{ce2} = 10^{-4} \, g_m$, we find

$$A_v \approx 5{,}000$$
$$R_i \approx 200 \text{ k}\Omega$$
$$R_0 \approx 5 \text{ M}\Omega$$

These results are somewhat high, owing to neglect of $g_{b'c}$. ////

Note that the larger gain and input resistance of the circuit using the active load are obtained at the expense of increased output resistance. Therefore, in order to realize the large gain it would be necessary that any circuit connected to the output of the amplifier have very large input resistance. Biasing must also be given careful consideration so that the bias current of the stage connected to the output does not cause saturation of one of the output transistors.

It should be emphasized that no resistors other than those needed for the bias current source are necessary in the circuit with active load; consequently this circuit would require much less area than its two-transistor counterpart.

9-6 GILBERT'S GAIN CELL[10]

In all the circuits we have discussed thus far, we have been concerned primarily with voltage gain. We consider now a circuit which provides current gain and which is quite similar to the differential amplifier. The basic circuit is shown in Fig. 9-28a; here I_E and I_B are provided by some means such as the bias current circuit of Fig. 9-6, and i is a small-signal input current. Because of the symmetry of the circuit we note that, as was the case for the differential amplifier, the small-signal voltage v_e is zero if all transistors are matched. We need therefore to analyze only the half circuit, the small-signal equivalent circuit for which is shown in Fig. 9-28b. Here it is assumed that $\alpha \rightarrow 1$; that is, β is very large. Since the transistors are matched, and Q_1 and Q_3 have identical bias current $I_E/2$, $g_{m1} = g_{m3}$.

It is also assumed that proper biasing in the collector circuits is employed so that all transistors are operating in the forward active region. Then the collector bias current I will be

$$I = I_B + \frac{I_E}{2}$$

We can easily calculate the small-signal current i_c from the equivalent circuit of Fig. 9-28b. For simplicity we assume that β is large enough that

$$g_{m2} \gg \frac{g_{m1}}{\beta}$$

Then

$$i_{e2} \approx -i$$

and

$$v = \frac{i}{g_{m2}}$$

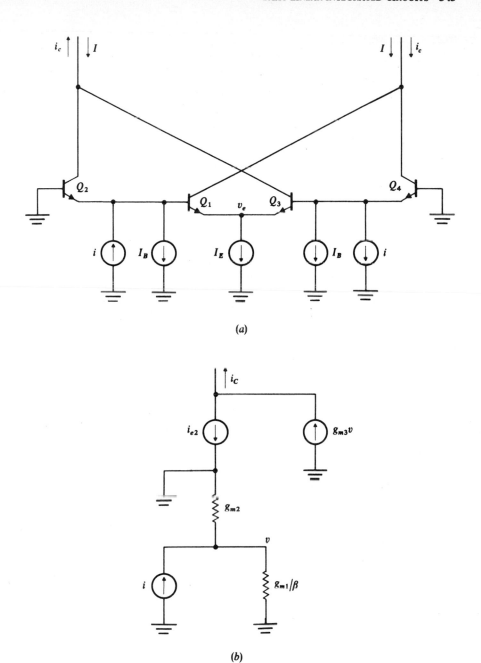

(a)

(b)

FIGURE 9-28
(a) Gilbert's gain cell; (b) small-signal equivalent half circuit.

from which we obtain

$$A_i \triangleq \frac{i_c}{i} = 1 + \frac{g_{m1}}{g_{m2}} = 1 + \frac{I_E}{2I_B} \qquad (9\text{-}24)$$

Equation (9-24) indicates that the gain can be chosen by selecting the proper ratio I_E/I_B. No resistors are used in the gain cell; the only resistors required are those for the bias circuits. Furthermore, the gain can be changed by varying either or both bias currents.

Exercise 9-16 Derive an expression for i_c/i in terms of I_E and I_B. Do not assume $\alpha \to 1$, and do not neglect g_{m1}/β.

Gain cells can be cascoded and resistors can be added to the last stage to produce an output voltage proportional to the small-signal current i_c. The small-signal input current i, as well as the bias current I_B, can be obtained very conveniently from a differential amplifier. A circuit with input differential amplifier, two cascoded gain cells, and output load resistors is shown in Fig. 9-29. In this case it has been assumed that the common-mode input voltage $V_{iCM} = 0$; base bias voltages V_1 and V_2 have been obtained by using forward-biased diodes. With the circuit shown, the collector-base bias voltage of Q_1, Q_2, Q_3, and Q_6 is V_D; if $V_{iCM} > V_D$, then either more diodes or a resistor must be added in series with D_1 and D_2 to prevent saturation of Q_1 and Q_3.

The current gain of gain cell 1 is

$$A_{i1} = 1 + \frac{I_{E1}}{2I_{B1}}$$

while that of gain cell 2 is

$$A_{i2} = 1 + \frac{I_{E2}}{2I_{B2}}$$

But since $I_{B2} \approx I_{B1} + I_{E1}/2$, we find

$$A_{i2} = 1 + \frac{I_{E2}}{2I_{B1} + I_{E1}}$$

The current gain of the cascoded cell is

$$A_i \triangleq \frac{i_L}{i} = A_{i1} A_{i2} = \left(1 + \frac{I_{E1}}{2I_{B1}}\right)\left(1 + \frac{I_{E2}}{2I_{B1} + I_{E1}}\right) \qquad (9\text{-}25)$$

The output voltage v_{oDM} for half the circuit is

$$v_{oDM} = i_L R_C = A_i i R_C$$

while the input small-signal current i is simply

$$i = v_{iDM} g_{m1} = \frac{v_{iDM} q I_{B1}}{mkT}$$

Thus the differential-mode voltage gain of the amplifier is

$$A_{DM} = \frac{v_{oDM}}{v_{iDM}} = \frac{A_i R_C q I_{B1}}{mkT} \qquad (9\text{-}26)$$

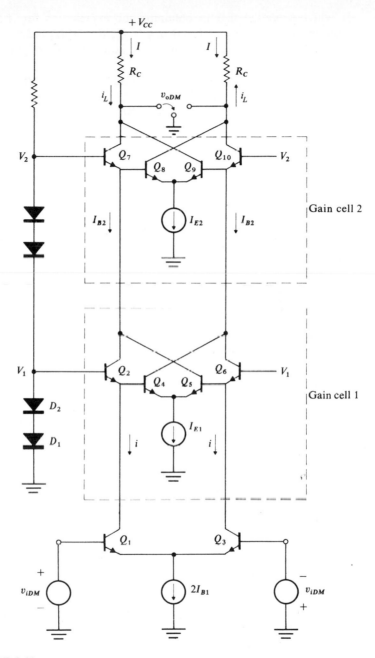

FIGURE 9-29
Cascode of two gain cells driven by a differential amplifier and having load resistors R_C.

It will be recalled that in obtaining the expressions for A_{i1} and A_{i2}, we have assumed that $g_{m2} \gg g_{m4}/\beta$ and $g_{m7} \gg g_{m8}/\beta$. If $I_{E1} \gg 2I_{B1}$ and $I_{E2} \gg 2I_{B2}$, this assumption will not be valid and more accurate expressions for A_{i1} and A_{i2} must be used.

It is particularly important to note that the voltage gain of the differential amplifier and of the first gain cell will be very small, owing to the small effective load resistance presented by the following stage. Therefore, Miller multiplication effects will be negligible in these stages, and the bandwidth of the circuit will be determined by the second gain cell with load resistors R_C.

Exercise 9-17 Suppose it is desired that no transistor have a collector-base voltage less than V_D. If the common-mode input voltage is V_{iCM}, what is V_1?

Exercise 9-18 For the circumstances of Exercise '9-17, find the maximum value of R_C in terms of V_{CC}, V_{iCM}, and the bias currents.

Exercise 9-19 For the circumstances of Exercise 9-17, what is the maximum possible A_{DM} if $I_{E2} = 2I_{E1}$, $I_{E1} = 20I_{B1}$, and $V_{CC} = 12$ V?

9-7 MULTIPLIERS[11,12]

Many applications require circuits which can perform analog multiplication, that is, produce an output signal which is proportional to the product of two input signals. This generally requires that the circuit have some parameter which can be made to vary with one of the input signals. The differential amplifier seems to be a likely candidate for such an application since the g_m of the transistors can be conveniently varied by varying the emitter current. Consider the circuit of Fig. 9-30a; if we focus attention only on the small-signal variations of collector current in Q_1 and use the model of Fig. 9-2b, we can write

$$i_{c1} = g_m v_{iDM} = \frac{q}{mkT}\left(\frac{I_E}{2\alpha_N} + \frac{i_e}{2\alpha_N}\right)v_{iDM}$$

or

$$i_{c1} = g_{m0} v_{iDM} + \frac{q}{2\alpha_N mkT} i_e v_{iDM} \tag{9-27}$$

where $g_{m0} = qI_E/2mkT\alpha_N$. The second term of (9-27) is the product term we are seeking. Thus we see that in principle the differential-amplifier circuit can perform multiplication. For practical applications, however, the circuit suffers from two major disadvantages:

> 1 *Dynamic range and distortion* The model of Fig. 9-2b applies only for small signals, that is, for $i_e \ll I_E$. If this inequality is not met, other terms will appear in (9-27) as a result of the nonlinearities of the emitter junctions. This can be seen by examining the large-signal transfer characteristic I_{C1} versus V_{iDM} of Fig. 9-30b. This transfer characteristic is reasonably linear

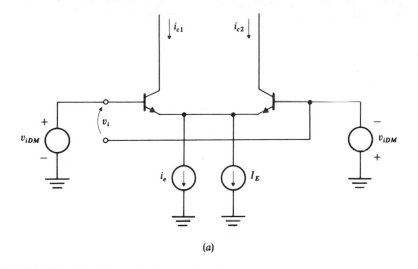

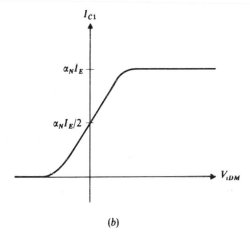

FIGURE 9-30
(a) Differential amplifier as a small-signal multiplier; (b) dc transfer characteristic for the differential amplifier.

in the vicinity of $I_{C1} = \alpha_N I_E/2$; it is quite nonlinear as I_{C1} approaches $\alpha_N I_E$.

2 Unwanted signals Note that even if $i_e = 0$, the first term of (9-27) is a differential-mode signal. An ideal multiplier would produce zero output signal when either of the input variables is zero. Thus even though no non-linearities are present, undesired signals appear when the simple differential amplifier is used as a multiplier.

The dynamic range limitation can be overcome by adding input diode-connected transistors and using an input current signal instead of an input voltage signal. These modifications are shown in Fig. 9-31. Here the input signals are

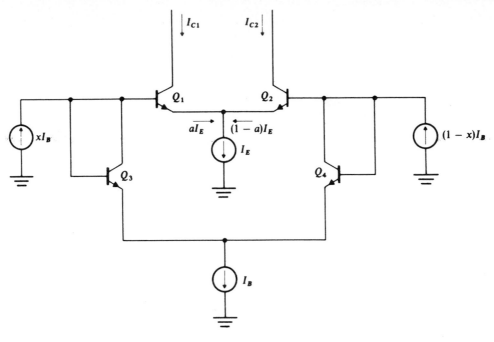

FIGURE 9-31
Use of input diodes to improve the dynamic range of the multiplier.

xI_B and $(1 - x)I_B$. If transistors Q_3 and Q_4 are identical to Q_1 and Q_2 the non-linearities of the emitter junctions of Q_3 and Q_4 will exactly compensate those of Q_1 and Q_2. This can be demonstrated as follows. From (9-1) the current-voltage relationship for the emitter junction is

$$I = I_{s1}(e^{q\phi_E/mkT} - 1)$$

where I is the forward current and ϕ_E is the junction voltage. (I_{s2} is assumed to be negligible.) For the circuit of Fig. 9-31, we can thus write, for large β,

$$\frac{mkT}{q} \ln \frac{xI_B}{I_{s1}} - \frac{mkT}{q} \ln \frac{aI_E}{I_{s1}} - \frac{mkT}{q} \ln \frac{(1 - x)I_B}{I_{s1}} + \frac{mkT}{q} \ln \frac{(1 - a)I_E}{I_{s1}} = 0 \qquad (9.28)$$

This reduces to, simply,

$$a = x$$

Therefore the collector currents are

$$I_{C1} = \alpha_N x I_E$$
$$I_{C2} = -\alpha_N (1 - x)I_E$$

These apply for $0 \leq x \leq 1$. The dynamic range of the circuit is therefore quite large. Multiplication can be performed by allowing I_E to vary with a second signal.

Two differential-amplifier circuits with a single pair of input diodes can be cross-coupled as shown in Fig. 9-32 to produce a four-quadrant multiplier circuit.

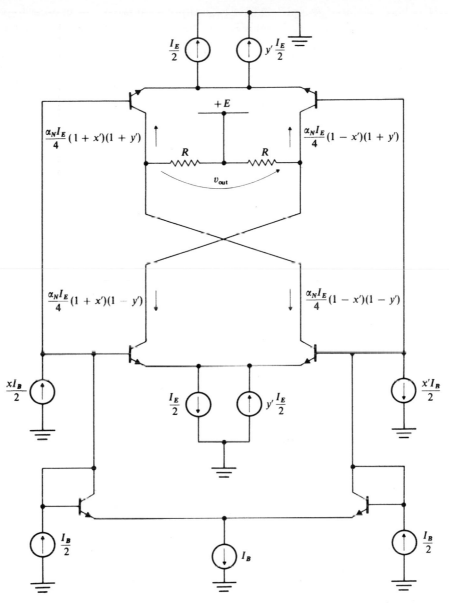

FIGURE 9-32
The four-quadrant multiplier.

Here bias currents $I_B/2$ and $I_E/2$ have been supplied at the inputs in order to permit bipolar input signals $x'I_B/2$ and $y'I_E/2$ to be used.   The values of x' and y' can thus be in the range

$$-1 \le x' \le 1$$
$$-1 \le y' \le 1$$

without nonlinearities being encountered in the multiplier circuit.   The collector currents are as shown in Fig. 9-32, and the output voltage V_{out} is easily found to be

$$V_{\text{out}} = RI_E x'y' \qquad (9\text{-}29)$$

Thus we see that the input is a true product with no unwanted terms.   Since x' and y' can be either positive or negative and (9-29) produces the correct sign, four-quadrant multiplier operation is realized by the circuit.

Exercise 9–20   Show that the collector currents of the transistors in the four-quadrant multiplier are as given in Fig. 9-32, and that V_{out} is given by (9-29).

PROBLEMS

9-1   A transistor with negligible $r_{bb'}$ is used as a common-emitter amplifier with load resistance R_L.   Use the model of Fig. 9-2e and find the ratio of collector voltage to base current.

9-2   Suppose it is desired to obtain a small bias current I.   Find the resistor values which should be used to consume the smallest total area if the small-value current-source circuit of Fig. 9-7 is to be used.

9-3   Investigate the behavior of the circuit of Fig. P9-3 to determine under what circumstances the approximation $I_C \approx I(R_1/R_2)$ is reasonable.   Assume large β.

9-4   Investigate the multiple small-value current-source circuit of Fig. P9-4 to determine under what circumstances the approximation $I_1/I_2 \approx R_2/R_1$ is reasonable.   Assume large β.

9-5   The circuit of Fig. P9-5 can be used to obtain a large-value current source from a smaller current.   Calculate the value of R required to produce $I_C = 10$ mA if $I = 1$ mA.   Assume large β.

9-6   In the circuit of Fig. P9-5, note that as $R \to \infty$, $I_C \to \beta I$, and the circuit is then very sensitive to changes of β.   Investigate the sensitivity of I_C to variations of β; in particular, find the maximum ratio I_C/I that can be allowed if a variation of β from 50 to 400 is to produce no more than a 10 percent variation of I_C.

9-7   For the circuit of Fig. P9-5, use electrothermal models and find $\partial I_{c2}/\partial T$.

9-8   Calculate the ouput conductance $\partial I_{c2}/\partial V_{c2}$ of the circuit of Fig. 9-7.

9-9   Find the output conductance $\partial I_c/\partial V$ of the circuit of Fig. 9-11.

9-10   Find V for the V_D multiplier circuit of Fig. 9-12 if the current source I is replaced by a resistor R connected to a positive supply voltage V_{cc}.

9-11   Find the output conductance of the circuit of Fig. 9-12.

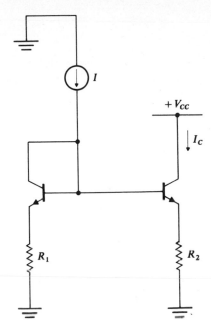

FIGURE P9-3

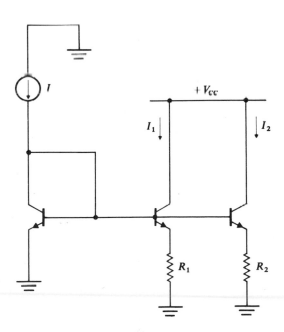

FIGURE P9-4

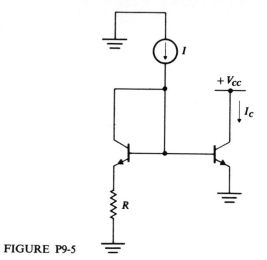

FIGURE P9-5

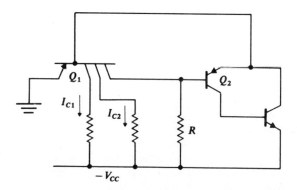

FIGURE P9-12

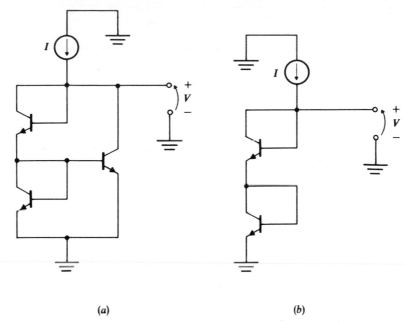

(a)   (b)

FIGURE P9-15

9-12 The circuit of Fig. P9-12 is a *p-n-p* multiple current source using a composite
p-n-p in the feedback loop to improve the performance. Q_1 and Q_2 are lateral
p-n-p transistors.   All collectors of Q_1 have equal area, and each has the same
area as the collector of Q_2. Q_2 has $\beta = 1$.   Find I_{c1} and I_{c2}.

9-13 Use the four-layer model to determine for the circuit of Fig. 9-14 under what
circumstances the approximation $V \approx V_D - IR$ is reasonable.

9-14 Lay out a V_D multiplier circuit using pinch resistors to produce an output voltage
$V = 2.5V_D$.   Assume 0.5-mil geometry.

9-15 Use electrothermal models to investigate the temperature behavior of the circuit
of Fig. P9-15a: compare it with that of Fig. P9-15b.   Assume large β.

9-16 Do not assume large β, and calculate I_C and I for the circuit of Fig. 9-16b.

9-17 Find the input resistance of the circuit of Fig. 9-16b.

9-18 Find the differential-mode input resistance of a differential amplifier with Darlington-connected input transistors.

9-19 Derive Eq. (9-21).

9-20 Design a three-stage amplifier, common-collector to common-base to common-
collector, using biasing similar to that of Fig. 9-16b.   Power supply voltage is $+10$,
$\beta = 100$, and all transistors are to operate at 1-mA bias current.   Give the values
of all dc voltages.

9-21 In the circuit of Fig. P9-21, i_s is a small-signal current source.   All *n-p-n* transistors
have $\beta = 100$, all *p-n-p* transistors have $\beta = 10$.   Find v_0/i_s, R_{in}, R_{out}.

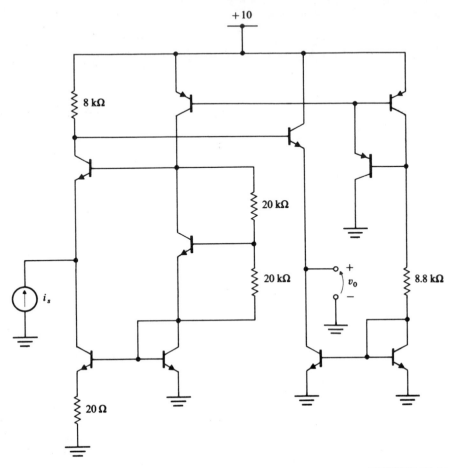

FIGURE P9-21

REFERENCES

1 SEARLE, C. L., A. R. BOOTHROYD, E. J. ANGELO, P. E. GRAY, and D. O. PEDERSON: "Elementary Circuit Properties of Transistors," vol. 3, chap. 3, "SEEC Notes," John Wiley & Sons, Inc., New York, 1964.

2 WIDLAR, R. J.: Some Circuit Design Techniques for Linear Integrated Circuits, *IEEE Trans. Circuit Theory*, vol. CT-12, pp. 586–590, 1965.

3 WILSON, G. R.: A Monolithic Junction FET-NPN Operational Amplifier, *IEEE J. Solid-State Circuits*, vol. SC-3, pp. 341–348, 1968.

4 GREBENE, A. B.: "Analog Integrated Circuit Design," Van Nostrand Reinhold Company, pp. 122–123, New York, 1972.

5 BERGER, H. H., and W. KAPPALLO: Superintegrated Voltage Clamp, *IEEE J. Solid-State Circuits*, vol. SC-8, pp. 231–232, 1973.

6 MIDDLEBROOK, R. D.: "Differential Amplifiers," John Wiley & Sons, Inc., New York, 1963.

7 GRAEME, J. G., G. E. TOBEY, and L. P. HUELSMAN: "Operational Amplifiers," McGraw-Hill Book Company, New York, 1971.

8 WIDLAR, R. J.: Super-gain Transistors for IC's, *IEEE J. Solid-State Circuits*, vol. SC-4, pp. 249–251, 1969.

9 ———: Design Techniques for Monolithic Operational Amplifiers, *IEEE J. Solid-State Circuits*, vol. SC-4, pp. 184–191, 1969.

10 GILBERT, B.: A New Wideband Amplifier Technique, *IEEE J. Solid-State Circuits*, vol. SC-3, pp. 353–365, 1968.

11 GREBENE: op. cit., chap. 7.

12 GILBERT, B.: A Precise Four-quadrant Multiplier with Subnanosecond Response, *IEEE J. Solid-State Circuits*, vol. SC-3, pp. 365–373, 1968.

10

INTEGRATED OPERATIONAL AMPLIFIERS

One of the most widely used integrated-circuit building blocks is the operational amplifier,[1] a circuit with large gain used principally with externally applied feedback to perform such functions as stable gain, integration, and filtering. The ideal operational amplifier is a voltage-controlled voltage source as shown in Fig. 10-1a. It has the following characteristics:

1 Large gain ($K \rightarrow \infty$)
2 Zero output voltage when $v_a - v_b = 0$
3 Zero input current
4 Zero output resistance
5 No frequency dependence
6 No temperature dependence
7 No distortion (large dynamic range)

Most practical operational amplifiers attempt to approach this ideal performance by employing the general structure shown in Fig. 10-1b. A differential input stage is used, and in some cases this is followed by an additional stage of gain. Because the transistors must operate in the forward active region, a bias offset exists between the input and the output of the gain stages. A level-shifting

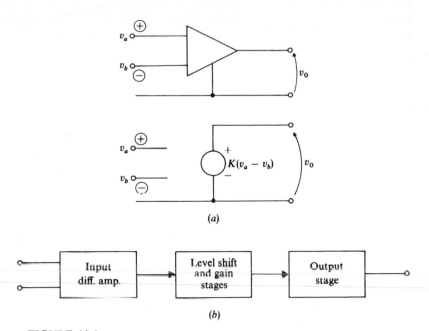

FIGURE 10-1
(*a*) Ideal operational amplifier; (*b*) general structure of practical operational amplifiers.

circuit is generally employed to shift the bias level so that the output will be zero for zero input voltage. Since low bias current is generally used in the input stage in order to produce large input resistance, the gain stages are seldom capable of supplying the signal current required at the output. Therefore an output stage is added which has little voltage gain but which provides low output resistance and large-signal current capability. Monolithic operational amplifiers have been designed with as much as 15-W output capability.[2]

In this chapter we examine some representative techniques which have been developed for realizing input and output stages. To focus attention on the basic concepts involved, we confine our analysis to first-order consideration of bias and low-frequency small-signal behavior. We then describe a novel current-differencing operational amplifier. Finally, we consider frequency behavior of typical operational-amplifier configurations.

10-1 CONVENTIONAL DIFFERENTIAL INPUT STAGES

In Chap. 9, we saw that the small-signal voltage gain of a single differential stage is limited to several hundred if conventional load resistors are used. In this case it is necessary to employ more than one stage in order to obtain the gains in excess of 10^3 required for operational amplifiers. Although the input stage may be a differential stage, the ouput of the operational amplifier is usually single-ended, and some sort of differential-to-single-end conversion must be employed at some stage in the amplifier.

The Motorola MC 1530, an early operational amplifier, is shown in Fig. 10-2a; the input stages are shown in Fig. 10-2b. A modified current source Q_5 with diodes for temperature compensation provides emitter bias current for the first stage. To analyze the performance of the circuit, we first calculate the bias currents. Assuming the diode voltage to be $V_D \approx 0.75$ and β to be large, we find

$$V_B = -5.1 \text{ V}$$

Then
$$I_{E1} = \frac{V_B - V_D + 9}{R_{E1}} = 1.43 \text{ mA}$$

With this bias current

$$g_{m1} = g_{m2} = 27.7 \times 10^{-3} \text{ U}$$

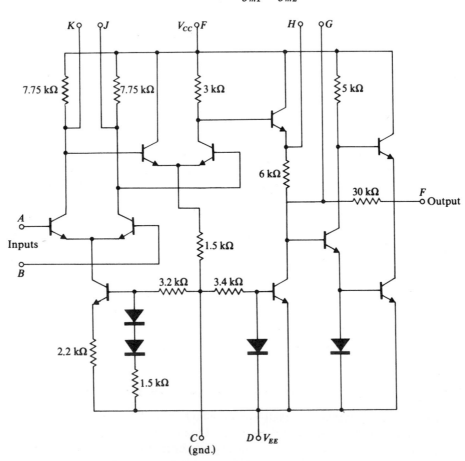

FIGURE 10-2
(a) The MC 1530 operational amplifier.

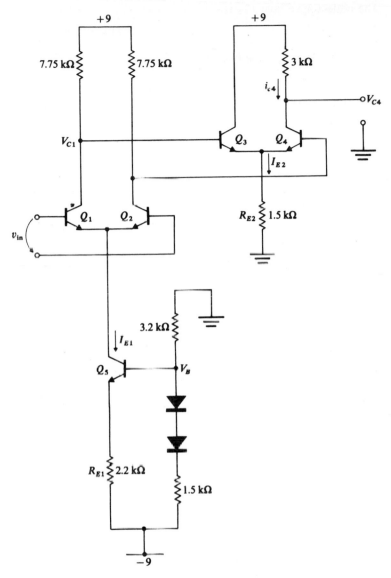

FIGURE 10-2
(b) Input stages of the MC 1530.

The collector voltage V_{C1} is

$$V_{C1} = 9 - (3.0 \text{ k}\Omega \times 0.89 \text{ mA}) = 3.4 \text{ V}$$

With this V_{C1}, the bias current for the second stage is then

$$I_{E2} = \frac{V_{C1} - V_D}{R_{E2}} = 1.76 \text{ mA}$$

From this we find

$$V_{C2} = 9 - (3 \text{ k}\Omega \times 0.86 \text{ mA}) = 6.4 \text{ V}$$

The small-signal gain of the two stages can now be determined. As far as collector currents are concerned, both stages are symmetrical, so we can calculate i_{c4} by considering the half circuits for each differential stage. We use the small-signal model of Fig. 9-2b; the small-signal equivalent circuit is shown in Fig. 10-3. It is assumed for concreteness that $\beta = 100$.

Since the bias currents are all known, we can find g_{m2} and g_{m4}, as well as $r_{b'e2}$ and $r_{b'e4}$:

$$g_{m2} = 27.7 \times 10^{-3}$$

$$r_{b'e2} \approx \frac{\beta}{g_{m2}} = 3600 \ \Omega$$

$$g_{m4} = 33 \times 10^{-3}$$

$$r_{b'e4} \approx \frac{\beta}{g_{m4}} = 2940 \ \Omega$$

Note that the collector load resistance R_{L2} must be taken into account in calculating the gain of the first stage:

$$\frac{v_4}{v_{in}/2} = \frac{-g_{m2}}{g_{b'e4} + G_{L2}}$$

$$= -61$$

The second-stage gain is

$$\frac{v_0}{v_4} = -g_{m4} R_{L4} = -100$$

Therefore the overall gain v_0/v_{in} is

$$\frac{v_0}{v_{in}} = \frac{6100}{2} = 3050$$

In this case the differential-to-single conversion has been accomplished simply by taking the signal from only one collector of the second stage. This procedure has two disadvantages:

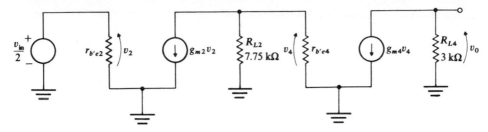

FIGURE 10-3
Small-signal equivalent circuit for the MC 1530 input stages.

1 The gain is only half that which would be obtained if a differential output were used.

2 The output signal contains not only a differential component but also a common-mode component.

Here the second disadvantage is unimportant because the first stage has current-source bias and therefore little common-mode signal will be applied to the second stage.

The input resistance can also be calculated from the circuit of Fig. 10-3; it is

$$R_{iDM} = 2r_{b'e2} = 7200 \ \Omega$$

We can make a consistency check on the applicability of the model used in Fig. 10-3, at least as far as input resistance is concerned. If $g_{b'c}$ is included in the model, it will appear at the input multiplied by $1 + g_{m2}/(g_{b'e4} + G_{L2})$. For $g_{b'c} \approx 10^{-7}$, this contribution will be negligible in comparison with $r_{b'e2}$, so the model is valid.

Exercise 10-1 If emitter followers are added at the input of the first stage, calculate the gain and the input resistance.

Common-mode Feedback

Although the input stage of the MC 1530 uses Q_5 as a bias current source to reduce the common-mode gain of the amplifier, finite values of $r_{b'c}$ and r_{ce} of Q_5 will lead to a nonzero common-mode gain. For the MC 1530, the common-mode rejection ratio is -75 dB, which is adequate for many applications. In cases where larger common-mode rejection is required, the two-stage differential amplifier can be modified slightly to provide negative feedback of the common-mode signal, thereby further improving the common-mode rejection ratio. Such a technique is employed in the MC 1533, the input stages for which are shown in Fig. 10-4.

Transistors Q_3 and Q_4 are emitter followers, Q_1 and Q_2 form the input differential amplifier, and Q_5 and Q_6 are the second stage of gain. Differential-

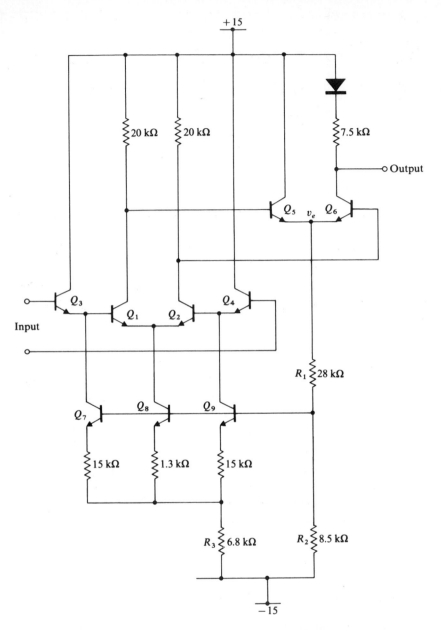

FIGURE 10-4
Input stages of the MC 1533.

to-single-end conversion is performed in the same way as in the MC 1530.   The small-signal voltage v_e appearing at the emitters of Q_5 and Q_6 is proportional to the common-mode input signal and of opposite polarity to it.   This common-mode voltage v_e is converted to a current signal by voltage divider R_1 and R_2, and current-source transistors Q_7, Q_8, and Q_9.   The result is negative feedback of a common-mode current to the differential amplifier and emitter followers.   This feedback causes a reduction of the common-mode collector signal current of Q_1 and Q_2, and an increase of the common-mode input resistance of the amplifier.
The MC 1533 has a common-mode rejection ratio of -100 dB.

Exercise 10-2   Calculate the bias currents for the circuit of Fig. 10-4.

10-2   A NOVEL DIFFERENTIAL-TO-SINGLE-END CONVERSION CIRCUIT

Differential-to-single-end conversion is accomplished without a factor-of-2 degradation of gain in the μA702, another early operational amplifier.   Although the μA702 has been superseded by other later designs, we describe its input stage because it is the precursor of the active load, a technique widely used in most later-generation amplifiers.   The circuit is shown in Fig. 10-5; to analyze it we first calculate the bias currents.   It is assumed that β is large and base bias currents are negligible.

The bias current I_{E1} can be found by noting that

$$V_{E6} \approx V_{E5}$$

Therefore

$$\frac{I_{E5}}{I_{E1}} = \frac{2.4 \text{ k}\Omega}{480}$$

But

$$I_{E5} = \frac{10 - V_D}{2880}$$

Thus we obtain

$$I_{E1} = 0.65 \text{ mA}$$

and

$$I_{C1} = I_{C2} = 0.32 \text{ mA}$$

Now since both Q_3 and Q_4 must operate in the forward active region, their base voltages are each V_D.   If the base currents are negligible, then

$$V_{C3} = V_D + I_{C1}R_1 = 1.39 \text{ V}$$

This means that I_3 must be

$$I_3 = \frac{10 - V_{C3}}{R_3} = 1.07 \text{ mA}$$

Then the collector current I_{C3} is

$$I_{C3} = I_3 - (I_{C1} + I_{C2}) = 0.43 \text{ mA}$$

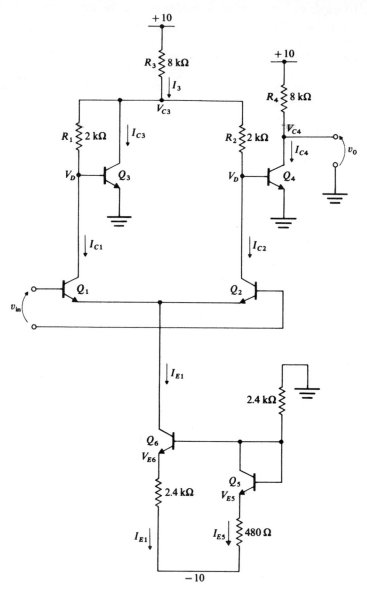

FIGURE 10-5
Input stage of the μA702.

Now Q_3 and Q_4 are identical transistors and have identical circuits connected to their bases; therefore they must have identical collector currents. Thus

$$I_{C4} = I_{C3} = 0.43 \text{ mA}$$

and

$$V_{C4} = 10 - I_{C4} R_4 = 6.6 \text{ V}$$

To analyze the small-signal behavior we use the zero-order model of Fig. 9-2a for Q_3 and Q_4, and the first-order model of Fig. 9-2b for Q_1 and Q_2, as shown in Fig. 10-6. The small-signal base current i_{b3} is

$$i_{b3} = g_m v_1 - v_{c3} G_1$$

while

$$v_{c3} = \frac{\beta i_{b3}}{G_1 + G_2 + G_3}$$

Solving for v_{c3} we obtain

$$v_{c3} = \frac{\beta_m g v_1}{(1 + \beta) G_1 + G_2 + G_3} \approx g_m v_1 R_1$$

Since v_{c3} is known, we can easily find i_{b4}:

$$i_{b4} = \frac{v_{c3}}{R_2} + g_m v_2 = g_m v_1 \frac{R_1}{R_2} + g_m v_2$$

With $R_1 = R_2$ this becomes

$$i_{b4} = g_m v_1 + g_m v_2$$

Note that the base current i_{b4} is the sum of the small-signal collector currents of Q_1 and Q_2. Thus there will be no loss of a factor of 2 in converting from the differential-stage Q_1 and Q_2 to the single-stage Q_4. Since $v_1 = v_2 = v_{in}/2$, we have

$$i_{b4} = 2g_m v_1 = g_m v_{in}$$

The output voltage is

$$v_0 = -\beta i_{b4} R_4 = -\beta g_m v_{in} R_4$$

For the bias current previously calculated,

$$g_m = 12 \times 10^{-3}$$

If $\beta = 100$, the gain is

$$\frac{-v_0}{v_{in}} = 12 \times 10^{-3} \times 100 \times 8 \times 10^3 = 9,840$$

The input resistance is

$$R_{iDM} \approx r_{b'e1} + r_{b'e2} = 2r_{b'e1} = 16.3 \text{ k}\Omega$$

Exercise 10-3 Calculate the voltage gain v_{c1}/v_1 and determine whether neglecting $r_{b'c}$ in the calculation of R_{iDM} was justified.

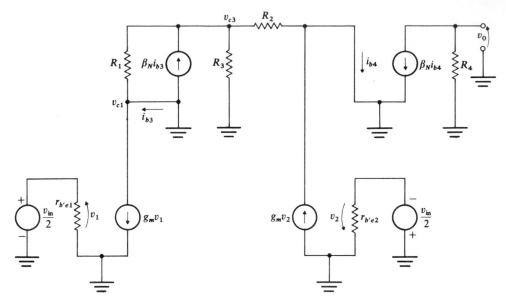

FIGURE 10-6
Small-signal equivalent circuit for the input stage of the μA702.

10-3   INPUT STAGE WITH ACTIVE LOAD

We saw in Chap. 9 that the gain of a differential amplifier can be made very large through the use of an active load circuit. Such a circuit is employed in the μA741 high-performance operational amplifier.[3] The input stage is shown in Fig. 10-7b. The functional behavior of the circuit as far as small signals are concerned is as follows. Transistors Q_7, Q_8, Q_9, Q_{10} form a differential amplifier in which Q_7 and Q_8 operate in the common-collector mode and Q_9 and Q_{10} in the common-base mode. In this respect, the differential amplifier is similar to a modified cascode circuit. An active load is formed by Q_{11}, Q_{12}, and Q_{13}, and the output is taken from the collectors of Q_{10} and Q_{12} as shown. All of the other transistors are used for biasing purposes. The p-n-p transistors in the circuit are all lateral p-n-p devices with β of approximately 4, while the n-p-n transistors are all high-β devices.

To analyze the circuit we consider first the biasing. Transistors Q_4 and Q_3 form a small-value current source which derives its bias current from R_1. Q_1 and Q_2 form a bias current source for use in other parts of the circuit. Feedback stabilization of the bias current of the differential-amplifier stage is provided by the bias current source comprising Q_5 and Q_6. To see this, we focus attention on the emitter bias currents I_{E7} and I_{E8} in the differential amplifier. If the differential input voltage is zero, $I_{E7} = I_{E8}$. We assume that $\alpha_N \to 1$ for the n-p-n transistors, and β of the p-n-p transistors is β_p. The current I_6 is given by

$$I_6 = 2I_{E7}$$

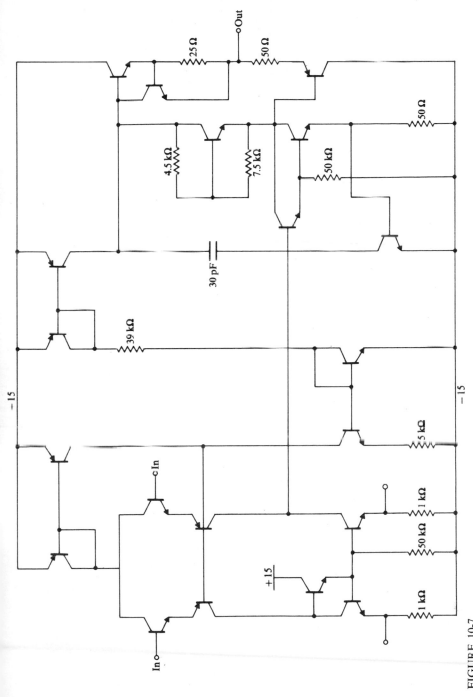

FIGURE 10-7

(*a*) The µA741 amplifier.

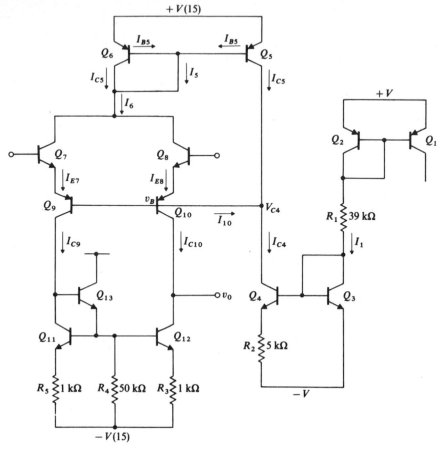

FIGURE 10-7

(b) Input stage of the μA741, using an active load circuit.

Since Q_5 and Q_6 are identical, they have equal collector currents.   Then

$$I_6 = I_{C5} + 2I_{B5} = I_{C5}\left(1 + \frac{2}{\beta_p}\right)$$

We also note that

$$I_{10} = \frac{2I_{E7}}{\beta_p + 1}$$

and that

$$I_{C4} = I_{C5} + I_{10} = I_{C5} + \frac{2I_{E7}}{\beta_p + 1}$$

Solving for I_{E7}, we obtain

$$I_{E7} = \frac{I_{C4}(\beta_p{}^2 + 3\beta_p + 2)}{2(\beta_p{}^2 + 2\beta_p + 2)}$$

In the $\mu A741$, I_{C4} has been chosen to be 27.5 μA. For $\beta_p = 4$, this yields

$$I_{E7} = 15 \ \mu A$$

while for $\beta_p \to \infty$, we obtain

$$I_{E7} \to 13.8 \ \mu A$$

Thus we see that the operating current of Q_7, Q_8, Q_9, and Q_{10} is well stabilized against variations of β_p.

For $I_{E7} = 15 \ \mu A$ and $\beta_p = 4$, we have

$$I_{C9} = I_{C10} = I_{C12} = 12.8 \ \mu A$$

Resistors R_3 and R_5 are provided so that an offset-adjustment can be made externally to compensate for differences in transistor characteristics which would result in $I_{C12} \neq I_{C10}$. This is done by connecting a potentiometer between the emitters of Q_{11} and Q_{12} with the wiper connector to $-V$. Resistor R_4 is used to provide emitter bias current for Q_{13}.

The small-signal analysis is quite complicated owing to the complexity of the circuit; in order to obtain an estimate of the voltage gain we therefore make many approximations at the outset. We ignore R_3, R_4, and R_5, and we neglect $r_{b'c}$ of all transistors and g_{ce} of all transistors except Q_{10} and Q_{12}. The approximate small-signal equivalent circuit is shown in Fig. 10-8.

If we assume that the effect of g_{ce10} on the emitter current of Q_{10} is negligible, then

$$i_{e8} = i_{e10} = i_{e7} = i_{e9}$$

and we can easily find i_{e7}:

$$i_{e7} = \frac{v_{in}}{r_{e7} + r_{e9} + r_{e8} + r_{e10}}$$

and since all four differential-amplifier transistors have equal emitter bias currents, we have

$$i_{e7} = \frac{v_{in}}{4r_{e7}} = \frac{v_{in} q I_{E7}}{4kT}$$

Analyzing the Q_{13}, Q_{11} combination, we find

$$v \approx \frac{\alpha_p i_{e9}}{g_{m11}}$$

Since $g_{m11} = g_{m12}$, we also find

$$g_{m12} v = \alpha_p i_{e9}$$

Then if $i_{e9} \approx i_{e10}$, we obtain

$$v_0 \approx -\frac{2\alpha_p i_{e9}}{g_{ce10} + g_{ce12}} = \frac{-v_{in}\alpha_p}{2r_{e7}(g_{ce10} + g_{ce12})}$$

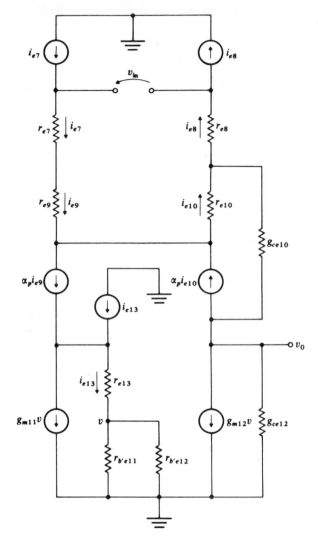

FIGURE 10-8
Approximate small-signal equivalent circuit for the input stage.

If g_{ce10} and g_{ce12} are approximately $10^{-5}/r_{e7}$, we finally obtain, with $\alpha_p \approx 0.8$,

$$\frac{v_0}{v_{in}} \approx -2 \times 10^5$$

The input resistance is

$$R_{in} \approx 4r_{e7}(\beta + 1) \approx 657 \text{ k}\Omega$$

10-4 INPUT STAGES WITH LOW INPUT BIAS CURRENT

Use of Super-gain Transistors[4]

We saw in Chap. 9 that the input resistance can be increased and the input bias current decreased by using super-gain transistors at the input of the differential stage. It will also be recalled that special circuit design is required in order to prevent punch-through of the super-gain transistors. In the Motorola MC 1556 amplifier, super-gain *n-p-n* transistors are used as emitter followers, and *p-n-p* transistors are employed in the differential amplifier. The use of complementary-type transistors in this manner provides the required protection for the super-gain devices.

The circuit of the MC 1556 is shown in Fig. 10-9a, and the basic input circuit is shown in Fig. 10-9b. Transistors Q_1 and Q_2 form a bias current source which supplies current to the differential-amplifier transistors Q_8 and Q_9. Super-gain transistors Q_6 and Q_7 are emitter-follower input transistors; their collectors are connected in such a way as to make the collector-emitter voltage slightly larger than V_D, so the collector-base junction is only slightly reverse-biased. An active load circuit quite similar to that of the μA741 is formed by Q_{10}, Q_{11}, and Q_{12}; offset adjustments are made in the same manner as in the μA741.

Because complementary transistors Q_6 and Q_8, as well as Q_7 and Q_9, are used, some provision must be made to provide bias current for the emitter of Q_6 and base of Q_8, and for the emitter of Q_7 and base of Q_9. This is done by means of the small-value current source comprising Q_3, Q_4, and Q_5.

To calculate the bias currents, we first calculate I_1:

$$I_1 = \frac{2V - 2V_D}{R_1 + R_2} = 0.5 \text{ mA}$$

If we now assume that Q_1 and Q_2 have sufficiently large β that base currents can be neglected, and if we use an analysis similar to that given in Chap. 9 for the small-value current source, we find

$$I_2 \approx \frac{I_1}{3.9} = 0.126 \text{ mA}$$

The currents I_4 and I_5 are found in a similar manner,

$$I_4 = I_5 \approx 0.019 I_1 = 9.5 \ \mu\text{A}$$

It is now a simple matter to calculate I_6 and I_7. If we assume Q_8 and Q_9 to have $\beta = 10$, we find

$$I_6 = I_7 = 59 \ \mu\text{A}$$

while

$$I_{E6} = I_{E7} = 4 \ \mu\text{A}$$

If the super-gain transistors have $\beta = 1000$, the input bias current is 4 nA.

To analyze the small-signal behavior on a first-order basis we must make many approximations, as was the case for the μA741. We assume $g_{b'e}$ to be

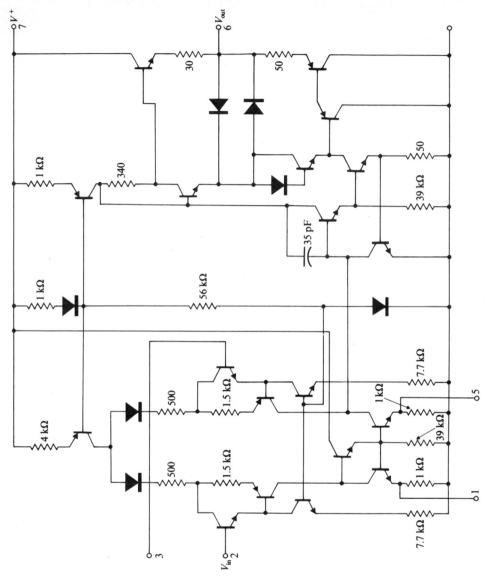

FIGURE 10-9
(a) Circuit of the MC 1556.

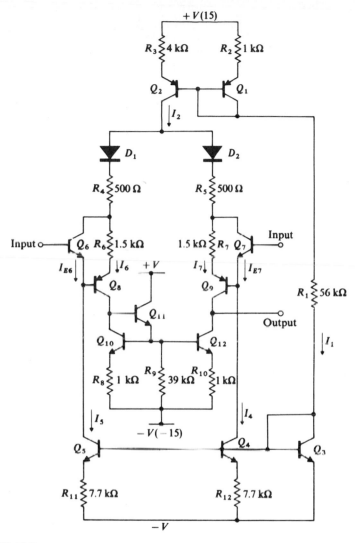

FIGURE 10-9
(b) Input stage for the MC 1556.

negligible for all transistors, and we neglect g_{ce} for all except Q_9 and Q_{12}. The approximate small-signal equivalent circuit is shown in Fig. 10-10. Making approximations similar to those used in analyzing the μA741, we also ignore the effects of the collector currents of Q_6 and Q_7 on i_{e8} and i_{e9}, and we assume g_{ce9} has negligible effect on i_{e9}. With the previously calculated bias currents, we determine

$$r_{e8} = r_{e9} = 433 \ \Omega$$
$$r_D = 415 \ \Omega$$

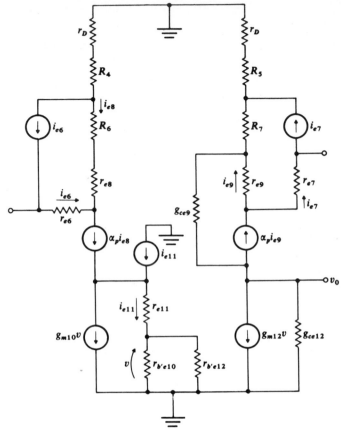

FIGURE 10-10
Small-signal equivalent circuit for the MC 1556 input stage.

Then
$$i_{e8} \approx \frac{v_{in}}{2(r_D + R_4 + R_6 + r_{e8} + r_{e6}/\beta_p)}$$

$$= \frac{v_{in}}{5.78 \times 10^3}$$

where v_{in} is the differential input voltage.

Using an analysis similar to that of Sec. 10-3, we find the output voltage to be

$$v_0 \approx \frac{-2\alpha_p i_{e8}}{g_{ce9} + g_{ce12}}$$

from which we find the gain to be

$$\frac{v_0}{v_{in}} = \frac{-2\alpha_p}{(g_{ce9} + g_{ce12})\, 5.78 \times 10^3}$$

If we assume $g_{ce12} \approx g_{ce9} \approx 10^{-5} g_{m12}$, we have

$$\frac{v_0}{v_{in}} = 0.136 \times 10^5$$

Exercise 10-4   Calculate the bias currents I_2, I_4, I_5.

Bias Cancellation[5-7]

In applications where integrated circuits are subjected to radiation, a degradation of material properties occurs which manifests itself as a reduction of transistor β and an increase of junction leakage current. In an operational amplifier, the input stage would be adversely affected, since both effects would cause an increase of the input bias current. To compensate for this increase of input bias current, or to reduce the input bias current in general, positive feedback can be used to supply the base current of the input stage; such a technique is called *bias cancellation*.

To evolve a bias-cancellation circuit, we make use of the fact that integrated *n-p-n* transistors having the same geometry will have equal base currents if their collector currents are equal, and that the bias circuits of Chap. 9 can be used to produce a current mirror. The basic circuit is shown in Fig. 10-11a; here Q_1 and Q_5 form a differential amplifier with cascode load transistors Q_2 and Q_6. Diodes D_1, D_2, D_3 are used to provide an offset voltage, and Q_3 and Q_4 form a current mirror, as do Q_7 and Q_8. If the base current of Q_1 is I_B, then

$$I_{C1} = \beta I_B$$

and

$$I_{B2} = \frac{\beta I_B}{\beta + 1}$$

The current mirror produces a current

$$I_{C4} = \frac{\beta \beta_p I_B}{(\beta + 1)(2 + \beta_p)}$$

where β_p is the *p-n-p* transistor current gain. The input current is then

$$I_{in} = I_B \left[1 - \frac{\beta_p \beta}{(\beta + 1)(2 + \beta_p)} \right]$$

The current mirror emitters are all connected to V_1, which is a common-mode point. With three diodes used for offset,

$$V_1 = V_E + 3V_D$$

The base voltage of Q_2 is

$$V_1 - V_D$$

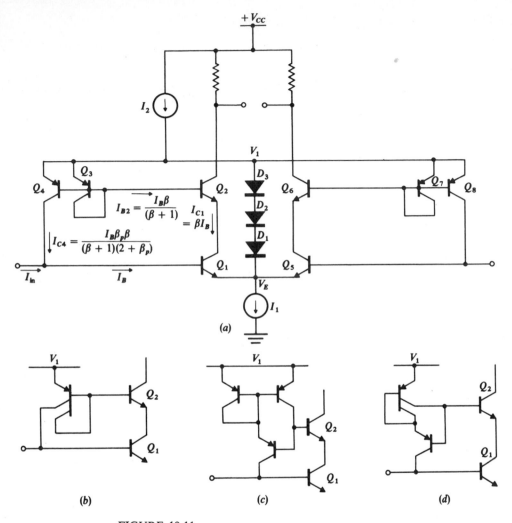

FIGURE 10-11
(a) Bias cancellation in an input stage; (b) use of a two-collector p-n-p current mirror; (c) the improved current mirror using p-n-p transistors; (d) implementation of the improved current mirror.

while the collector voltage of Q_1 is

$$V_1 - 2V_D$$

The base voltage of Q_1 is $V_E + V_D$, so the collector-base voltage of Q_1 is

$$V_1 - 2V_D - (V_E + V_D) = 0$$

It is not necessary that two separate transistors be used for the current mirror; a single multiple-collector device can be used, as shown in Fig. 10-11b.   The input current can be further reduced by improving the current mirror, as shown in Fig. 10-11c; this can be implemented with a multiple-collector transistor, as shown in Fig. 10-11d. For the circuits of Fig. 10-11c and d it is necessary to add another diode to make

$$V_1 = V_E + 4V_D$$

Operational amplifiers with input stages using the bias-cancellation schemes of Fig. 10-11d have input bias currents of the order of 3 nA.

While the above analysis has been carried out only for bias currents, it applies on a first-order basis to small-signal currents as well. Thus the input resistance of the amplifier is markedly increased by the bias-cancellation circuits.

Although bias cancellation can improve the input bias and input resistance, we do not wish to leave the impression that it is a panacea for all input stages. It has two principal disadvantages:

1   The input noise current is increased.
2   For a given β mismatch, the input offset current is increased. In many applications input offset current is more important than input bias current.

10-5   OUTPUT STAGES

The principal functions of the output stage are to provide low output resistance and large load current capability. Usually output stages have voltage gains less than 10. In cases where large output currents are required, class B or class AB output stages are used to conserve bias power. In such cases, complementary output transistors are usually used so that output currents of either polarity can be accommodated. If large output current is not required, class A operation can be used, and negative feedback can be employed to reduce the output resistance.

The MC 1530

An output stage with negative feedback is used in the MC 1530; the circuit diagram is shown in Fig. 10-12. An emitter-follower input transistor Q_6 is used to drive the level-shift resistor R_1. Transistors Q_7 and Q_8 form a bias current source whose current I is used to produce the level-shift voltage across R_1. Transistors Q_{10} and Q_{11} provide a bias current source for emitter-follower Q_{12}, while Q_9 functions as a common-emitter amplifier. Feedback is provided by R_f.

The circuit is designed so that $V_0 \approx 0$. For this case

$$V_{C9} = V_0 + V_D = V_D$$

If we assume that the base current of Q_{12} is negligible, then $I_2 = I_{C9} \approx I_1$. Therefore R_2 is chosen to produce the desired value of I_1:

$$\frac{V - V_D}{R_2} = I_1$$

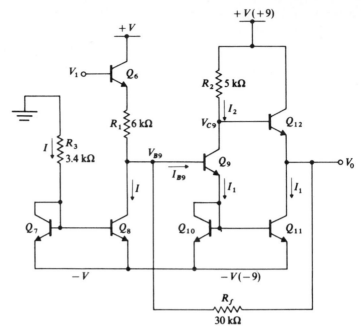

FIGURE 10-12
Output and level-shift stages of the MC 1530.

Note that $V_{B9} = -V + 2V_D$. Then

$$I_{B9} = \frac{V_1 - V_D + V - 2V_D}{R_1} - I + \frac{V - 2V_D}{R_f}$$

But

$$I_{B9} = \frac{I_1}{\beta + 1}$$

Thus we see that I, R_1, and R_f must be chosen so that

$$\frac{V - V_D}{R_2(\beta + 1)} = \frac{V_1 + V - 3V_D}{R_1} - I + \frac{V - 2V_D}{R_f}$$

For the values shown in Fig. 10-12,

$$I_1 = 1.66 \text{ mA}$$

and

$$I = 2.4 \text{ mA}$$

To analyze the small-signal behavior, we assume that the voltage gain of Q_6 is unity, and we use a zero-order model for transistors Q_9 and Q_{12} and a first-order model for Q_{10} and Q_{11}. The small-signal equivalent circuit is shown in Fig. 10-13.

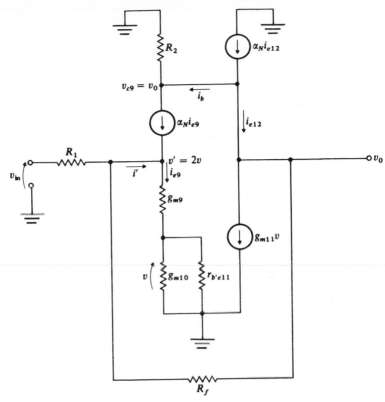

FIGURE 10-13
Small-signal equivalent circuit for MC 1530 output stage.

Since Q_{10}, Q_9, and Q_{11} have the same bias current,

$$g_{m9} = g_{m10} = g_{m11}$$

and

$$g_{m10} \gg \frac{1}{r_{b'e11}}$$

Therefore

$$i_{e9} \approx g_{m10}v \approx g_{m11}v$$

and

$$v' \approx 2v$$

Since $i_{e9} = (\beta + 1)i'$, we can write

$$(\beta + 1)[(v_{in} - 2v)G_1 + (v_0 - 2v)G_f] = g_{m10}v \qquad (10\text{-}1)$$

If we assume $\alpha_N \approx 1$, at the collector of Q_9 we have

$$v_0 G_2 + i_{e9} = i_b$$

At the emitter of Q_{12}

$$i_{e12} = g_{m11}v + (v_0 - 2v)G_f \qquad (10\text{-}2)$$

Since

$$i_{e12} = -(\beta + 1)i_b \qquad \text{and} \qquad i_{e9} = g_{m10}v = g_{m11}v$$

Equation (10-2) becomes

$$-(\beta + 1)(v_0 G_2 + g_{m10}v) = g_{m10}v + (v_0 - 2v)G_f \qquad (10\text{-}3)$$

Eliminating v from (10-3) and (10-1), we obtain the following relation between v_0 and v_{in}:

$$-v_0 \left[\frac{G_f + (\beta + 1)G_2}{g_m(\beta + 2) - 2G_f} \right] - v_0 \left[\frac{(\beta + 1)G_f}{g_m + 2(\beta + 1)(G_f + G_1)} \right]$$

$$= \frac{v_{in}(\beta + 1)G_1}{g_m + 2(\beta + 1)(G_f + G_1)} \qquad (10\text{-}4)$$

The first term on the left side of (10-4) will be small in comparison with the second term; the gain v_0/v_{in} then becomes

$$\frac{v_0}{v_{in}} \approx \frac{-G_1}{G_f} = \frac{-R_f}{R_1}$$

For the values shown in Fig. 10-12, we obtain

$$\frac{v_0}{v_{in}} = \frac{-30 \text{ k}\Omega}{6 \text{ k}\Omega} = -5$$

The output resistance is found by analysis of the circuit of Fig. 10-13 to be

$$R_0 \approx \frac{R_f}{2 + 2\beta + \beta^2}$$

Exercise 10-5   Repeat the analysis of Fig. 10-13 for v_0/v_{in}, but make the following simplifying approximations at the outset:

$$v' \ll v_{in}$$
$$v' \ll v_0$$

Show that this leads to

$$\frac{v_0}{v_{in}} = \frac{G_1[(\beta + 1)^2 + (\beta + 1)]}{G_f[(\beta + 1)^2 + (\beta + 1) + 1] + G_2(\beta + 1)} \approx \frac{-R_f}{R_1}$$

Output Stage of the μA741

The μA741 amplifier uses an output stage in which complementary emitter-follower transistors are biased for class AB operation; the circuit is shown in Fig. 10-14. Here Q_{18} and Q_{19} function only as overload protection for the circuit; if the output

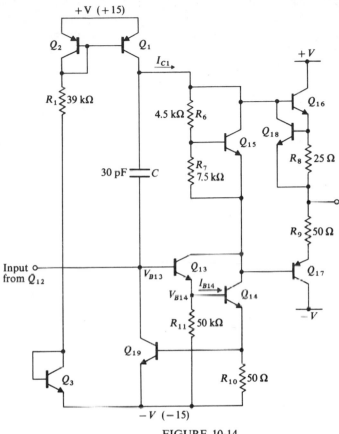

FIGURE 10-14
Output stage of the μA741 amplifier.

is shorted to either of the supply voltages, current in either R_8 or R_{10} will become large enough to turn on Q_{18} or Q_{19}, which will in turn cause base current of Q_{16} or Q_{13} to be reduced, thereby limiting the output current. Under normal operating conditions, Q_{18} and Q_{19} are cut off.

Class AB operation of Q_{16} and Q_{17} is provided by Q_{15}, which functions as a V_D multiplier. If both Q_{16} and Q_{17} were operating class A, the voltage between their bases would be approximately $2V_D$. However, the voltage provided by Q_{15} is

$$V_{CE15} = V_D\left(1 + \frac{R_6}{R_7}\right) = 1.6V_D$$

This voltage produces only a small emitter bias current in Q_{16} and Q_{17}; yet it is sufficient to eliminate the crossover distortion that would result from class B operation.

Transistors Q_{13} and Q_{14} form a Darlington-connected common-emitter amplifier, and Q_1 acts both as a bias current source and as a small-signal load for the Darlington stage. The base of Q_{13} is connected to the collector of Q_{12} in the input stage.

The μA741 has internal compensation in order to obviate the need for external capacitors. This is accomplished by using a 30-pF capacitor on the chip, and connecting it as shown in Fig. 10-14. The capacitance is multiplied by the Miller effect resulting from the gain of the Darlington stage. In this way the gain roll-off can be made to begin at about 10 Hz and continue at 20 dB per decade; thus the 0-dB point occurs at about 1 MHz, ensuring that the amplifier will be stable under closed-loop conditions.

The input resistance of this output stage is of interest, since the output stage acts as a load on the input stage. Since the latter uses an active load, its output resistance is large and the loading effects of Q_{13} become important in the determination of the gain. To calculate the input resistance, we must first calculate the bias currents for the Darlington stage. The collector current of Q_1, the bias current source, is

$$I_{C1} = \frac{2V - 2V_D}{R_1} = 0.73 \text{ mA}$$

Since this is also the collector current of Q_{14}, the base current of Q_{14} is

$$I_{B14} = 0.73 \frac{\text{mA}}{\beta} = 7.3 \ \mu\text{A} \qquad \text{for } \beta = 100$$

The emitter current of Q_{13} is then

$$I_{E13} = I_{B14} + \frac{V_{B14} + V}{R_{11}}$$

and since $V_{B14} + V \approx V_D$, we find

$$I_{E13} = 22.5 \ \mu\text{A}$$

Then
$$g_{m14} = 28.1 \times 10^{-3}$$

and
$$g_{m13} = 0.87 \times 10^{-3}$$

If we neglect $g_{b'c13}$, the input resistance is

$$R_{in} \approx \beta \left(\frac{1}{g_{m13}} + \frac{\beta}{g_{m14}} \right) = 470 \text{ k}\Omega$$

Note that $g_{b'c13}$ will be multiplied by Miller effect, and therefore the neglect of $g_{b'c13}$ is probably a poor approximation.

The maximum output current capability of the amplifier can be estimated by noting that Q_{18} will turn on when the voltage across R_8 is approximately V_D. This will occur at

$$I_{out(max)} \approx \frac{V_D}{R_8} = 30 \text{ mA}$$

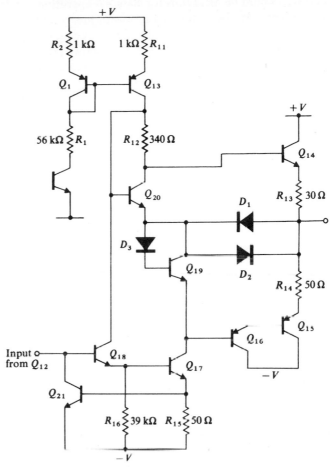

FIGURE 10-15
Output stage of the MC 1556.

Output Stage of the MC 1556

The output stage of the MC 1556 is quite similar to that of the μA741; the circuit is shown in Fig. 10-15. Basically, the circuit employs complementary emitter-follower output transistors Q_{14} and Q_{15}, biased for class AB operation. These are driven by Q_{18} and Q_{17}, which, for small signals, function as a Darlington stage. Bias current and small-signal load are provided by Q_{13}.

The output emitter follower Q_{15} is driven by emitter follower Q_{16} to provide increased current gain, since the β of the substrate p-n-p transistors will not be as large as that of the n-p-n devices. However, this also means that a base-to-base bias voltage larger than $2V_D$ is necessary to ensure class AB operation for Q_{14} and Q_{15}. Note that for class A operation the voltage between the bases of Q_{14}

and Q_{16} would be $3V_D$; thus for class AB operation a voltage slightly less than this is required. In the circuit of Fig. 10-15, this is done by using Q_{19} and D_3 to provide $2V_D$, and using Q_{20} to provide an additional voltage slightly less than V_D. This is accomplished by using R_{12} as shown; since the collector of Q_{20} is not directly connected to its base, Q_{20} will saturate, causing $V_{CE20} < V_D$.

Short-circuit protection is provided by D_1, D_2, and Q_{21}.

10-6 A CURRENT-DIFFERENCING OPERATIONAL AMPLIFIER[8]

Most of the operational amplifiers we have thus far described are generally termed high-performance amplifiers and have the following salient properties:

1 Large gain ($\approx 10^5$, or 100 dB)
2 High input resistance
3 Low input bias current
4 DC output level ≈ 0 for zero differential input voltage
5 Two supply voltages
6 Large output current capability

There are many applications in which such high performance is not a requisite, but in which other factors such as low cost and single-supply operation are of prime importance. The MC 3401P makes use of a very simple circuit configuration, making possible the fabrication of four identical amplifiers on a single chip. Since all four amplifiers are on the same chip, a single biasing circuit on the chip suffices. Only one power supply voltage is required. An example of an application for which this device is ideal is an active RC filter which may require four operational amplifiers, but in which high-performance amplifiers are not required. A single MC 3401P provides all of the amplifiers required for this application.

The basic amplifier circuit is shown in Fig. 10-16a; all of the voltage gain in the amplifier is provided by a single common-emitter stage Q_1. Bias current sources are used for I_1 and I_2, and the current source I_1 also serves as a small-signal load for Q_1. To prevent an external load from reducing the gain, an emitter follower Q_4 is used at the output. Base current for Q_1 is the difference between an external current I_{in1} and a current I_{in2} derived from an external current by a bias current source comprising Q_2 and Q_3. The currents I_1 and I_2 are provided by the common bias circuit in the chip; currents I_{in1} and I_{in2} must be supplied by external resistors. The output voltage level is determined by the difference $I_{in1} - I_{in2}$.

Fig. 10-16b shows the complete circuit of one amplifier, together with the bias circuit which supplies current for all four amplifiers. Transistors Q_6 and Q_7 provide bias current, with Q_6 functioning also as the load for Q_1. To ensure large voltage gain, an additional emitter follower Q_5 is inserted between Q_1 and the current source Q_6; this helps to further isolate Q_1 from any output load.

In order to make all bias currents independent of supply voltage variations, forward junction voltages are used to set the voltage drops across current-deter-

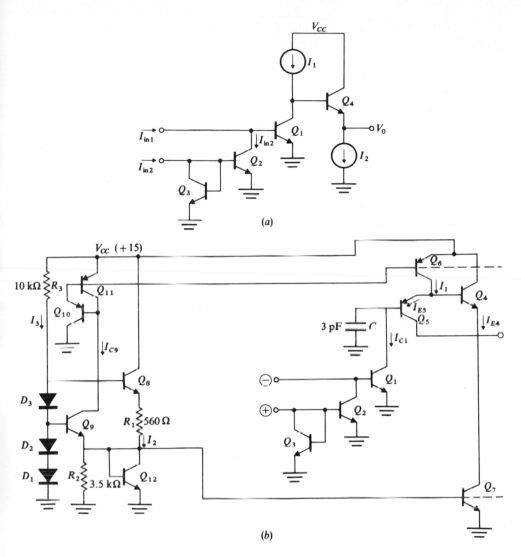

FIGURE 10-16
(a) Basic circuit of the MC 3401P; (b) bias circuit and one amplifier.

mining resistors R_2 and R_1. A current I_3 is established by R_3 to forward-bias D_1, D_2, and D_3; this produces a voltage V_D across R_2 and across R_1. Thus the bias currents are

$$I_{C9} = I_1 = \frac{V_D}{R_2} = 0.2 \text{ mA}$$

$$I_2 = \frac{V_D}{R_1} = 1.3 \text{ mA}$$

Diode-connected transistors Q_{11} and Q_{12} provide base bias for current-source transistors Q_6 and Q_7, as well as for similar transistors in the other three amplifiers on the chip.

Figure 10-17 shows how external resistors are used to establish the bias currents I_{in1} and I_{in2} in a typical inverting-amplifier application. Using this circuit, together with that of Fig. 10-16b, we find

$$I_{in1} = (V_0 - V_D)G_f$$

and

$$I_{in2} = (V_{CC} - V_D)G_B$$

At the output,

$$I_{E4} = I_{in1} + I_2$$

Then the collector current of Q_1 is

$$I_{C1} = \frac{I_1 - I_{E4}/(\beta + 1)}{\beta_p + 1}$$

But I_{C1} must also be given by

$$I_{C1} = \beta(I_{in1} - I_{in2})$$

Combining these relations and solving for V_0, we find

$$V_0 \approx \frac{I_1/(\beta_p + 1) - I_2/(\beta_p + 1)(\beta + 1) + V_D \beta(G_f - G_B) + V_{CC} \beta G_B}{\beta G_f}$$

For the values of I_1 and I_2 calculated above, if G_f and G_B are both larger than 10^{-6} ℧, this expression reduces to

$$V_0 \approx V_{CC} \frac{G_B}{G_f} = V_{CC} \frac{R_f}{R_B}$$

To bias the amplifier for maximum dynamic range, R_f and R_B would be chosen so that

$$R_B = 2R_f$$

thus producing

$$V_0 \approx \frac{V_{CC}}{2}$$

The input resistance of the amplifier itself is rather low, particularly at the noninverting input. However, since it is a current-differential amplifier, input voltages are converted to input currents by external series resistors. Thus the input resistance seen by the input signals is determined by these external resistors, which can be of the order of 1 MΩ. In the circuit of Fig. 10-17, the voltage gain is

$$A_v \approx - \frac{R_f}{R_1}$$

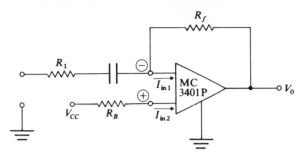

FIGURE 10-17
Typical inverting-amplifier application.

If R_f and R_1 are each chosen to be 1 MΩ, and R_B is 2 MΩ, the gain is -1 and the input resistance is 1 MΩ.

External frequency compensation is not required for the MC 3401P, as this is provided internally by C. The open-loop voltage gain is approximately 65 dB, and C causes roll-off to start at 1 kHz and continue at 20 dB per decade, producing a unity-gain frequency of about 5 MHz.

Exercise 10-6 Show that the small-signal voltage gain of the circuit of Fig. 10-17 is approximately $-(R_f/R_1)$.

10-7 FREQUENCY BEHAVIOR OF OPERATIONAL AMPLIFIERS

Compensation[9]

Operational amplifiers are seldom used in an open-loop circuit configuration; rather, large amounts of feedback are used in order to obtain precise performance. The amplifier must therefore be stable in the closed-loop condition; the most stringent stability requirements occur when the amplifier is connected as a unity-gain voltage follower. For unconditional stability it is necessary that the open-loop phase shift be less than 180° at the frequency at which the open-loop gain is unity, or 0 dB. In a typical uncompensated amplifier, excess phase resulting from the use of more than one stage of gain, as well as from devices such as lateral *p-n-p* transistors, causes the phase shift to be greater than 180° at the unity-gain frequency, and it is necessary to add means for compensation. In early amplifiers this was done by using external components, but in more advanced designs it is done on the chip.

To see how compensation can be accomplished, we consider the circuit of Fig. 10-18; here the current source represents the input stage of the amplifier with active load, and the gain block represents the remaining stages. The capacitor

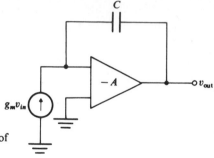

FIGURE 10-18
Equivalent circuit for the analysis of compensation.

C is added for compensation. If we assume for the moment that the gain block has no bandwidth limitations and has zero input current, that is, is an ideal integrator, we find

$$\frac{V_{out}(s)}{V_{in}(s)} = g_m/sC$$

The gain will be unity at a frequency

$$\omega_1 = \frac{g_m}{C}$$

and the rate of change of gain with frequency will be 20 dB per decade. Now suppose that excess phase in A begins to appear, from whatever causes, at a frequency ω_0. If C is chosen large enough so that the overall open-loop gain is unity at a frequency below ω_1, the amplifier will be stable under closed-loop conditions. Thus it is required that

$$\omega_1 < \omega_0$$

That is,

$$C > \frac{g_m}{\omega_0}$$

EXAMPLE 10A *Calculation of the Open-loop Gain Corner Frequency.* For large values of open-loop gain, the compensated gain will fall at 20 dB per decade for several decades before reaching 0 dB. If the zero-frequency gain is known we can easily find the -3-dB corner frequency of the open-loop compensated amplifier. Consider a case such as the μA741, in which the gain is 10^5 and a 30-pF compensating capacitor is used. We have already found the bias current of the input transistors to be 13 μA. Since a common-collector common-base input stage is used with an active load, the effective g_m for the input stage is

$$g_m = \frac{qI_{C9}}{2kT}$$

The compensated unity-gain frequency is then

$$\omega_1 = \frac{qI_{C9}}{2kTC} = 8.3 \times 10^6 \text{ rad/s} = 1.33 \text{ MHz}$$

Since the amplifier has a gain of 100 dB, the -3-dB corner frequency must occur approximately 5 decades below ω_1, that is, at 13.3 Hz. This indicates that properly compensated high-gain amplifiers will have very low open-loop corner frequencies. ////

Slew-rate Limitations

In discussing compensation, we have considered only small-signal behavior. We now show that the requirement for compensation also degrades the large-signal performance as a result of nonlinear behavior of the input stage. If we consider a simple differential input stage with active load, and if we represent the remainder of the amplifier by a gain block with compensating capacitor, as shown in Fig. 10-19a, we see that the rate of change of the output voltage is

$$\frac{dv_0}{dt} = \frac{i_1}{C}$$

The current i_1 will be maximum when Q_2 cuts off, for which case

$$i_1 = I_0$$

and

$$\frac{dv_0}{dt} = \frac{I_0}{C}$$

If a sinusoidal input is applied, producing an output $V \sin \omega t$, the maximum amplitude V_p of the output before distortion occurs is thus

$$V_p = \frac{I_0}{\omega C}$$

Conversely, if an amplifier has a maximum peak rated output voltage V_p, the maximum frequency at which an undistorted sinusoidal output can be obtained with peak voltage V_p is

$$f_{max} = \frac{I_0}{2\pi V_p C}$$

The frequency f_{max} is sometimes called the *full-power bandwidth* of the amplifier.

Suppose that the amplifier is connected as a unity-gain voltage follower, and that a step function of large magnitude is applied to the input. Q_2 will cut off, and the amplifier will act as an integrator, producting a ramp output whose slope S is

$$S = \frac{dv_0}{dt} = \frac{I_0}{C}$$

as shown in Fig. 10-19b. S is called the *slew rate* of the amplifier; it is clearly related to the maximum sinusoidal output amplitude by

$$V_p = \frac{S}{\omega}$$

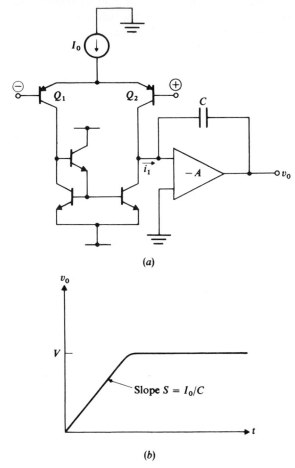

(a)

(b)

FIGURE 10-19
(a) Typical input stage followed by compensated amplifier; (b) response of voltage follower to input step function of amplitude V.

EXAMPLE 10B *Slew Rate of the μA741.*   For the μA741 amplifier, the input-stage bias current is 16 μA for each transistor, making $I_0 = 32$ μA.   With $C = 30$ pF, we find

$$S \approx 1 \times 10^6 = 1 \text{ V}/\mu\text{s}$$

Slew rates of 1 V/μs are typical for amplifiers of the same general structure as the μA741.   ////

Improving the Slew Rate[10]

It appears at first glance that the slew rate can be increased by increasing I_0. This is not the case, because g_m also depends directly on I_0; increasing I_0 increases

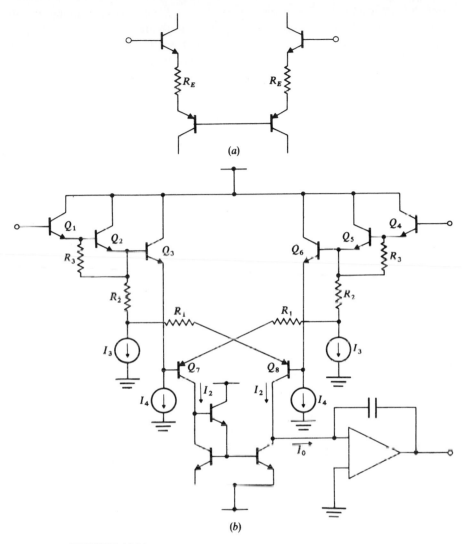

FIGURE 10-20

(*a*) Use of emitter degeneration resistors to improve slew rate; (*b*) cross-coupled input stage with fast slew rate (after Hearn[10]).

g_m, causing larger open-loop gain with a consequent requirement for larger C to ensure stability. The slew rate remains the same.

A straightforward method of increasing slew rate is to increase I_0 and decrease the gain; this can be done by using emitter degeneration resistors in the input stage, as shown in Fig. 10-20*a*. Even greater improvement can be made by arranging the circuit so that the current available for charging C during large-signal operation is different from the bias current for the input stage. This is done by operating the input transistors class AB and cross-coupling the input stage, as shown in Fig. 10-20b. Here I_4 is set at about 1 mA, and I_3, R_1, and R_2 are closed to produce about 50-μA collector current I_2 in Q_7 and Q_8. For small signals, the cross-coupling produces an effective transconductance for the circuit of

$$g_m = \frac{2}{R_1 + R_2 + kT/qI_2}$$

For large signals, if Q_7 cuts off, for example, the emitter current of Q_8 is determined not by the bias current but by the input signal and R_1 and R_2. In this mode of operation, dI_0/dv_{in} is easily seen to be $1/(R_1 + R_2)$. The dynamic range of the input stage is therefore considerably increased by this technique. Typical slew rates for an operational amplifier using an input stage of this type are of the order of 30 V/μs, more than an order of magnitude greater than that of a conventional stage.

Exercise 10-7 Calculate the value of R_E required for emitter degeneration in the μA741 to double the slew rate.

PROBLEMS

10-1 Use the model of Fig. 9-2b for Q_3 and Q_4 in the circuit of Fig. 10-6 and calculate the voltage gain.

10-2 Use electrothermal models to find the temperature coefficient of the current I_{E1} in the circuit of Fig. 10-2.

10-3 Calculate the common-mode rejection ratio of the MC 1533.

10-4 Calculate the input bias current for the circuit of Fig. 10-11d.

10-5 Calculate the output resistance of the MC 1530, using the circuit of Fig. 10-13. Do this first by using the approximations suggested in Exercise 10-4.

10-6 Take into account $g_{b'c13}$ of Q_{13} in Fig. 10-14, and calculate R_{in} of the output stage of the μA741.

10-7 Show that the small-signal voltage gain of the circuit of Fig. 10-17 is approximately $-(R_f/R_1)$.

10-8 Design an input stage of the type shown in Fig. 10-20b with $R_1 + R_2 = kT/qI_2$. Calculate the slew rate if $C = 30$ pF.

REFERENCES

1 GRAEME, J. G., G. E. TOBEY, and L. P. HUELSMAN: "Operational Amplifiers," McGraw-Hill Book Company, New York, 1971, offers a complete treatment of operational amplifiers, both circuit design and applications.

2 GRAY, P. R.: A 15 Watt Monolithic Power Operational Amplifier, *IEEE J. Solid-State Circuits*, vol. SC-7, pp. 474–480, 1972.

3 FULLAGAR, D.: A New High-performance Monolithic Operational Amplifier, Fairchild Semiconductor Corporation Application Brief, May, 1968.

4 WIDLAR, R. J.: Design Techniques for Monolithic Operational Amplifiers, *IEEE J. Solid-State Circuits*, vol. SC-4, pp. 184–191, 1969.

5 GRAEME et al.: op. cit., pp. 74–76.

6 NISHIKAWA, Y., and J. E. SOLOMON: A General Purpose Wideband Operational Amplifier, *ISSCC Digest Tech. Papers*, pp. 144–145, 1973.

7 KUIJK, K. E.: A Fast Integrated Comparator, *IEEE J. Solid-State Circuits*, vol. SC-8, pp. 458–462, 1973.

8 FREDERIKSEN, T. M., W. F. DAVIS, and D. W. ZOBEL: A New Current-differencing Single-supply Operational Amplifier, *IEEE J. Solid-State Circuits*, vol. SC-6, pp. 340–347, 1971.

9 GREBENE, A. B.: "Analog Integrated Circuit Design," pp. 167–172, Van Nostrand Reinhold Company, New York, 1972.

10 HEARN, W. E.: Fast Slewing Monolithic Operational Amplifier, *IEEE J. Solid-State Circuits*, vol. SC-6, pp. 20–24, 1971.

11

APPLICATIONS EMPLOYING LINEAR
INTEGRATED CIRCUITS

11-1 MICROPOWER CIRCUITS[1]

In some applications, it is essential that the power consumption of electronic circuits be very low, on the order of tens of microwatts. Two examples of such applications are implantable circuits for biomedical telemetry, and telemetry circuits for space probes. In such circuits, power supply voltages of 1 to 2 V at current drains of 10 to 100 μA are common. With these current levels, large-value resistors must be used to develop voltages of the order of 1 V. We saw in Chap. 4 that large values of sheet resistance can be obtained by using pinch-resistor techniques, thus permitting large-value resistors to be realized with relatively small surface areas. In micropower circuits nearly all resistors used are either base-pinch resistors or collector-pinch resistors.[2] Some processing modifications can also be made for micropower circuits. Since currents are very small, series collector resistance in the transistors is of little importance, and epitaxial material of several ohm-centimeter resistivity can be used. With the low voltages encountered, the epitaxial-layer thickness can be reduced to about 5 μm without danger of punch-through to the substrate; the combination of high resistivity and thin material produces a high value of sheet resistance for collector resistors and collector-pinch resistors. Moreover, the base width of the *n-p-n* transistors can be decreased without danger of punch-through; this reduction leads to improved β for the transistors and also increased sheet resistance for the base-pinch resistors.

Exercise 11-1 Calculate the sheet resistance of collector resistors if an epitaxial layer is used having thickness 5 μm and impurity concentration 5×10^{14} cm^{-3}. Neglect depletion regions.

Biasing

A circuit which can be used for biasing in micropower circuits is shown in Fig. 11-1. Here Q_2, Q_3, and Q_4 form a current source similar to that of Fig. 9-9. A similar *p-n-p* current source is formed by Q_5, Q_6, and Q_7, except that Q_7 is designed to have twice the emitter and collector area of Q_6. Therefore the current I_1 will be

$$I_1 = 2I$$

If we assume for the moment that Q_1 is not present we can analyze the combination of Q_2, Q_3, Q_4, and R as a small-value current source supplied by a bias current I_1. With R in the emitter of Q_4, we have

$$IR = V_{BE2} - V_{BE4} = \frac{mkT}{q} \ln \frac{I_{C2}}{I}$$

Since $I_{C2} \approx I_1 = 2I$, we obtain

$$IR = \frac{mkT}{q} \ln 2$$

For the low currents generally encountered in micropower circuits, m will have a value of approximately 2. For a given current I, the resistor value is

$$R = \frac{mkT}{qI} \ln 2$$

For $I = 10$ μA, $R = 3.6$ kΩ.

Other transistors can now be connected to Q_6 and Q_4 as shown; the collector currents of Q_8 and Q_9 will be I, while those of Q_{10} and Q_{11} will be $2I$.

Transistor Q_1 together with R_1 and the three diodes are added to provide a "starting" circuit. The thermally generated currents in Q_2, Q_4, Q_6, and Q_7 can be sufficiently low that negligible I is produced when the voltage V_{CC} is applied to the circuit. Therefore R_1 is added so that emitter current will flow in Q_1, producing a collector current in Q_4. Once this current is established, operation becomes self-sustaining; for the self-sustaining condition $V_{CE2} \approx 2V_D$. If I_0 is chosen to be only a few microamps, the voltage V_s is somewhat less than $3V_D$ and Q_1 is almost cut off when the self-sustaining condition is reached.

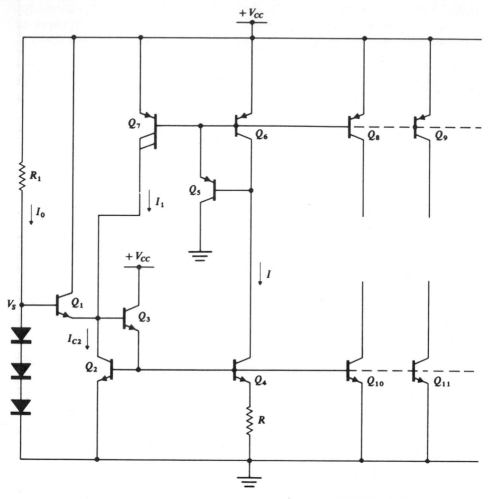

FIGURE 11-1
A micropower biasing circuit.

A Micropower Gain Cell[3]

A micropower self-biasing gain cell is shown in Fig. 11-2a; note its similarity to the circuit of Fig. 9-16b. Transistors Q_1 and Q_2 provide gain, while Q_3 and Q_4 are diode-connected and are used to establish stable biasing. The bias current is set by R_1, and the output voltage is developed across R_2.

A first-order small-signal low-frequency model for the circuit is shown in Fig. 11-2b. To determine the element values, we first note that the emitter current of Q_3 is also the base current of Q_1, and the emitter current of Q_4 is the base

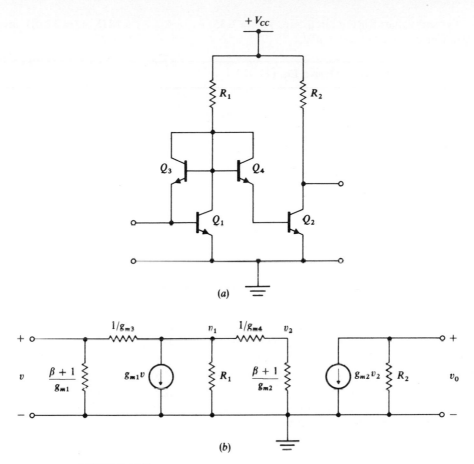

FIGURE 11-2
(a) A self-biasing micropower gain cell; (b) first-order small-signal equivalent circuit.

current of Q_2. If all transistors are identical, Q_1 and Q_2 will have identical collector currents with value

$$I_c \approx \frac{V_{cc} - V_{CE1}}{R_1}$$

Since very low currents are used, V_{CE1} will be less than $2V_D$.

With the currents as described above,

$$g_{m3} = \frac{g_{m1}}{(\beta + 1)} = g_{m4} = \frac{g_{m2}}{(\beta + 1)}$$

The voltage gain of the cell is then

$$A_v = \frac{v_0}{v_{in}} = \frac{\beta g_{m1}^2 R_2}{3g_{m1} + 2G_1(\beta + 1)} \qquad (11\text{-}1)$$

Typical values for the circuit are $V_{CC} = 1.35$, $R_1 = R_2 = 1.5$ MΩ, $A_v = 32$ dB, and power dissipation $= 1.1$ μW.

Exercise 11-2 Derive Eq. (11-1).

11-2 HIGH-POWER CIRCUITS[4,5]

In the operational-amplifier circuits described in Chap. 10, output current capabilities of the order of 10 mA were common. With these current levels, no special precautions were necessary in device design, and because the resulting maximum power dissipations were of the order of 200 mW, no significant thermal gradients were produced on the chip. The designer needed only observe some simple layout rules such as not locating input transistors near output devices.

In power operational amplifiers where peak power of the order of 15 W is encountered, device design, circuit design, and thermal gradients must all be considered. The μA791 amplifier, shown in Fig. 11-3, provides an excellent illustration of these aspects. As can be seen from the block diagram of Fig. 11-4, the circuit contains two gain stages and an output stage, and in addition includes provisions for load current limiting and thermal overload protection.

Referring to Fig. 11-3, we see that the structure of the first and second stages is quite similar to the μA741 circuit. Emitter-follower input transistors drive p-n-p transistors having an active load consisting of Q_6, Q_7, and Q_5. Bias current sources are provided by Q_8, Q_9, and Q_{10}. The second gain stage is provided by emitter follower Q_{13} and common-emitter transistor Q_{14}. Note that the circuit connected to the collector of Q_6 is almost identical to that connected to Q_7; this helps to balance the collector currents. Collector bias current for Q_{14} is provided by p-n-p current source Q_{11}, which is in turn coupled to Q_{10}.

Level shifting at the output of Q_{14} is accomplished by emitter follower Q_{17} with R_{15}. The current established in R_7 for the n-p-n current source also flows in p-n-p current source Q_{12}, and is used to bias the output stage.

Voltage biasing for the output stage is provided by Q_{25} and V_D multiplier Q_{24}; the total voltage across Q_{24} and Q_{25} is $2.7V_D$. Load current limiting is provided by Q_{26} and Q_{27}. The output stage uses no p-n-p transistors except for current limiting; its operation can best be understood by considering the simplified circuit of Fig. 11-5a. To analyze the bias conditions for the circuit we make the following assumptions:

1 The zero-signal output voltage is approximately zero.
2 The diode D_2 is not conducting.
3 Base currents I_{B29} and I_{B28} are negligibly small.

With these assumptions, the bias currents can easily be estimated. The current I_3 is

$$I_3 = \frac{V_2 - 2V_D}{R_{11}}$$

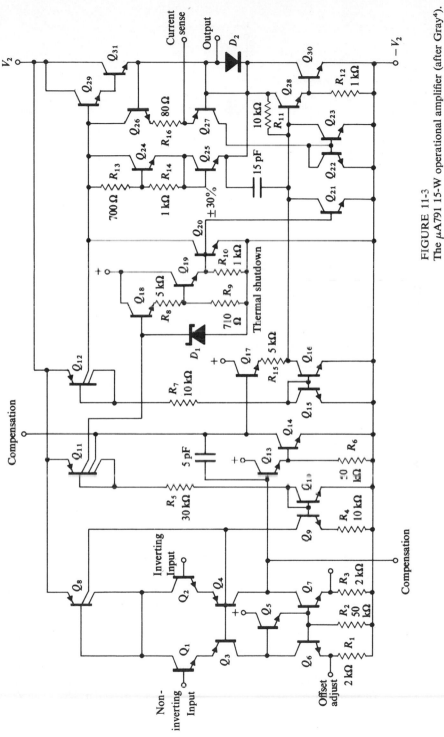

FIGURE 11-3
The μA791 15-W operational amplifier (after Gray[4]).

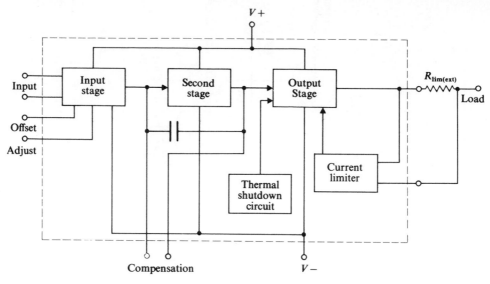

FIGURE 11-4
Block diagram of the μA791.

while the emitter current of Q_{31} is

$$I_E \approx I_3$$

and the collector current of Q_{30} is

$$I_C \approx I_1$$

From Fig. 11-3, I_1 is easily found to be

$$I_1 = \frac{2V_2 - 2V_D}{R_7} = 2.86 \text{ mA}$$

This is also the collector current of Q_{16}. The bias current I_2 is

$$I_2 \approx I_3$$

    We can check the consistency of assumption (2) as follows. The base volt-age of Q_{29} is $2V_D$, so the cathode of D_2 is at $-0.7V_D$; this forward bias of D_2 will be insufficient to produce significant current in D_2. The bias currents I_E and I_C will be somewhat larger than calculated above owing to the presence of R_{12}, which was neglected in the simplified analysis.

    The simplified analysis shows that the output stage is biased for class AB operation, with just sufficient bias current in the output transistors to prevent crossover distortion. To see how the circuit functions when a signal is applied, we return to Fig. 11-3, and imagine a positive-going signal applied to the base of Q_{17}; this causes an increase of current in R_{15}, with a consequent decrease of I_2 in Fig. 11-5. I_{B28} now increases, causing a large collector current to flow in Q_{30}.

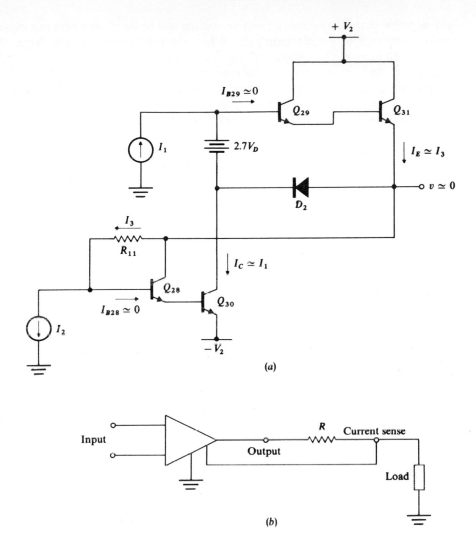

FIGURE 11-5
(*a*) Simplified output stage; (*b*) use of an external resistor for current limiting.

The diode is now forward-biased and v becomes negative; Q_{29} is cut off, as is Q_{31}. Thus for negative excursions of the output voltage, Q_{30} supplies the load current.

If a negative-going signal is applied to the base of Q_{17}, the current I_2 increases, causing Q_{28} to tend to cut off. Collector current I_C in Q_{30} is reduced, D_2 is cut off, and I_{B29} increases. This produces a large emitter current in Q_{31}; for positive excursions of the output voltage, Q_{31} supplies the load current.

Current limiting is provided by connecting an external resistor as shown in Fig. 11-5*b*. Referring to Fig. 11-3, we see that if overload current flows during the negative output excursion a positive voltage will be developed between current sense terminal and output terminal. When the load current is too large, this voltage will turn on Q_{27}, which will cause current to flow into current mirror Q_{22} and Q_{23}. This in turn reduces the base current of Q_{28} and hence the collector current of Q_{30}. If the output excursion is positive, excessive current produces a negative voltage between current sense and output terminals, causing Q_{26} to turn on. The base current of Q_{29}, and hence the emitter current of Q_{31}, is thus reduced to limit the load current.

The thermal shutdown provision operates as follows. Zener diode D_1, Q_{18}, and voltage divider R_8 and R_9 provide a base voltage for Q_{19} which is almost temperature-independent. As the temperature increases, the base-emitter voltages of both Q_{19} and Q_{20} decrease, causing an increase of base current in Q_{20} and Q_{21}. The collector current of Q_{20} reduces the base current of Q_{29} and the collector current of Q_{21} reduces the base current of Q_{28}; thus the output transistors tend to be cut off.

In the layout of a power circuit, the usual method employed is to make the layout as symmetrical as possible as far as power devices are concerned in order to make isotherms easy to locate. Critical components such as input transistors can be placed on isotherms to avoid thermal feedback effects. In cases where there are several power devices, it would be necessary to increase the chip area significantly in order to make the layout symmetrical. This was avoided in the μA791 by performing a careful thermal analysis of the chip to determine where the power devices should be located in order to reduce thermal feedback effects to an acceptable level. The result is sketched in Fig. 11-6a; note that five power devices are involved. The two output transistors, which must be capable of 15-W dissipation, each require an area 36 × 18 mils; the total chip size is 100 × 70 mils.

We have seen in Chap. 7 that special attention must be given to the design of high-current transistors in order to reduce the effects of current crowding and high-level injection. If large power is involved, thermal effects must also be taken into account. In order to obtain acceptable device characteristics, a number of emitter "fingers" are used, and provisions are made for so-called "ballasting" resistors in each finger in order to equalize currents and prevent hot spots from occurring. In the μA791, the thermal analysis was used to determine that 12 Ω was required at each emitter site in order to maintain neighboring site currents within 10 percent of each other. The emitter layout is also designed so that an effective pinch resistor R_x occurs in areas adjacent to the base contact; this helps to confine emitter injection to the periphery of the emitter region adjacent to the base contacts. The μA791 has the following characteristics:

> Open-loop gain: 50,000
> Input offset voltage: 3 mV
> Input bias current: 100 nA
> Power supply rejection ratio: 50 μV/V
> Maximum output current: 1.2 A

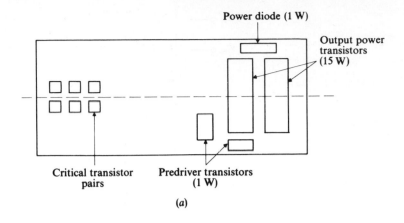

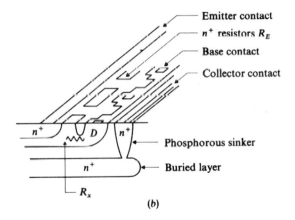

FIGURE 11-6
(*a*) Location of power devices to minimize thermal feedback; (*b*) structure of the output transistors.

Maximum power dissipation: 15 W
Maximum operating voltage: ± 22 V
Unity-gain bandwidth: 750 kHz
Zero-signal current: 10 mA
Open-loop output resistance: 2 Ω
Chip size: 100 × 70 mils

11-3 COMPARATORS

Comparators are used to detect when an input signal has reached a predetermined reference level.   An ideal comparator would have zero output for $v_{in} < V_{ref}$ and maximum output for $v_{in} \geq V_{ref}$, as shown in Fig. 11-7*a*.   A high-gain operational

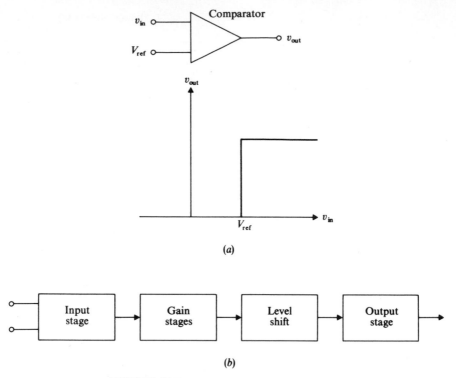

FIGURE 11-7
(a) The ideal comparator; (b) block diagram of the Kuijk comparator.

amplifier can be used as a comparator by connecting one input to V_{ref}; in fact, nonregenerative comparators are very similar to operational amplifiers. In contrast to normal operational-amplifier applications, comparators are usually operated open-loop; since fast response time is required in a comparator, the bandwidth of the comparator is generally greater than that of an ordinary operational amplifier. A comparator circuit designed by Kuijk[6] employs some of the techniques we have described in previous chapters, and in addition makes use of a novel method for obtaining wide-band gain. The comparator employs an input differential amplifier, several succeeding gain stages, level shift, and an output stage, as shown by the block diagram of Fig. 11-7b.

The input stage is shown in Fig. 11-8; for convenience we discuss only one side. Transistor Q_1 functions as half of a differential amplifier; diode-connected Q_7 prevents breakdown of the emitter junction of Q_1 when large negative input voltage is applied to the base of Q_1. Base current cancellation of the type described in Chap. 10 is provided by Q_2, Q_4, Q_3, and Q_5; Q_6 is cascode-connected to provide the load for Q_1. A load resistor is used for Q_6.

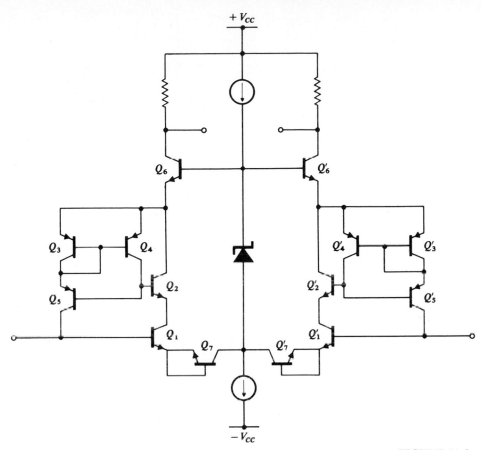

FIGURE 11-8
The input stage.

Since the comparator will only be operated open-loop, the closed-loop gain and phase margin requirements for operational amplifiers do not apply, and overall compensation is not required. This means that the excess phase problems attendant with several gain stages will not be important; what is required is good bandwidth. A novel "zig-zag" method is used for succeeding differential gain stages; the basic scheme is shown in Fig. 11-9. Here stage 1 bias current is used also for stage 3, and stage 2 bias current is used for stage 4. All stages are direct-coupled; although only four stages are shown in Fig. 11-9, it is easily seen that any number can be used, depending only upon the power supply voltage available.

In order to maintain operation in the active region, the collector-base voltage of each transistor must always be greater than 0. Referring to Fig. 11-9, we suppose that the input signal is large enough to cause Q'_3 to almost cut off. Then

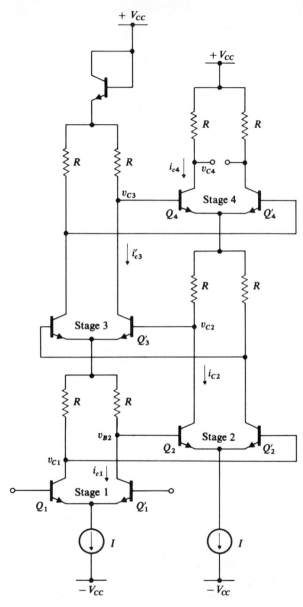

FIGURE 11-9
Zig-zag gain stages.

$i'_{c3} \approx 0$, and since this condition will also cause Q'_4 to cut off, we have $i_{c4} \approx I$. Then

$$v_{C4} = V_{CC} - IR$$

$$v_{C3} = V_{CC} - V_D$$

$$v_{cb4} = v_{C4} - v_{C3}$$

Therefore if $v_{cb4} \geq 0$, we see that

$$IR \leq V_D$$

We can easily calculate the maximum differential voltage gain that can be obtained from each stage.

$$A_{DM} = g_m R$$

$$A_{DM(\text{max})} = g_m R_{\text{max}}$$

$$= g_m \frac{V_D}{I}$$

$$= \frac{qI}{2kT} \frac{V_D}{I}$$

$$= \frac{qV_D}{2kT}$$

$$\approx 14$$

With small-value load resistors, the bandwidth of each stage will be large. Measurements for a five-stage zig-zag amplifier show a unity-gain frequency of 124 MHz.

Exercise 11-3 Suppose that the input voltage of the first stage has $+7$-V reference. If the power supply voltage is $+14$, what is the maximum voltage gain that could be obtained from a zig-zag scheme?

Since the dc bias level at the output of the nth stage is $V_{CC} - V_D$, some means of level shifting is necessary in order to couple into the output stage. This is accomplished by using a p-n-p differential stage as shown in Fig. 11-10. Here the p-n-p transistors are used as a common-base differential amplifier in order that their frequency response not degrade the amplifier performance; the level-shift stage is also zig-zag-connected. The p-n-p transistors are bypassed at high frequency by 5-pF capacitors C.

The output stage is shown in Fig. 11-11. Transistors Q_8 and Q_9 act as differential amplifiers, and Q_{11}, Q_{12}, and Q_{13} are used as a current mirror. An output voltage reference level is provided by V_B, and the load resistor for the differential stage is R_0. Offset voltage is provided by Q_{14} so that the dc level of v_{out} is the same as V_B. Differential to single-ended output conversion is provided by

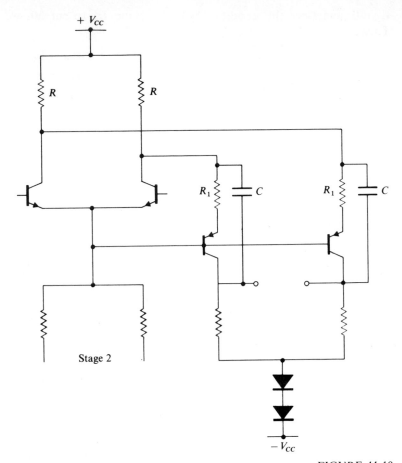

FIGURE 11-10
Level shifting.

the way in which R_0 is connected.   It is interesting to note that the current mirror in this case is not being used as an active load, since no signal current flows in it.
The comparator has the following performance:

Input bias current: 0.4 mA
Input offset current: <0.3 mA
Input offset voltage: <3 mV
Sensitivity: 0.3 mV for full output swing
Common-mode voltage range: +9.5 to − 12.5
Maximum differential input voltage: 25
Speed: delay, 40 ns; rise time, 65 ns; fall time, 40 ns
Common-mode rejection ratio: 100 dB
Supply current: 5.8 mA

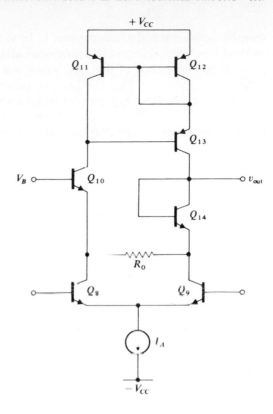

FIGURE 11-11
Output stage.

11-4 *RC* ACTIVE FILTERS

It is difficult to fabricate frequency-selective amplifiers in monolithic form because of the lack of conveniently fabricated inductors of reasonable quality. For frequencies in the range of tens of MHz, the methods described in Chap. 4 can be used to make spiral inductors with deposited ferrite, if the designer is willing to use the extra processing steps required. For lower frequencies, the large values of inductance and capacitance required make monolithic fabrication nearly impossible, even with extra processing steps. It then becomes necessary to use off-the-chip components. Since capacitors of high quality are more easily and cheaply obtained than inductors, techniques have been developed for using monolithic gain blocks with discrete capacitors and resistors to produce frequency-selective circuits. Many of these techniques involve Wien-bridge circuits, Sallen-Key networks,[7] or twin-T networks.[8] These methods have two principal disadvantages:

1 Their performance is highly sensitive to variations of active device gain and passive component values.

2 In some cases they can become unstable.

One *RC*-active method which appears to be superior is the state-variable synthesis method,[9] in which integrated operational amplifiers are used as integrators and gain blocks. A circuit of this type which can produce both bandpass and notch functions is shown in Fig. 11-12*a*. Amplifiers 1 and 2 function as integrators, amplifier 3 serves as a unity-gain inverting amplifier, and amplifier 4 is a summing amplifier. If we assume that the amplifiers are ideal infinite-gain operational amplifiers, we can easily see that bandpass and notch functions are performed. Let *s* be the complex-frequency variable; then

$$V_4(s) = -V_3(s)$$

$$V_3(s) = -V_2(s)\frac{G}{sC}$$

and
$$-V_2(s)(G_1 + sC) = GV_4(s) + G_2 V_1(s)$$

Eliminating $V_4(s)$ and $V_3(s)$ from the above equations, we can solve for the transfer function $V_2(s)/V_1(s)$ to obtain

$$\frac{V_2(s)}{V_1(s)} = \frac{-(G_2/C)s}{[s^2 + (G_1/C)s + (G/C)^2]}$$

This transfer function has a pair of complex poles if $G > G_1/2$, and can therefore provide the bandpass function. The center frequency ω_0 is

$$\omega_0 = \frac{G}{C} = \frac{1}{RC}$$

and for a narrow-band filter the Q is

$$Q = \frac{G}{G_1} = \frac{R_1}{R}$$

A notch is obtained by adding v_2 to the input signal; this produces an output v_5 with transfer function $V_5(s)/V_1(s)$ given by

$$\frac{V_5(s)}{V_1(s)} = \frac{-G_2}{G_3}\frac{[s^2 + (G/C)^2]}{[s^2 + (G_1/C)s + (G/C)^2]}$$

This transfer function has the same poles as $V_2(s)/V_1(s)$, but it also has zeros on the imaginary axis at

$$j\omega = \pm j\frac{G}{C}$$

There a notch will be produced at $\omega = (G/C) = \omega_0$.

It can be shown that by using four operational amplifiers, two of which are connected as integrators, one can realize a general second-order transfer function

$$F(s) = \frac{as^2 + bs + c}{ds^2 + es + f}$$

By proper choice of element values, the various coefficients can be selected to produce low-pass, high-pass, bandpass, or notch functions. For example, if

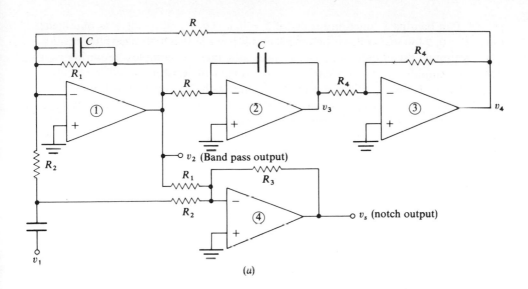

(a)

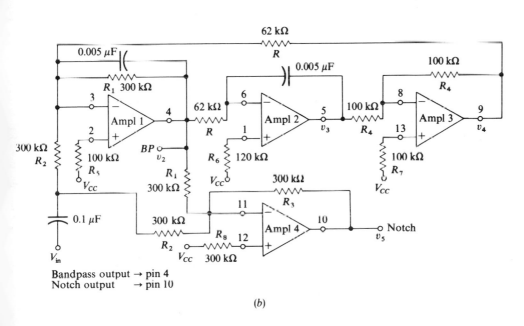

Bandpass output → pin 4
Notch output   → pin 10

(b)

FIGURE 11-12
(a) Basic state-variable circuit for bandpass and notch transfer functions;
(b) realization of the circuit with MC 3401P amplifiers.

$a = c = 0$, a bandpass function results; if $b = 0$, $a = d$, $c = f$, a notch is obtained; if $a = b = 0$, a low-pass filter is obtained; and if $b = c = 0$, a high-pass filter results.

Although the circuit of Fig. 11-12a seems somewhat complex, requiring as it does four amplifiers, two capacitors, and a number of resistors, it will be recalled that in Chap. 10 we described a monolithic chip with four operational amplifiers: the MC 3401P. This chip is ideally suited to state-variable synthesis applications, and the filter of Fig. 11-12a can be realized with the four amplifiers of the MC 3401P, as shown in Fig. 11-12b.[10] For the element values shown, the center frequency is $f_0 \approx 500$ Hz, the Q is 4.8, and the bandpass gain is 1.

In using the MC 3401P in such an application as this, we must be concerned not only with realizing the proper transfer function but also with ensuring the proper bias conditions. Fortunately this is easily done, since dc negative feedback is used around the first three amplifiers. It will be recalled that the MC 3401P amplifiers are current-differencing amplifiers; because of the negative feedback we can find the bias conditions by assuming that the two input currents are approximately equal for each amplifier. For amplifier 2, we can write

$$\frac{V_2}{62 \text{ k}\Omega} = \frac{V_{cc}}{120 \text{ k}\Omega}$$

from which we obtain

$$V_2 = 0.515 V_{cc}$$

For amplifier 1,

$$\frac{V_4}{62 \text{ k}\Omega} + \frac{V_2}{300 \text{ k}\Omega} = \frac{V_{cc}}{100 \text{ k}\Omega}$$

from which we find

$$V_4 = 0.515 V_{cc}$$

Finally, for amplifier 3,

$$\frac{V_4}{100 \text{ k}\Omega} + \frac{V_3}{100 \text{ k}\Omega} = \frac{V_{cc}}{100 \text{ k}\Omega}$$

or

$$V_4 + V_3 = V_{cc}$$

Therefore

$$V_3 = 0.485 V_{cc}$$

The summing amplifier 4 is not part of the closed loop of amplifiers 1, 2, and 3, but it has local feedback applied and we can write

$$\frac{V_5}{300 \text{ k}\Omega} + \frac{V_2}{300 \text{ k}\Omega} = \frac{V_{cc}}{300 \text{ k}\Omega}$$

from which we obtain

$$V_5 = 0.485 V_{cc}$$

All bias voltages are now determined and we see that the dc output voltage for each stage is approximately $V_{cc}/2$, indicating that each amplifier is biased for maximum dynamic range.

11-5 VERY-LOW-FREQUENCY FILTERS

The *RC*-active method described in the preceding section is useful for frequencies from the audio range to about 1 MHz; for lower frequencies the capacitor values become quite large.   For frequencies from less than 1 Hz to well into the audio range it is possible to make use of thermal effects in silicon and thereby avoid entirely the use of capacitors or inductors.   Moreover, since only transistors and resistors are needed, and since there are no stringent requirements on their characteristics, they can be made with the standard fabrication process.   A complete filter of this type, called an *electrothermal* filter or ETC filter, can be monolithic, with no external components required.

For simplicity, we describe only low-pass filters.   The basic structure of the essential filter elements is shown in Fig. 11-13a.[11,12]   Heater and sensor transistors are diffused into the chip in the usual manner; although only the heaters and sensors are shown in Fig. 11-13a, other components can also be placed on the chip if care is given to their location.   After the chips have been scribed and separated, they are mounted on insulators as shown.

The heaters H_1 and H_2 are differentially driven by transistors Q_1 and Q_2 as shown in Fig. 11-13b, and the electrical input signal is converted to a heat signal by power dissipation in the collector junctions of the heaters.   The temperature is sensed differentially by the emitter junctions of sensor transistors S_1 and S_2; these are connected as a differential amplifier.   Thus the sensor output is the difference between emitter junction voltages (caused by the temperature difference) multiplied by the gain of the differential amplifier.   Use of differentially connected heaters and sensors minimizes the dependence of circuit performance on ambient temperature variations.   It is important to note that there are no electrical connections between heaters and sensors; only thermal coupling exists.

The distributed thermal resistance and thermal capacitance of the silicon chip act as a low-pass filter to the heat signals; indeed, one can make a direct analogy between the electrothermal structure and a distributed *RC* structure, in which current corresponds to power (heat flow) and voltage to temperature.

That the circuit produces a low-pass filter can be seen from Fig. 11-14a; in this case the chip size was 75 × 25 mils and the cutoff frequency ω_{co} is about 10 Hz.   As the chip is made smaller, the cutoff frequency increases.   Fig. 11-14b shows response curves for various relationships between the distances L_1 and L_2; each curve in Fig. 11-14b is normalized to its zero-frequency value.

Other components, such as transistors to provide additional gain, can be included on the chip if they are located properly.   If the gain stages are differentially connected they can be placed on isotherms and will be uninfluenced by thermal signals.   Nondifferential circuits can be located along the chip centerline since this is a line of zero thermal signals.

More complex filters can be synthesized in ETC form, and high-pass, bandpass, and notch filters can also be realized.   Figure 11-15 shows a particular filter response which was synthesized without the use of feedback by using three heaters each with a different weighting factor.   Analysis shows that for simple, low-pass

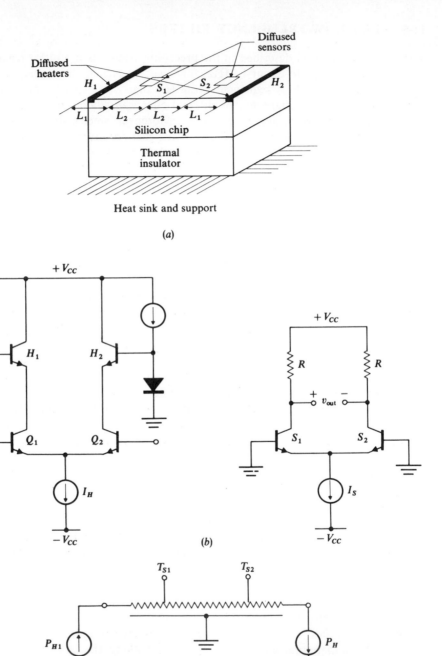

FIGURE 11-13
(a) Basic ETC low-pass filter structure; (b) realization of differentially connected
heaters and sensors; (c) distributed RC transmission line analog for the ETC filter.

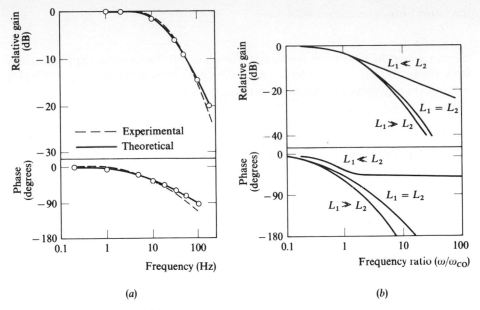

FIGURE 11-14
(a) Calculated and measured response curves for the low-pass ETC filter;
(b) response curves for various conditions of L_1 and L_2.

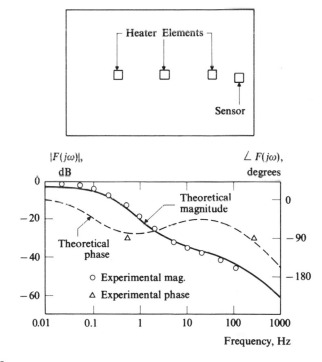

FIGURE 11-15
Calculated and measured response of the filter synthesized for an AGC applica-
tion. The sketch shows only the location of emitters of transistors used in the
ETC filter. Chip size is 150×125 mils.

filters realized in ETC form the limitations on cutoff frequency are approximately 0.03 Hz to 100 kHz. In order to reach the upper limit it is necessary to employ a fabrication process such as Isoplanar II.

11-6 USE OF OPERATIONAL AMPLIFIERS FOR TESTING OPERATIONAL AMPLIFIERS

The most common application of the operational amplifier is as the gain element in a negative-feedback loop; we have seen examples of this in Chap. 10. Figure 11-16a shows an operational amplifier used to provide stable negative gain; the gain of this configuration is

$$\frac{v_2}{v_1} = \frac{-R_2}{R_1}$$

This can be easily seen by making use of the "virtual ground" principle as follows. If the amplifier has a large gain (typically 10^5), for any reasonable output voltage the differential input voltage must be nearly zero. When negative feedback is used around the amplifier, as in Fig. 11-16a, point A is therefore maintained at approximately zero volts relative to point B. Since negligible input current can flow in the amplifier, the currents in R_1 and R_2 must have the same magnitude. The output voltage of the amplifier assumes the value required to cause this current balance to occur. Point A is thus a virtual ground.

EXAMPLE 11A *Gain of a Follower.* Consider the positive-gain follower circuit of Fig. 11-16b. The positive input to the amplifier is held at v_1 by the input. Following the virtual ground principle, the negative input, point A, must also be at v_1. Thus the voltage across R_1 must be

$$i_1 = \frac{v_1}{R_1}$$

Since the amplifier input impedance is large compared to other circuit resistances, $i_1 \approx i_2$. Thus

$$v_2 = i_1 R_2 + i_1 R_1$$

$$= \frac{v_1(R_1 + R_2)}{R_1}$$

from which

$$\frac{v_2}{v_1} = \frac{R_2 + R_1}{R_1} \qquad\qquad ////$$

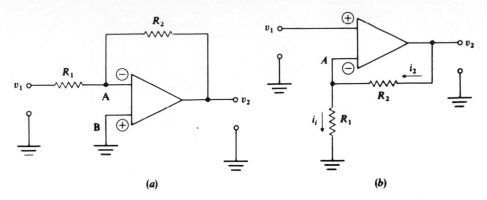

(a)   (b)

FIGURE 11-16
(a) Negative-gain configuration; (b) positive-gain, or "follower," configuration.

The virtual-ground principle is useful in analyzing a nonideal operational amplifier; a model for such an amplifier, including the most important dc nonideal aspects, is shown in Fig. 11-17.   Here:

$$V_{IO} = \text{input offset voltage}$$
$$IIB_+, IIB_- = \text{input bias currents}$$
$$R_{in} = \text{input resistance}$$
$$R_{out} = \text{output resistance}$$
$$A_{VS} = \text{open-loop voltage gain}$$

Since V_{IO} is often of the order of several millivolts, IIB is several nano-amperes, and A_{VS} is of the order of 10^6, direct measurement of these nonideal parameters is extremely difficult.

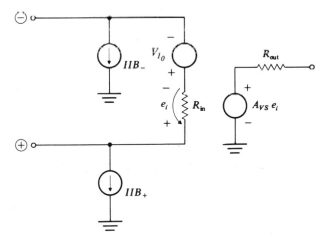

FIGURE 11-17
Model for a nonideal operational amplifier.

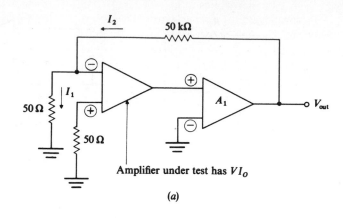

Amplifier under test has VI_o

(a)

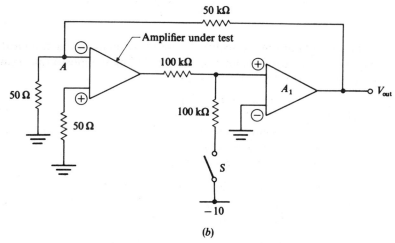

(b)

FIGURE 11-18
(a) Circuit for measuring V_{Io}; (b) circuit for measuring A_{VS}.

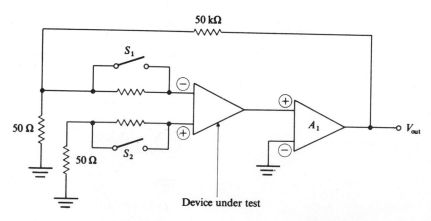

Device under test

FIGURE 11-19
Circuit for measuring IIB.

A simple circuit for determination of operational-amplifier offset voltage is shown in Fig. 11-18*a*. Since the loop has overall negative feedback, the input of A_1 must be at zero volts, ignoring offset in A_1. The input voltage of the device under test must then be V_{IO} (the internal offset voltage plus zero for the internal ideal amplifier). It now follows that

$$I_1 = \frac{V_{IO}}{50} = I_2$$

Finally,
$$V_{out} = (50 \text{ k}\Omega)I_2 + V_{IO}$$

$$= \frac{50{,}000}{50}V_{IO} + V_{IO} = 1001V_{IO}$$

Measurement of V_{out} in volts is therefore equal to V_{IO} in millivolts.

Exercise 11-4 Compute the error due to

$$IIB_{+,-} = 0.5 \text{ μA}$$
$$A_{VS} = 50{,}000$$
$$R_{out} = 5 \text{ }\Omega$$
$$R_{in} - 1 \text{ M}\Omega$$
$$V_{IO} = 5.0 \text{ mV}$$

for both operational amplifiers.

A similar circuit for determination of A_{VS} is given in Fig. 11-18*b*. With switch S open, V_{out} is $1000V_{IO}$, just as with the circuit of Fig. 11-18*a*. With switch S closed, the negative input to A_1 is still zero volts, and the output of the device under test is $+10$ V. The voltage of point A must therefore be:

$$V_A = V_{IO} + \frac{10}{A_{VS}}$$

Thus
$$V_{out/S(closed)} = 1000(V_{IO} + 10/A_{VS})$$

Therefore
$$V_{out/S(closed)} - V_{out/S(open)} = 1001\left(V_{IO} - V_{IO} + \frac{10}{A_{VS}}\right)$$

$$= \frac{10{,}010}{A_{VS}}$$

and
$$A_{VS} = \frac{10{,}010}{V_{out/S(closed)} - V_{out/S(open)}}$$

Exercise 11-5 Compute the errors in the measurement of A_{VS} if both amplifiers have the parameters of Exercise 11-4.

Exercise 11-6 Compute the input currents $IIB_- \, IIB_+$ and input offset current $IIO - IIB_+ - IIB_-$ by comparing circuit performance with S_1 and S_2 in all possible positions in Fig. 11-19.

11-7 DIGITAL-TO-ANALOG CONVERSION

Conversion of binary digital information to analog form is carried out by a digital-to-analog converter. Consider a typical binary digital word:

$$1\ 0\ 1\ 1\ 0\ 1$$

In decimal format this becomes:

$$1 \times 32 + 0 \times 16 + 1 \times 8 + 1 \times 4 + 0 \times 2 + 1 \times 1 = 45$$

Digital-to-analog converters contain the weighting factors (2, 4, 8, 16, ...) which can be summed or not depending upon the state of the input digit.

Weighting factors can be either voltages or currents. In this discussion, currents will be used because of the ease of summing currents without offset problems in bipolar transistor circuits. The fundamental block diagram of a simple digital-to-analog converter is shown in Fig. 11-20a, where

$$I_{out} = A_1 \times 32I + A_2 \times 16I + A_3 \times 8I + \cdots$$

and
$$A_i = 1 \text{ or } 0 \text{ depending on the state of the } i\text{th digit}$$

Creation of the current weighting factors can be carried out several ways. Two common methods are resistor weighting and R-$2R$ ladders; a weighted resistor circuit is shown in Fig. 11-20b. The ratio of resistor values determines the ratio of currents for the same voltage applied across all resistors. This circuit has the advantage of being able to reproduce any weighting scheme, but requires large ratios of resistor values.

The R-$2R$ ladder circuit is shown in Fig. 11-20c. The R-$2R$ ladder is limited in the current ratios it can reproduce, but has a maximum resistor ratio of 2 : 1, thus simplifying the resistor fabrication problem.

Exercise 11-7 Verify the current ratios of the R-$2R$ ladder of Fig. 11-20c by reducing the circuit to series and parallel resistances beginning at the right-hand end.

Both current weighting circuits require that all resistors be terminated at identical voltages. In addition, a current drive circuit is required to produce the total ladder current.

Termination Circuit

The termination circuit for the current weighting resistor ladder should have the following three circuit properties:

1 Low termination impedance to avoid producing series error resistance in the weighting circuit.
2 A fixed terminating voltage determined by an external bias voltage.
3 Unity current gain to the current from the current weighting network.

A block diagram is shown in Fig. 11-21a.

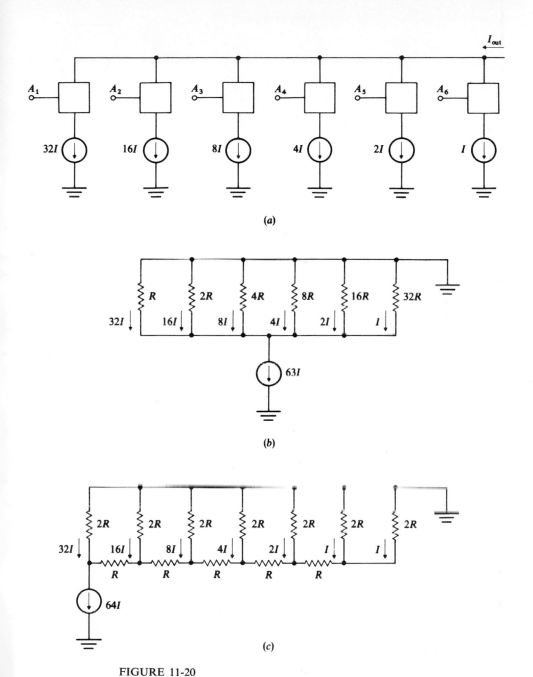

FIGURE 11-20
(a) Basic current weighter; (b) resistance current weighter; (c) R-2R current weighter.

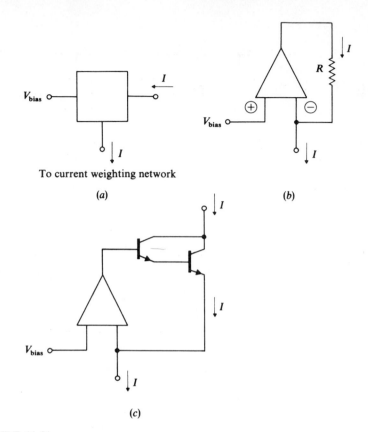

To current weighting network

(a)

(b)

(c)

FIGURE 11-21
(a) Block diagram of terminator; (b) terminator using operational amplifier;
(c) terminator in which negligible current flows at the output of the amplifier.

The properties required by the termination circuit are precisely those of a high-gain negative-feedback system having a virtual ground point. The circuit of Fig. 11-21b shows an operational-amplifier feedback loop where the current I flows from the amplifier output. In order to make the output current available for summation, the feedback resistor must be replaced by a feedback transistor as shown in Fig. 11-21c.

Exercise 11-8 Given that transistor $\beta = 50$ and that the termination circuit must add an error of no more than 1/8 of the least significant bit, compare the use of a single-feedback transistor, a two-transistor Darlington, and a three-transistor Darlington in the circuit of Fig. 11-21c for 6-, 8-, and 10-bit digital-to-analog converters.

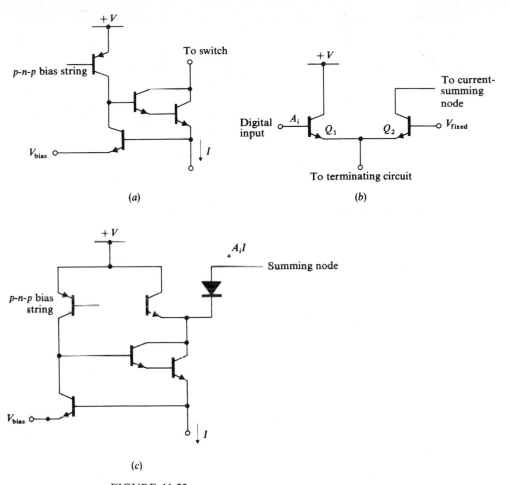

FIGURE 11-22
(a) Simplified terminator; (b) emitter-coupled current switch; (c) switch using a diode.

Rather than using a complete operational amplifier in the terminating circuit, a simpler circuit has been devised, with a single n-p-n transistor and a p-n-p current source, as shown in Fig. 11-22a. The switching function can most easily be carried out using a differential switch similar to an emitter-coupled logic gate, as shown in Fig. 11-22b. The base current in Q_2 represents an error. To eliminate the error, Q_2 is converted to a diode, as shown in Fig. 11-22c. Note that this circuit responds to negative logic. The total digital-to-analog circuit using an R-$2R$ ladder now appears as shown in Fig. 11-23.

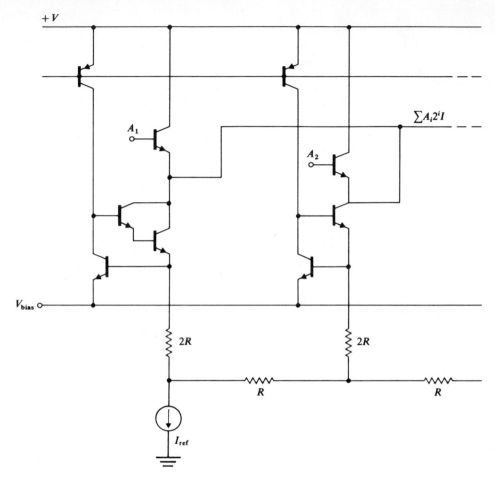

$+V$

A_1

A_2

$\sum A_i 2^i I$

V_{bias}

$2R$

$2R$

R

R

I_{ref}

FIGURE 11-23
Two sections of a digital-to-analog converter using an R-$2R$ ladder.

Current Drive Circuit

The current drive circuit serves to translate a reference voltage into a reference current which is then used to drive the ladder network. In addition, it must perform a level translation from the grounded reference to a negative voltage to which current can flow from the ladder network.

Here again, a virtual-ground node in a negative-feedback loop is useful. A system to generate the reference current is shown in Fig. 11-24a. Since point A is maintained at zero volts, being the virtual ground, the reference voltage V_{ref} appears across the reference resistor R_{ref} to produce reference current $I_{ref} = V_{ref}/R_{ref}$. In the circuit of Fig. 11-24a, a feedback resistor is used to close the loop;

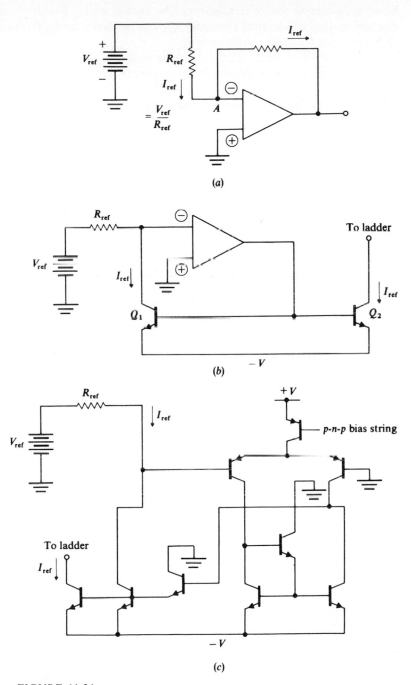

FIGURE 11-24
(a) Reference circuit with amplifier; (b) use of transistors in the feedback loop;
(c) simplified amplifier in a reference circuit.

however, a current-source transistor can be used as well.   Recall, from the biasing discussion, that an identical transistor Q_2 can be slaved to the feedback transistor Q_1 to generate the ladder drive current.

The amplifier circuit should combine both the gain needed to close the loop and the driver-end level translation.   A simple circuit to accomplish these two objectives is the input stage from the μA741 circuit discussed earlier.   The complete circuit is shown in Fig. 11-24c.   The complete digital-to-analog circuit, including current weighting ladder, terminating circuits, and bias and current drive circuit, is shown in Fig. 11-25.

11-8   VOLTAGE REGULATORS[13,14]

A voltage regulator can be thought of as a negative-feedback amplifier having very low output resistance, in which the input is a reference voltage and the output is the desired regulated voltage; supply voltage for the amplifier is provided by the unregulated voltage.   Regulators usually employ the general structure shown in

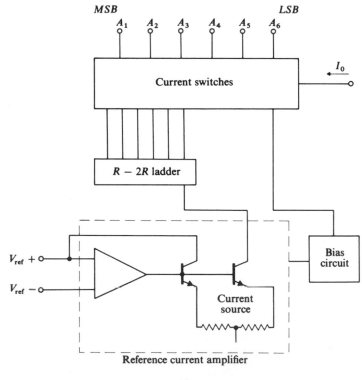

(a)

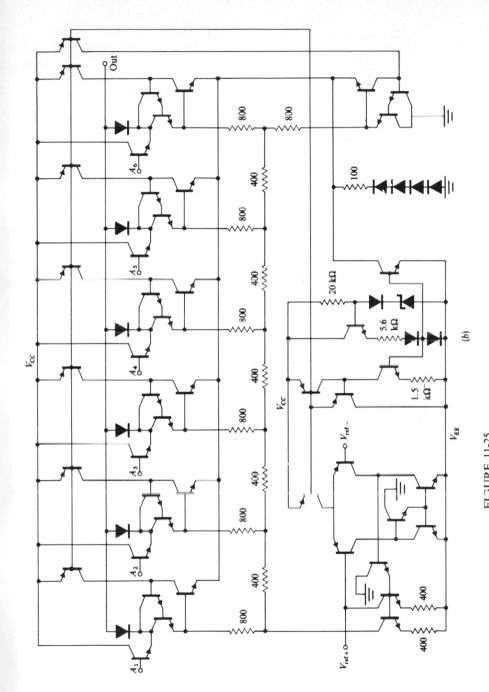

FIGURE 11-25

Complete circuit of the MC 1506L digital-to-analog converter. (*a*) Block diagram; (*b*) circuit diagram.

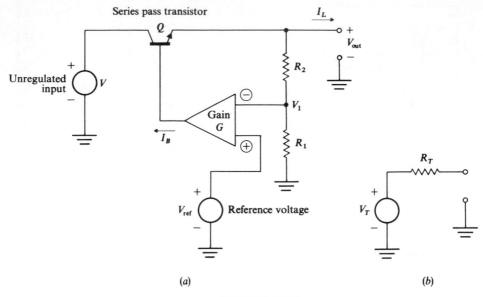

FIGURE 11-26
(a) Basic regulator structure; (b) Thévenin equivalent.

the block diagram of Fig. 11-26a, in which a fraction of the regulated output is compared with a reference voltage V_{ref}. The difference is amplified by the voltage-controlled current source with transconductance gain G; the output of this amplifier is used as base current for the series pass transistor Q. In Fig. 11-26b is shown the Thévenin equivalent for the regulator; the voltage V_T is easily calculated by assuming I_L to be zero in Fig. 11-26a.

$$V_T = (\beta + 1)I_B(R_1 + R_2) = (\beta + 1)(R_1 + R_2)G\left(V_{\text{ref}} - \frac{V_T R_1}{R_1 + R_2}\right)$$

Solving for V_T we obtain

$$V_T \approx \frac{GV_{\text{ref}}(R_1 + R_2)(\beta + 1)}{1 + G(\beta + 1)R_1}$$

If G is large, this becomes

$$V_T \approx \frac{V_{\text{ref}}(R_1 + R_2)}{R_1}$$

To find R_T, we let $V_{\text{out}} = 0$ and calculate I_L:

$$I_L = G(\beta + 1)V_{\text{ref}}$$

Then

$$R_T = \frac{V_T}{I_L} = \frac{(R_1 + R_2)/R_1}{G(\beta + 1)}$$

It will be noted that for the structure of Fig. 11-26a, the reference voltage must be chosen to be less than the desired regulated output voltage and less than the unregulated input. A series combination of a Zener diode and forward-biased diodes can be used to obtain a reference voltage with nearly zero temperature coefficient; however, it will be recalled that the breakdown voltage of the emitter junction is approximately 7 V. For output voltages less than this, some other scheme must be used for the reference voltage.

The so-called *band-gap reference method* provides a reference voltage proportional to V_{GO}, the band-gap voltage of silicon, which is 1.23 V. This method is best understood by first recalling that the temperature coefficient of the emitter-base voltage is negative and has a value of 2 mV/°C. If two transistors are operated with different emitter current densities, the difference between their emitter-base voltages is

$$\Delta V_{BE} = \frac{kT}{q} \ln \frac{J_1}{J_2}$$

This voltage difference is seen to have a positive temperature coefficient for fixed J_1 and J_2. By combining the positive temperature coefficient of ΔV_{BE} with the negative temperature coefficient of V_{BE}, one can achieve a low-voltage reference with nearly zero temperature coefficient.

The basic circuit for the reference voltage is shown in Fig. 11-27a. Here Q_1 and Q_2 are connected as a small-value current source, with the collector current of Q_1 set at about 10 times that of Q_2. The voltage ΔV_{BE} is

$$\Delta V_{BE} = \frac{kT}{q} \ln \frac{I_{C1}}{I_{C2}}$$

and this voltage appears across R_3. If we assume the base current of Q_3 to be negligible, we can write the voltage across R_2 as

$$V_2 = \frac{R_2}{R_3} \Delta V_{BE}$$

if α of Q_2 is assumed to be unity. The reference voltage is then

$$V_{ref} = V_{BE} + \frac{R_2}{R_3} \Delta V_{BE} \qquad (11\text{-}2)$$

The temperature behavior of V_{BE} can be shown to be given by

$$V_{BE} = V_G \left(1 - \frac{T}{T_0}\right) + V_{BEO} \frac{T}{T_0} + \frac{nkT}{q} \ln \frac{T_0}{T} + \frac{kT}{q} \ln \frac{I_C}{I_{CO}} \qquad (11\text{-}3)$$

where V_G = band-gap voltage
T = temperature, °K
I_C = collector current at T
I_{CO} = collector current at T_0
V_{BEO} = base-emitter voltage at T_0
n = factor depending on processing methods, having a typical value of 1.5

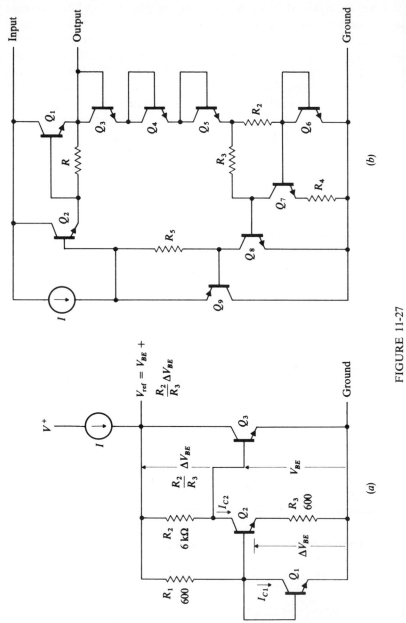

FIGURE 11-27

(a) The band-gap reference; (b) simplified circuit of the LM 109 5-V regulator.

The last two terms of (11-3) are small in comparison with the first two; neglecting them and inserting (11-3) in (11-2), we obtain

$$V_{ref} = V_G\left(1 - \frac{T}{T_0}\right) + V_{BE0}\frac{T}{T_0} + \frac{R_2}{R_3}\frac{kT}{q}\ln\frac{I_{C1}}{I_{C2}} \qquad (11\text{-}4)$$

If we now set $\partial V_{ref}/\partial T = 0$, we find

$$V_G = V_{BE0} + \frac{R_2}{R_3}\frac{kT_0}{q}\ln\frac{I_{C1}}{I_{C2}} \qquad (11\text{-}5)$$

The base-emitter voltage V_{BE0} depends upon the emitter current of Q_3; in the circuit of Fig. 11-27a the supply current I and the resistors R_1, R_2, and R_3 can be chosen so that (11-5) is satisfied. The reference voltage is then

$$V_{ref} = V_G$$

A band-gap reference of this type is used in the National LM 109 voltage regulator, a simplified circuit for which is shown in Fig. 11-27b. Here a series combination of diodes is used to raise the output voltage to the desired level. The negative temperature coefficients of Q_8, Q_5, Q_4, and Q_3 are balanced by the positive temperature coefficient of the ΔV_{BE} of Q_6 and Q_7 multiplied by R_3/R_4. If we assume that Q_8, Q_5, Q_4, and Q_3 all have the same V_{BE}, we see that the output voltage is

$$V_{out} = 4V_{BE} + \frac{R_3}{R_4}\Delta V_{BE}$$

For zero temperature coefficient of V_{out}, the resistors R_2, R_3, and R_4, and the current in the diode-connected transistors, would be chosen so that

$$V_G = V_{BE0} + \frac{R_3}{R_4}\frac{kT_0}{q}\ln\frac{I_{C6}}{I_{C7}}$$

In the circuit of Fig. 11-27b, negative feedback is used to maintain this current constant as the load current varies. The output of Q_8 is applied to the base of vertical p-n-p Q_9, whose output is applied to the Darlington combination Q_2 and Q_1. Series-pass transistor Q_1 is connected to the output. To see qualitatively how regulation occurs, suppose that a load is applied which tends to reduce V_{out}. This would lower the emitter voltage of Q_5 which would in turn cause a reduction of V_{BE8}. The collector current of Q_8 would then decrease, causing more of the bias current I_0 to be diverted into the base of Q_2; amplification of the base current by Q_2 and Q_1 would then tend to raise the output voltage to its correct value.

PROBLEMS

11-1 A micropower circuit is to be fabricated in an epitaxial layer of thickness 5 μm with 10 $\Omega \cdot$ cm resistivity. The substrate resistivity is 10 $\Omega \cdot$ cm, and the base diffusion has junction depth 3.0 μm and surface concentration 10^{17} cm^{-3}. Take into account depletion regions and calculate the sheet resistance of collector pinch resistors.

11-2 For the base diffusion of Prob. 11-1, calculate the minimum epitaxial thickness that could be used without collector punch-through to the substrate if the power supply voltage in the circuit is 2.0 V.

11-3 In the circuit of Fig. 11-2a, if $R_1 = R_2 = 1$ MΩ, and if the collector bias current in each of Q_1 and Q_2 is 0.5 μA, calculate the voltage gain.

11-4 In the circuit of Fig. 11-5a do not assume $I_{B28} \approx 0$ and $I_{B29} \approx 0$; calculate I_C and I_E.

11-5 In the circuit of Fig. 11-8, if the emitter current of Q_1 is 0.6 mA, calculate the input current of the input stage. Assume $\beta = 5$ for *p-n-p* transistors and $\beta = 100$ for *n-p-n* transistors. Calculate the differential-mode voltage gain of the circuit if 600-Ω load resistors are used.

11-6 Calculate the gain and output resistance of the output stage of Fig. 11-11.

11-7 For the circuit of Fig. 11-12b let the amplifiers have a finite transimpedance K, and calculate the transfer functions $V_2(s)/V_1(s)$ and $V_s(s)/V_1(s)$.

11-8 For an electrothermal filter with $L_1 = 0$, derive the transfer function $[T_{S1}(s) - T_{S2}(s)]/[P_{H1}(s) - P_{H2}(s)]$.

REFERENCES

1 MEINDL, J. D.: "Micropower Circuits," John Wiley & Sons, Inc., New York, 1969.

2 HUDSON, P. H., J. S. KESPERIS, and J. D. MEINDL: Large-value Monolithic Resistors for Micropower Integrated Circuits, *IEEE J. Solid-State Circuits*, vol. SC-7, pp. 160–168, 1972.

3 HUDSON, P. H., and J. D. MEINDL: A Monolithic Micropower Command Receiver, *IEEE J. Solid-State Circuits*, vol. SC-7, pp. 125–135, 1972.

4 GRAY, P. R.: A 15-W Monolithic Power Operational Amplifier, *IEEE J. Solid-State Circuits*, vol. SC-7, pp. 474–480, 1972.

5 LONG, E. L., and T. M. FREDERIKSEN: High-gain 15-watt Monolithic Power Amplifier with Internal Fault Protection, *IEEE J. Solid-State Circuits*, vol. SC-6, pp. 35–45, 1971.

6 KUIJK, K. E.: A Fast Integrated Comparator, *IEEE J. Solid-State Circuits*, vol. SC-8, pp. 458–462, 1973.

7 SALLEN, R. P., and E. L. KEY: A Practical Method of Designing RC Active Filters, *IRE Trans. Circuit Theory*, vol. CT-2, pp. 74–85, 1955.

8 MEYER, C. S., D. K. LYNN, and D. J. HAMILTON: "Analysis and Design of Integrated Circuits," pp. 460–470, McGraw-Hill Book Company, New York, 1968.

9 KERWIN, W. J., L. P. HUELSMAN, and R. W. NEWCOMB: State-variable Synthesis for Insensitive Integrated Circuit Transfer Functions, *IEEE J. Solid-State Circuits*, vol. SC-2, pp. 87–92, 1967.

10 MC 3401P Specifications and Applications Information, Motorola, Inc., 1972.

11 FRIEDMAN, M. F.: An Integrated High-Q Bandpass Filter, *ISSCC Digest Tech. Papers*, pp. 162–163, 1970.

12 GRAY, P. R., and D. J. HAMILTON: Analysis of Electrothermal Integrated Circuits, *IEEE J. Solid-State Circuits*, vol. SC-6, pp. 8–14, 1971.

13 WIDLAR, R. J.: New Developments in IC Voltage Regulators, *IEEE J. Solid-State Circuits*, vol. SC-6, pp. 2–8, 1971.

14 GREBENE, A. B.: "Analog Integrated Circuits," pp. 189–202, Van Nostrand Reinhold Company, New York, 1972.

BASIC SATURATING LOGIC CIRCUITS

12-1 DC BEHAVIOR OF THE SATURATING INVERTER

Several basic bipolar integrated logic circuit families employ a saturating inverter as an integral part of the circuit. It is this inverter which provides the current gain, the logical inversion property enabling the implementation of the NAND or NOR logic function, and the well-determined output voltage when the inverter is saturated. Since the inverter is such a basic part of saturating logic circuits, we first consider some of its general properties.

The Saturated Case

Consider the basic inverter circuit of Fig. 12-1 in which we have shown the parasitic substrate p-n-p transistor to emphasize the importance of the substrate. To determine the saturation voltage $\phi_{CE(\text{sat})}$ for the intrinsic transistor with no bulk resistances, we can make use of the four-layer nonlinear model of Chap. 7. If we define the transistor to be *just-saturated* when $\phi_C = 0$, we see that for this case the

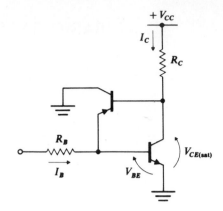

FIGURE 12-1
The basic saturating inverter.

parasitic *p-n-p* has zero emitter-base voltage, and the effect of the *p-n-p* will be negligible. The emitter-base voltage is

$$V_{BE} \approx \phi_E \approx V_D$$

and

$$\phi_{CE(\text{sat})} = \phi_E$$

For these conditions the currents are related by

$$I_C \approx \beta_N I_B$$

If the base current is increased while I_C is held fixed, $\phi_{CE(\text{sat})}$ will decrease because the collector junction becomes forward-biased. Note that the effective current gain of the inverter is now less than β_N. In fact, we have already calculated the effective current gain $\beta_N' = I_C/I_B$, sometimes called the *circuit β* or the *forced β*, in Chap. 7 and noted that β_N' decreases as $\phi_{CE(\text{sat})}$ is decreased. This means that more base drive must be supplied to produce a lower $\phi_{CE(\text{sat})}$ for a given I_C. The result obtained in Chap. 7 is

$$\beta_N' = \frac{\alpha_N(I_{s1}/I_{s2})e^{q\phi_{CE(\text{sat})}/mkT} - (1 - \alpha_s)}{(1 + \alpha_I) + (I_{s1}/I_{s2})(1 - \alpha_N)e^{q\phi_{CE(\text{sat})}/mkT}} \qquad (12\text{-}1)$$

In addition to the junction voltages of the intrinsic transistor, there is a voltage resulting from the bulk collector resistance $r_{C(\text{sat})}$: thus the total saturation voltage is

$$V_{CE(\text{sat})} = \phi_{CE(\text{sat})} + I_C r_{C(\text{sat})}$$

EXAMPLE 12A *Base Drive for a Typical Case.* Suppose a circuit requires a saturating inverter with $V_{CE(\text{sat})} \leqq 0.300$ V with $I_C = 2.0$ mA. The integrated-circuit transistor has

$$\frac{I_{s1}}{I_{s2}} = 0.25 \qquad \alpha_N = 0.98 \qquad \alpha_I = 0.10$$

$$\alpha_s = 0.80 \qquad r_{C(\text{sat})} = 100 \ \Omega$$

Note that the contribution from $r_{C(sat)}$ alone is 200 mV; thus it is necessary that $\phi_{CE(sat)} \leqq 100$ mV. Inserting the above values in (12-1), we find

$$\beta_N' \approx 11$$

For $I_C = 2$ mA, we therefore require

$$I_B \geqq 0.182 \text{ mA}$$

In this case the effective current gain of the transistor is only 11, while the actual active-region current gain is $\beta_N = 49$.

This example shows that logic circuits which require low $V_{CE(sat)}$ will have low effective overall current gain. It also emphasizes the importance of using a buried-layer diffusion for saturating inverters, in order that the $I_C r_{C(sat)}$ term be kept as small as possible. If $r_{C(sat)}$ decreases to 75 Ω, then for $V_{CE(sat)} = 0.300$, we have

$$\beta_N' \approx 31$$

while if $r_{C(sat)}$ increases to 125 Ω, we obtain

$$\beta_N' \approx 1.72 \qquad ////$$

The Cutoff Case

Integrated logic circuits nearly always use only a single power supply voltage; therefore an inverter is seldom truly cut off since negative base-emitter voltage cannot usually be applied. Fortunately, because of the exponential nature of the emitter-base junction characteristic, a reduction of the base-emitter voltage of several hundred millivolts is sufficient to cause a reduction of the collector current by several orders of magnitude. Thus the transistor can be effectively cut off by reducing the forward voltage of the emitter junction.

To quantify this statement, we can use the four-layer model to calculate the collector current when the transistor is in the forward active region. If the collector current is always larger than I_{s2} or I_{s3}, we obtain

$$I_C \approx \alpha_N I_{s1}(e^{q\phi_E/mkT} - 1)$$

Now in the saturating inverter with a load resistor R_C and with $V_{CC} \gg V_D$, the collector current changes very little as operation changes from just-saturated (the edge of the forward active region) to fully saturated; let this current be I_{C1}. Let the collector current when the transistor is designated cut off be $I_{C(off)}$. If the voltages are ϕ_{E1} and $\phi_{E(off)}$ for these two cases, we have

$$I_{C1} \approx \alpha_N I_{s1}(e^{q\phi_{E1}/mkT} - 1)$$

and

$$I_{C(off)} \approx \alpha_N I_{s1}(e^{q\phi_{E(off)}/mkT} - 1)$$

Taking the ratio and solving for $\phi_{E(off)}$, we obtain

$$\phi_{E(off)} = \phi_{E1} - \frac{mkT}{q} \ln \frac{I_{C1}}{I_{C(off)}} \qquad (12\text{-}2)$$

Since the junction currents can be very small, we should be conservative and use the asymptotic value $m = 2$ (see Fig. 7-15).

EXAMPLE 12B *Base-emitter Voltage for a Typical Case.* Suppose that the saturating inverter of Example 12A with $I_{C1} = 2$ mA must be cut off so that the collector current is less than 0.1 percent of I_{C1}. Then $I_{C(off)} = 2$ μA, and the junction voltage at this current is

$$\phi_{E(off)} = \phi_{E1} - \frac{mkT}{q} \ln 1000$$

$$= \phi_{E1} - 360 \text{ mV}$$

To estimate $\phi_{E(off)}$ we recall that ϕ_{E1} will be approximately V_D; thus

$$\phi_{E(off)} \approx 340 \text{ mV}$$

We would therefore design for $\phi_{E(off)} < 340$ mV to ensure cutoff.   ////

12-2 TRANSIENT BEHAVIOR OF THE INVERTER

The transient analysis of even a very simple logic circuit is quite difficult, for the following reasons:

1 The driving waveforms are not simple.
2 Since the inputs come from logic circuits and the outputs drive other logic circuits, nonlinearities exist in both source and load circuits.
3 Since large-signal behavior is involved as the inverter operation moves from cutoff to saturation and vice versa, nonlinearities of junctions and junction capacitances must be taken into account.

As a result, if an accurate analysis of a simple logic circuit is required, use of a computer is advisable; for a complex circuit it is mandatory. Large-scale computer analysis programs which take into account the nonlinearities encountered are available.

For an understanding of circuit behavior, an accurate, detailed analysis is not required; in fact it is generally not desired, for the details often obscure the salient relationship between the important aspects of the transient behavior and the circuit and device parameters. For this purpose one does not require the detailed waveshapes, but rather an estimate of such important aspects as delay time, rise and fall times, and storage times. To obtain such an estimate, we use the following guidelines for our analysis.

1 A fairly general inverter circuit is considered; the results can then be adapted to any of the saturating logic families by use of proper values for the parameters.
2 Driving waveforms are idealized so that they can be approximated by step functions.
3 Nonlinear capacitors are replaced by linear capacitors whose values are the averages of the nonlinearities over the voltage excursions present.
4 Piecewise-linear methods of analysis are employed.

5   Junction nonlinearities are replaced by ideal diodes, and first-order models are used for the transistor.

6   First-order system approximations are used wherever possible in order to simplify the analysis.

The circuit to be analyzed is shown in Fig. 12-2a.   In this circuit, R_B and v represent the Thévenin equivalent of the driving circuit, and R_L and V_L the Thévenin equivalent of the load circuit.   Both these equivalents may have to be piecewise-linear approximations of the actual source and load, depending upon the particular logic circuit. C_1 is the emitter junction capacitance, together with any additional capacitance between base and ground resulting from the particular circuit configuration; C_2 is the collector junction capacitance plus additional circuit capacitance between collector and base; and C_3 is the substrate capacitance plus any circuit capacitance between collector and ground.

The idealized output waveform is shown in Fig. 12-2b.   It is assumed that a turn-on signal is applied at $t = 0$; the delay time t_D is the time elapsed until the output begins to change, and the transient time t' is the time from the end of the delay until the transistor enters the saturation region.   (For actual waveforms, no clear-cut demarcation exists as in the idealized waveforms.   Therefore it is often more desirable to define the delay as the time required to reach the 50 percent point of the excursion, while the transient can be defined as the time required for a 90 to 10 percent excursion.)

When a turnoff signal is applied, a storage time t_s elapses before the transistor enters the active region; this is followed by a transient time of t'' for 90 percent of the excursion to be completed.   The total turn-on time is

$$t_{on} = t_D + t'$$

while the total turnoff time is

$$t_{off} = t_s + t''$$

The average propagation delay can be defined as

$$\bar{t}_D \triangleq \frac{t_{on} + t_{off}}{2}$$

It is important to note that during the times t_D, t', and t'', the transient behavior will probably be determined primarily by parasitic capacitances rather than by the intrinsic frequency limitations of the transistor.   Most integrated transistors have sufficiently high cutoff frequencies that parasitics are dominant.   This is not the case in the saturation region, however, since here there is very little change of voltage and hence very little capacitive current.   Thus the storage time t_s will be determined basically by device performance.

The Delay Time t_D

To calculate the delay time t_D we assume the driving waveform v to be a step function between the limits V_1 and V_2 as shown in Fig. 12-3a; it is necessary that V_1 be

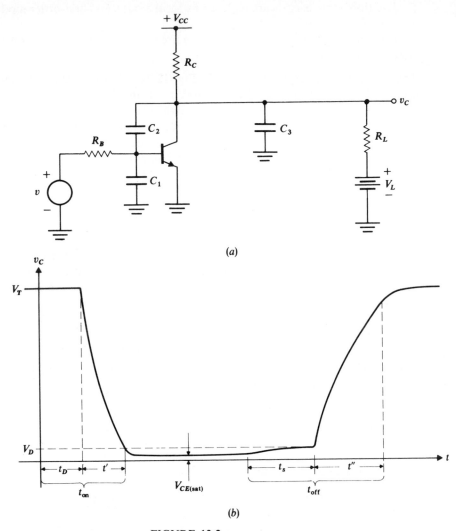

(a)

(b)

FIGURE 12-2
(a) The general inverter circuit; (b) idealized output waveform.

$\phi_{E(\text{off})}$ or less.   The first-order model for the transistor is shown in Fig. 12-3b; note that until $t = t_D$, $v_B < V_D$ and $i_B = 0$.   Thus the equivalent circuit for the inverter is as shown in Fig. 12-3c.   This is a second-order circuit; however, it becomes first-order under two asymptotic conditions: $R_T \to 0$ and $R_T \to \infty$.   The case $R_T \to 0$ will overestimate t_D, and we therefore use it as an approximation.   For this case the time constant T is

$$T = R_B(C_1 + C_2)$$

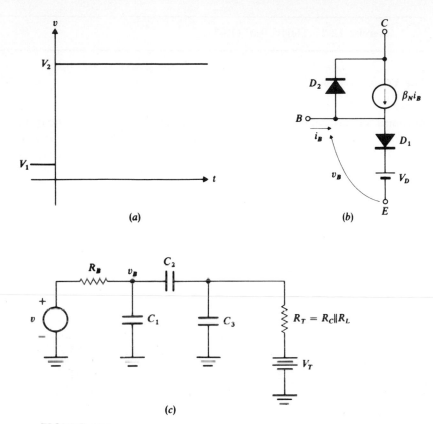

FIGURE 12-3
(a) Input waveform; (b) first-order transistor model; (c) inverter equivalent circuit.

We next write an expression for $v_B(t)$, and set

$$v_B(t_D) = V_D$$

Solving for t_D we obtain

$$t_D = T \ln \frac{V_1 - V_2}{V_D - V_2} \qquad (12\text{-}3)$$

EXAMPLE 12C *Calculation of the Delay Time.* Consider a case for which $R_B = 4\,\text{k}\Omega$, $C_1 = C_2 = 3$ pF, and input voltage limits are $V_1 = V_D/2$ and $V_2 = 6V_D$. For this case $T = 24 \times 10^{-9}$ s. Inserting the values given in (12.3), we obtain

$$t_D = T \ln \frac{5.5}{5} = 2.3 \times 10^{-9} \text{ s} \qquad ////$$

Exercise 12-1   Derive Eq. (12-3).

The Transient Time t'

When the base voltage has reached V_D, the diode D_1 in the model of Fig. 12-3b becomes forward-biased and no further change of v_B occurs.   The transistor is in the forward active region and the equivalent circuit is shown in Fig. 12-4a.   To simplify the circuit, we note that no current flows in C_1, but that C_2 is multiplied by Miller effect as far as v_C is concerned.   We also form the Thévenin equivalent of V_{CC}, R_{C1}, V_L, and R_L to obtain V_T and R_T; the resulting equivalent circuit is shown in Fig. 12-4b.   In accordance with the approximation that parasitic capacitances dominate the transient behavior, no frequency dependence is associated with the dependent source.

Since this is a first-order circuit, we can easily solve for the time t' elapsed as v_C changes from V_T to V_D; the result is

$$t' = R_T C' \ln \frac{\beta_N I_B R_T}{V_D - V_T + \beta_N I_B R_T}$$

$$= R_T C' \ln \frac{\beta_N (R_T/R_B)(V_2 - V_D)}{(V_D - V_T) + \beta_N (R_T/R_B)(V_2 - V_D)} \qquad (12\text{-}4)$$

If the load circuit changes during the excursion of v_C, then piecewise-linear methods must be used, and the problem must be treated as a composite of several parts in which V_T and R_T take on different values in each part.

EXAMPLE 12D   *Calculation of t'.*   Consider a hypothetical case for which

$$C_1 = C_2 = 3\,\text{pF} \qquad C_3 = 10\,\text{pF}$$
$$R_T = 2\,\text{k}\Omega \qquad V_T = 6V_D \qquad \beta_N = 49$$
$$V_2 = 6V_D \qquad R_B = 4\,\text{k}\Omega$$

We first calculate $C' = C_3 + C_2(\beta_N + 1) = 160\,\text{pF}$.   Then we obtain $R_T C' = 320 \times 10^{-9}$ s and

$$t' = 320 \times 10^{-9} \ln \frac{24.5}{23.5} = 13.3 \times 10^{-9}\,\text{s}$$

Since this is the same device as was used in Example 12C, we can use the result of that example to calculate the total turn on time t_{on}

$$t_{\text{on}} = t' + t_D = 15.6 \times 10^{-9}\,\text{s} \qquad ////$$

Exercise 12-2   Derive Eq. (12-4).

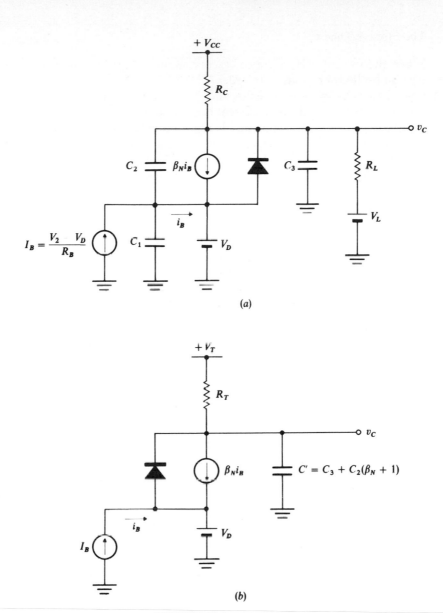

(a)

(b)

FIGURE 12-4
(a) Circuit for the calculation of t'; (b) equivalent circuit.

The Storage Time t_s

When the transistor is saturated, excess minority carrier charge is stored in the base and collector regions. In order for the transistor to reenter the forward active region, this charge must be removed. If a turnoff signal is applied to the base as shown in Fig. 12-5a, the base current is reduced to approximately zero if $V_3 \approx V_D$. The excess base charge in this case is removed principally by recombination.

   To obtain a first-order estimate of the storage time, we use the transistor model of Fig. 12-5b, in which $1/\omega_s$ is the time constant associated with the removal of excess base charge. Since negligible voltage changes occur at the transistor terminals during saturation, the parasitic capacitances have no effect, and the intrinsic device dominates the behavior. The equivalent circuit for the inverter is shown in Fig. 12-5c. We wish to calculate the time t_s elapsed until the diode current i_D is zero. At this time the transistor enters the forward active region.

   The circuit of Fig. 12-5c is a first-order system with step-function excitation; we can therefore write the current i_D as

$$i_D(t) = i_D(\infty) + [i_D(0) - i_D(\infty)]e^{-\omega_s t}$$

where $i_D(0)$ is the initial value and $i_D(\infty)$ is the *virtual* final value. Once these two constants are evaluated, we can set $i_D(t_s) = 0$ and solve for t_s. By inspection of the circuit we see that

$$i_D(0) = \frac{\beta_N(V_2 - V_D)}{R_B} - \frac{V_T - V_D}{R_T}$$

and that

$$i_D(\infty) = \frac{\beta_N(V_3 - V_D)}{R_B} - \frac{V_T - V_D}{R_T}$$

Note that if the transistor is to enter the forward active region, V_3 must be small enough that

$$i_D(\infty) < 0$$

Using these values and setting $i_D(t_s) = 0$, we obtain

$$t_s = \frac{1}{\omega_s} \ln \frac{\beta_N(V_2 - V_3)/R_B}{(V_T - V_D)/R_T - \beta_N(V_3 - V_D)/R_B} \tag{12-5}$$

EXAMPLE 12E *Calculation of t_s.* Consider the same device which was used in Example 12D and let

$$\omega_s = 5 \times 10^7 \qquad R_B = 4\,k\Omega \qquad R_T = 2\,k\Omega \qquad V_2 = 6V_D \qquad V_3 = V_D$$

For these values we obtain

$$t_s = 20 \times 10^{-9} \ln 24.5 \approx 64 \times 10^{-9} \text{ s}$$

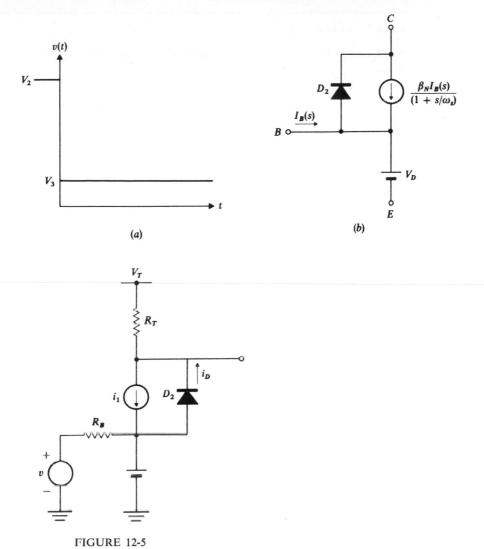

FIGURE 12-5
(a) Turnoff input waveform; (b) model for the transistor in saturation; (c) equivalent circuit for the inverter during t_s.

The value of ω_s chosen is typical of devices with $f_T = 500$ MHz and $\beta_N = 50$. Note that for the value of V_3 used, the base current is zero during the storage time interval. This example clearly illustrates that logic circuits which employ a saturating inverter can be expected to exhibit long storage times. To reduce the storage time it is necessary to provide some means for producing reverse base current during the storage time interval. ////

Exercise 12-3   Derive Eq. (12-5).

Exercise 12-4   How much improvement in storage time is obtained if $V_3 = V_D/2$?

The Transient Time t''

When, after t_s, operation enters the forward active region, we again assume that parasitic effects are dominant and that the effects of intrinsic cutoff frequency are negligible.   Two cases will be considered here: the case for which $V_3 = V_D$ and the case for which $V_3 < V_D$.   In the former, the base current applied by the driving circuit is zero, and operation asymptotically approaches the cutoff condition.   In the latter, the driving circuit causes reverse base current to flow and operation reaches the cutoff condition at approximately the same time that v_C reaches V_T.

The equivalent circuit for $V_3 = V_D$ is shown in Fig. 12-6a.   At the end of the storage time interval, $v_C = V_D$.   After that, v_C is given by

$$v_C(t) = V_T + (V_D - V_T)e^{-t/R_T[C_3 + C_2(\beta_N + 1)]}$$

If we define t'' to be the time required to complete 90 percent of the total excursion, then

$$v_C(t'') = V_T - 0.1(V_T - V_D)$$

Solving for t'' we obtain

$$t'' = 2.3R_T[C_3 + C_2(\beta_N + 1)] \qquad (12\text{-}6)$$

If $V_3 < V_D$, we must first calculate the time t'' elapsed until $i_E = 0$.   Note that the initial value of i_E is

$$i_E(0) = \frac{V_T - V_D}{R_T} + \frac{V_3 - V_D}{R_B}$$

and the virtual final value is

$$i_E(\infty) = \frac{(V_3 - V_D)(\beta_N + 1)}{R_B}$$

When $i_E = 0$, t'' is

$$t'' = R_T C' \ln \frac{(V_T - V_D)/R_T - \beta_N(V_3 - V_D)/R_B}{-(V_3 - V_D)(\beta_N + 1)/R_B} \qquad (12\text{-}7)$$

where $C' = C_3 + C_2(\beta_N + 1)$.

The collector voltage at this time is approximately V_T.

EXAMPLE 12-F   *Calculation of t''*.   We again consider the case for which:

$$\beta_N = 49 \qquad V_T = 6V_D \qquad R_B = 4 \text{ k}\Omega \qquad R_T = 2 \text{ k}\Omega$$
$$C_1 = C_2 = 3 \text{ pF} \qquad C_3 = 10 \text{ pF}$$

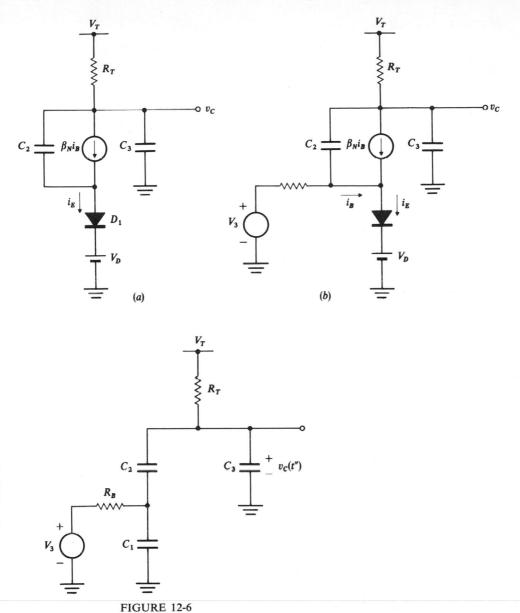

FIGURE 12-6
(a) Equivalent circuit for $V_3 = V_D$; (b) equivalent circuit for $V_3 < V_D$; (c) equivalent circuit for cutoff operation.

If $V_3 = V_D$, we find

$$t'' = 2.3 \times 320 \times 10^{-9} = 736 \times 10^{-9} \text{ s}$$

If $V_3 = V_D/2$, we obtain

$$t'' = 320 \times 10^{-9} \ln \frac{34.5}{25} = 103 \times 10^{-9} \text{ s}$$

The total turnoff time is then

$$t_{\text{off}} = t_s + t'' = 167 \times 10^{-9} \text{ s} \qquad \qquad ////$$

Although our transient analysis of the saturating inverter is crude because of the many simplifying approximations involved, it can be expected to yield order-of-magnitude results. Moreover, it indicates that the average propagation delay of the inverter is dominated by the turnoff effects unless significant reverse base drive or a nonlinear load is provided. Thus logic circuits using saturating inverters in which no provision is made for reverse base drive or nonlinear load can be expected to be slow.

12-3 LOGIC OPERATIONS[1]

Logic systems generally deal with binary variables and binary functions, that is, variables and functions which can have only two possible values. The most common logic operations performed with binary variables are complement, or NOT, AND, OR, NAND, NOR; these operations are defined in Table 12-1, in which X_1, X_2, X_3, etc., are binary variables which may have the values 1 or 0. The function F is a binary function of the binary variables; that is, F may also have only the values 1 or 0. Although Table 12-1 shows the above operations for only three variables, it is clear that the definitions can be extended to n variables.

Table 12-1 DEFINITION OF LOGIC OPERATIONS

Operation	Expression	Definition
Complement or NOT	$\bar{X}$ = complement of X	$\bar{X} = 0$ when $X = 1$ $\bar{X} = 1$ when $X = 0$
AND	$F = X_1 X_2 X_3$ or $F = X_1 \cdot X_2 \cdot X_3$	$F = 1$ for X_1, X_2, X_3 all 1 $F = 0$ otherwise
OR	$F = X_1 + X_2 + X_3$	$F = 0$ for X_1, X_2, X_3 all 0 $F = 1$ otherwise
NAND	$F = \overline{X_1 X_2 X_3}$	$F = 0$ for X_1, X_2, X_3 all 1 $F = 1$ otherwise
NOR	$F = \overline{X_1 + X_2 + X_3}$	$F = 1$ for X_1, X_2, X_3 all 0 $F = 0$ otherwise

Since binary variables and functions have only two possible values, for a function involving n variables there are only 2^n possible combinations of these variables. A binary function can thus be specified by tabulating the value of the function for each combination of variables. Such a specification is called a *truth table*; truth tables for AND, OR, NAND, and NOR functions of two variables are given in Table 12-2.

Exercise 12-5 Make up truth tables for AND, OR, NAND, NOR for three binary variables.

Logic functions involving AND, OR, complement, etc., can be manipulated according to the rules of Boolean algebra. Briefly, Boolean algebra has the same associative, distributive, and commutative properties as ordinary algebra. Thus, for example, the function

$$F = (X_1 + X_2)(X_1 + \bar{X}_2)$$

can also be written

$$F = X_1 X_1 + \bar{X}_1 X_2 + X_1 \bar{X}_2 + X_2 \bar{X}_2$$

But by using the definition of the AND operation, we see that

$$X_1 \bar{X}_1 = 0 \quad \text{and} \quad X_2 \bar{X}_2 = 0$$

The function F is therefore

$$F = X_1 \bar{X}_2 + \bar{X}_1 X_2$$

This particular function is called the EXCLUSIVE OR function.

Table 12-1 defines the complement of the variable X; a function F also has a complement. To find the complement of a function in terms of the variables and their complements, we use DeMorgan's theorem:

Given a function $F(X_1, X_2, X_3, \ldots, X_n, +, \cdot)$, the complement is $\bar{F} = F(\bar{X}_1, \bar{X}_2, \bar{X}_3, \ldots, \bar{X}_n, \cdot, +)$

Table 12-2 TRUTH TABLES FOR SEVERAL FUNCTIONS OF TWO VARIABLES

X_1	X_2	$X_1 X_2$	$X_1 + X_2$	$\overline{X_1 X_2}$	$\overline{X_1 + X_2}$
0	0	0	0	1	1
0	1	0	1	1	0
1	0	0	1	1	0
1	1	1	1	0	0

EXAMPLE 12G   *Complement of the Function* $F = X_1\overline{X}_2 + X_2\overline{X}_1$.   To find the complement replace all variables by their complements, all AND operations by OR, and all OR operations by AND.   We then obtain

$$\overline{F} = (\overline{X}_1 + X_2)(\overline{X}_2 + X_1) = \overline{X}_1\overline{X}_2 + X_2\overline{X}_2 + X_2X_1 + \overline{X}_1X_1 = \overline{X}_1\overline{X}_2 + X_2X_1$$

//////

Exercise 12-6   Make up a truth table for $F = X_1\overline{X}_2 + X_2\overline{X}_1$.

Exercise 12-7   Use DeMorgan's theorem and the properties of Boolean algebra to express the function $F = (X_1 + X_2)(\overline{X_1X_2})$ in a different way.

Our discussion thus far has dealt with binary variables only as abstract quantities.   In a logic system, these variables are represented by electrical quantities, the most common being voltage.   Two voltage levels are selected, v_1 representing the binary value 1 and v_0 the binary value 0.   This representation is defined as *positive logic* if $v_1 > v_0$, and *negative logic* if $v_0 > v_1$; in this book we shall usually deal with positive logic.   In addition to v_1 and v_0, a third quantity is often of interest in characterizing logic circuits; it is the *logic swing* v_l, defined by

$$v_l \triangleq |v_1 - v_0|$$

In the design of logic systems, the actual logic circuits themselves do not have to be known, provided the functions they realize are known, and provided it is known that all circuits are electrically compatible, that is, v_1 and v_0 are the same for all circuits.   The logic circuits can be represented symbolically, as shown in Fig. 12-7a, and the logic system can be drawn as the interconnection of the proper symbols.   The function $F = X_1\overline{X}_2 + X_2\overline{X}_1$, for example, can be represented as shown in Fig. 12-7b.

We have thus far discussed logic functions which involve the operations AND, OR, and NOT.   It can be shown that any logic function can also be realized with only the NAND operation, or with only the NOR operation.   This result has great significance in terms of logic circuit design, because it means that, in principle, for any logic system only one type of logic circuit need be designed.   Such a situation is ideally suited to the large-volume mass-production aspects of integrated-circuit fabrication.   Moreover, in terms of circuit design, it means that if only one type of circuit is used, what is driving the inputs and what the outputs must drive are always known; only the number of inputs and the number of circuits connected to the output are dependent upon the particular logic function being realized.   Also, if only one type of circuit is used, the output levels must be compatible with the input levels.   It is therefore convenient to characterize the circuit performance in terms of the number m of inputs it can accept and the number n of identical circuits which it can drive at its output.   The number m is called the fan-in number, and n the fan-out number.   Clearly a logic circuit must be designed so that it has m and n of 2 or greater; otherwise it will be of no use in a logic system.

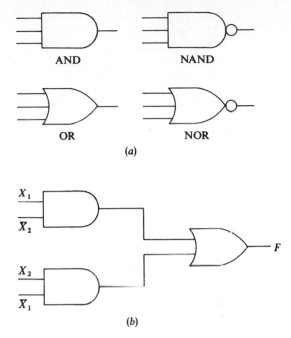

(a)

(b)

FIGURE 12-7
(a) Symbolic representation of logic circuit; (b) the function $F = x_1 \bar{x}_2 + x_2 \bar{x}_1$.

EXAMPLE 12H *Realization of a Logic Function by Only NOR Circuits.* To see how a single type of logic circuit can be used, we consider the implementation of the NOT and the AND operations by NOR circuits. The NOT operation is easily obtained by using a two-input NOR circuit and connecting the inputs together as shown in Fig. 12-8a. To implement $F = X_1 X_2$ we use 3 two-input NOR circuits as shown in Fig. 12-8b. ////

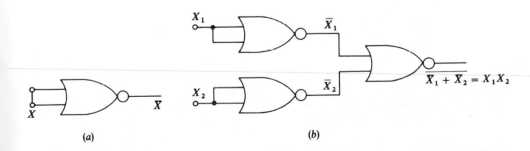

(a)

(b)

FIGURE 12-8
(a) Use of a NOR circuit to perform NOT; (b) use of NOR circuits to perform AND.

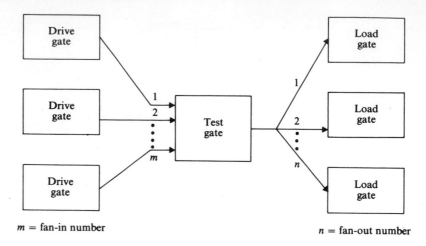

m = fan-in number   n = fan-out number

FIGURE 12-9
Method of characterizing gate performance.

Exercise 12-8   Show how NOR can be implemented with only NAND circuits.

Exercise 12-9   Show how the function $F = X_1 \overline{X}_2 + X_2 \overline{X}_1$ can be implemented (*a*) with NOR circuits and (*b*) with NAND circuits.

Since all logic circuits, or gates, in a system can be of the same type, it is convenient for circuit design and for measurement of performance to focus attention on a test gate whose inputs are supplied by m drive gates and whose output is connected to n load gates.   This is shown in Fig. 12-9.

12-4   DIODE-TRANSISTOR LOGIC (DTL)[2]

Perhaps the most straightforward way to obtain a NAND circuit is to use a diode-resistor AND gate followed by a saturating inverter.   If all the electrical compatibility requirements can be satisfied, such a circuit would perform the NAND operation.   This method is shown in block-diagram form in Fig. 12-10*a*, and is called diode-transistor logic, or DTL.   The basic circuit configuration is shown in Fig. 12-10*b*.   Diodes D_1, D_2, D_3, ..., D_m form the diode-resistor AND gate. The output of this gate is connected to the saturating inverter Q_2 by two diodes D_{01} and D_{02} whose only function is to provide offset voltage to ensure compatibility between the output of the diode-resistor gate and the base input of Q_2.

Case 1   *All drive-gate inverters are cut off.*   For this case, all of the input diodes will be reverse-biased, and D_{01} and D_{02} will be forward-biased.   If R_L is sufficiently small, I_B will be large enough to saturate Q_2, and the output will be

$$v_{C2} = V_{CE(\text{sat})}$$

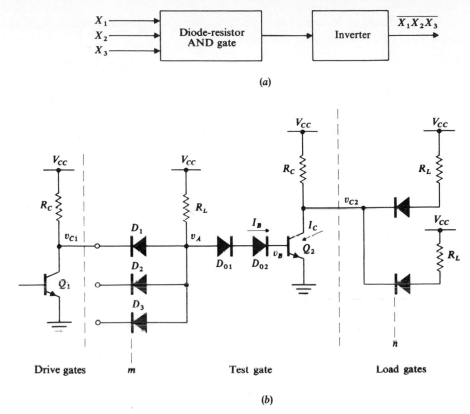

FIGURE 12-10
(a) Block diagram of DTL; (b) basic DTL circuit.

For this condition note that all inputs are V_{CC}. Thus in this case we have

$$v_1 = V_{CC}$$

$$v_0 = V_{CE(\text{sat})}$$

Note that $V_A = 3V_D$, so it is necessary that

$$V_{CC} > 3V_D$$

Case 2 Q_1 *is saturated.* For this case $v_{C1} = V_{CE(\text{sat})}$, and $v_A = V_{CE(\text{sat})} + V_D$. If D_{01}, D_{02}, and Q_2 all have identical emitter junctions, they will all have equal voltages, and

$$v_B = \frac{V_A}{3}$$

But as we saw in our analysis of the saturating inverter, a base-emitter voltage of

$V_D/3$ is insufficient to produce any collector current, so Q_2 is effectively cut off and $v_{C2} = V_{CC}$. For this case we thus have

$$v_0 = V_{CE(sat)}$$
$$v_1 = V_{CC}$$

Thus the input and output are compatible for both cases, and the circuit functions properly as a NAND circuit.

It is important to note that even if Q_1 is only just-saturated, $v_A = 2V_D$ and $v_B = \frac{2}{3}V_D$. This may be a sufficiently low v_B to keep Q_2 cut off. The logic swing for the circuit is

$$v_l = V_{CC} - V_{CE(sat)}$$

Design Considerations

If we assume Q_2 to be saturated, then

$$I_B = \frac{V_{CC} - 3V_D}{R_L}$$

and

$$I_C = \frac{V_{CC} - V_{CE(sat)}}{R_C} + \frac{n[V_{CC} - V_D - V_{CE(sat)}]}{R_L}$$

Since Q_2 can be operated just-saturated,

$$V_{CE(sat)} = V_D$$

and the base current required is

$$I_B = \frac{I_C}{\beta_N}$$

Note that the transistor β_N, and not the forced β'_N, can be used because the transistor is just-saturated. Combining these results we obtain

$$\frac{R_L}{R_C} = \frac{\beta_N(V_{CC} - 3V_D) - n(V_{CC} - 2V_D)}{V_{CC} - V_D} \qquad (12\text{-}8)$$

If it is known that the circuit drives only identical gates, then R_C is not really necessary, since collector current can flow through the R_L of each of the n outputs. If we let $R_C \to \infty$ in (12-8) we can solve for n, which will be the maximum fan-out number that could be expected from this circuit. The result is

$$n_{max} = \frac{\beta_N(V_{CC} - 3V_D)}{V_{CC} - 2V_D}$$

Equation (12-8) is not sufficient to determine R_L and R_C; for this other criteria must be involved. As R_L and R_C are made smaller, parasitic capacitances can be charged and discharged faster, but the power consumed by the circuit increases. Thus one needs to invoke some constraints regarding speed of operation or power dissipation in order to complete the design. Note that as the fan-out number is increased, the parasitic capacitance at the output increases. It is therefore unlikely that a practical circuit will have a fan-out capability approaching n_{max}, since speed requirements will usually limit n to a much lower value. It is generally true for saturating logic circuits that speed, rather than dc considerations, determines n_{max}.

Power Dissipation

The maximum power dissipation in the gate occurs when Q_2 is saturated. It is given by

$$P_{D(max)} = \frac{V_{CC}(V_{CC} - 3V_D)}{R_L} + \frac{(V_{CC} - V_D)V_{CC}}{R_C} = I_C V_{CC} + I_B(sat) V_{CC}$$

Layout Considerations

For the input diodes, we need fast recovery time and low capacitance to the substrate since the substrate capacitance is connected to node A of the diode cluster. We therefore choose type a diodes (collector-base short).

The offset diodes should have low substrate capacitance but their storage time is not critical; in fact, long storage time is desirable. Type d diodes (collector open) are used. If space is at a premium and speed is not a primary concern, one type b diode and one type a diode could be used, since a single isolation region would suffice for the two. All the input diodes can share a common isolation region since they have common collectors. The number of isolation regions required is then

Resistors:	1
Input diodes:	1
Q_2:	1
Offset diodes:	2
Total:	5

Since the input diodes all have common collectors and collector-base shorts, they can actually be the emitters of a multiple-emitter transistor with collector-base short, as shown in Fig. 12-11a. Parasitic capacitances for the circuit are shown in Fig. 12-11b. Once the layout dimensions are known, the values for the capacitors can be calculated and the circuit can be reduced to approximately the form shown in Fig. 12-2a. The transient analysis of Sec. 12-2 can then be used to estimate the various times involved.

Modifications of the Basic DTL Circuit

The speed of the DTL circuit can be increased by decreasing R_L and R_C to provide more current to charge parasitic capacitances and more base drive for the inverter. As R_L is decreased, of course, the drive current required from the drive gates is also

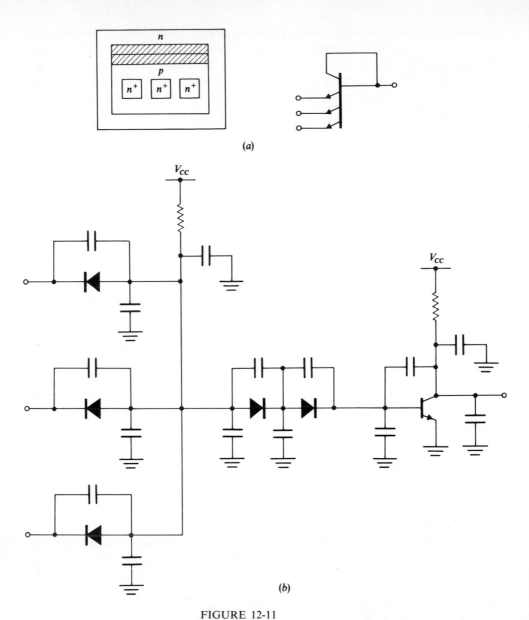

FIGURE 12-11
(*a*) Multiple-emitter input transistor; (*b*) parasitic capacitances.

increased. This disadvantage can be overcome by taking advantage of the fact that D_{01} is a diode-connected transistor. If D_{01} is used as a transistor in which the emitter junction provides the required offset, increased base current for the inverter can be obtained without an increase of the required current from the drive gates. This is accomplished in the MC 961 circuit, as shown in Fig. 12-12a. Here the emitter junction of Q_D provides the necessary offset voltage. When all inputs are in the high state, causing all input diodes to be reverse-biased, Q_D is in the forward active region and the current I_E is

$$I_E = \frac{(1/\alpha_N)(V_{CC} - 3V_D)}{R_1[1 + (1/\beta_N)(1 + R_2/R_1)]} \approx \frac{V_{CC} - 3V_D}{R_1} \qquad (12\text{-}9)$$

If any input is low, Q_D is cut off, and the maximum current the drive gate must absorb is

$$I_D = \frac{V_{CC} - V_D - V_{CE(sat)}}{R_1 + R_2}$$

Thus we see that the effective R_L as far as base drive is concerned is $R_L = R_1$, while the effective R_L seen by the drive gates is $R_L = R_1 + R_2$. The resistor R_3 is added to ensure better cutoff of Q_2.

When a large fan-out number is employed, the parasitic capacitive loading effects on the inverter become large and the turnoff speed is drastically reduced. If R_C is reduced to provide more current for the parasitic capacitances, the power dissipation becomes prohibitive. What is needed is a nonlinear load resistor which would provide large current for turnoff, but would have no current when the inverter is saturated. This is done in the MC 932 circuit as shown in Fig. 12-12b.

Diode D_{02} is used as transistor Q_3 with its emitter junction providing the required offset voltage. Its collector drives emitter follower Q_4. When all inputs are in the high state, Q_D and Q_3 are both on, and Q_3 saturates, as does Q_2. Resistors R_6 and R_7 produce sufficient voltage at the emitter of Q_4 to keep Q_4 cut off. Thus the effective collector resistor for Q_2 is

$$R_C = R_6 + R_7$$

The only increase of power dissipation results from R_5.

Now when any input changes to the low state, Q_D, Q_3, and Q_2 all are cut off, but Q_4 is on. A large transient current is now available at the output to charge the capacitance; but when the output rises to V_{CC}, Q_4 is again cut off. Thus Q_4 is only on during the turnoff transient. This use of a transistor to provide large transient current is called an *active pull-up* technique.

In the circuit of Fig. 12-12b, R_3 is purposely made small in order to provide significant reverse base current to reduce the storage time of Q_2. To ensure cutoff of Q_3, R_4 is added.

Typical values for power dissipation and average propagation delay for the MC 932 are 43 mW and 35 ns, respectively.

Exercise 12-10 Derive Eq. (12-9).

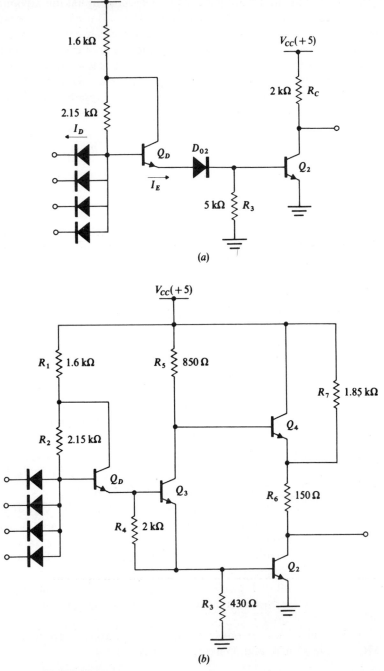

FIGURE 12-12
(a) The MC 961 DTL circuit; (b) the MC 932 circuit with active pull-up.

12-5 TRANSISTOR-TRANSISTOR LOGIC (TTL)

Perhaps the most popular saturating logic circuit is transistor-transistor logic (TTL) because of its high speed and low power dissipation. The basic TTL circuit is easily evolved from the DTL circuit using a multiple-emitter input transistor as shown in Fig. 12-13a. In the DTL circuit, the offset diodes adversely affect the transient performance because they contribute parasitic capacitance and they prevent reverse base drive from flowing in the inverter to reduce storage time. Inspecting the circuit of Fig. 12-13a, we see that an offset voltage is available which is not being used. If the collector-base short is removed from Q_D, and if Q_D can be made to saturate, the forward-biased collector junction would itself provide an offset voltage V_D. This would eliminate one diode. Let us now remove the other diode as well, and see if the circuit will function properly as a logic circuit with compatible input and output. The basic TTL circuit which results is shown in Fig. 12-13b.

Case 1 *One or more inputs are in the low state.* If one or more drive-gate inverters are saturated with $v_{C1} = V_{CE(sat)}$, the emitter of Q_2 connected to that inverter will be forward-biased. This will result in a base current

$$I_B = \frac{V_{CC} - V_{CE(sat)} - V_D}{R_B}$$

which will tend to make a collector current I_{B3} flow. But note that because the collector of Q_2 is connected to the base of Q_3, I_{B3} cannot be negative since the base current of Q_3 would have to be reverse base current. Therefore I_{B3} must be either approximately zero or positive. If it is zero, Q_3 must be cut off. To see if this is possible, we investigate the voltages in the circuit. If $I_{B3} = 0$, then Q_2 is heavily saturated, with its collector junction forward-biased by approximately V_D. Then V_{B3} is given by

$$V_{B3} = V_{CE(sat)} + V_D - V_D = V_{CE(sat)}$$

But if Q_1 is saturated to make $V_{CE(sat)}$ sufficiently less than V_D, Q_3 will be cut off.
It is important to note that Q_1 cannot be operated in the just-saturated condition; if it were, v_{B3} would be V_D, which would cause Q_3 to turn on rather than be cut off. Thus the TTL gate must be designed to provide sufficient base drive to the inverter to ensure a low $V_{CE(sat)}$. The logic 0 state is $v_0 = V_{CE(sat)}$.

Case 2 *All inverters are cut off.* With all inverters cut off, we assume that all emitters of Q_2 are reverse-biased. Now the collector junction of Q_2 will be forward-biased, as will the emitter junction of Q_3. Then $v_{B3} = V_D$ and $v_{B2} = 2V_D$. It is important to note that Q_2 is now operating in the *inverse-active* region. This means that

$$I_{B3} = I_B(1 + \beta_I)$$

and

$$I_B = \frac{V_{CC} - v_{B2}}{R_B} = \frac{V_{CC} - 2V_D}{R_B}$$

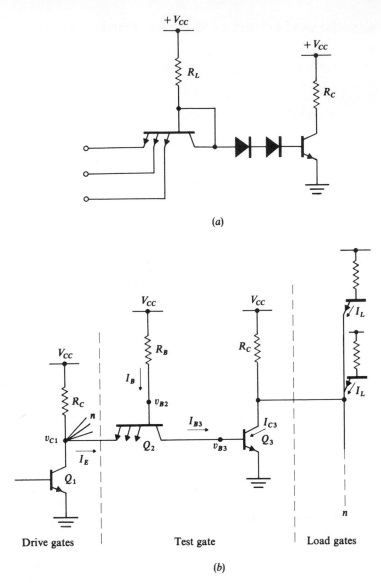

FIGURE 12-13
(a) DTL circuit with multiple-emitter input transistor; (b) the basic TTL circuit.

where β_I is the inverse active region β of Q_2. Note that, in contrast to DTL, when the emitters of Q_2 are all reverse-biased their currents are not zero. Rather, since Q_2 is in the inverse active region, a current I_E will flow in each emitter, in the direction shown in Fig. 12-13b. If there are m inputs, the current in each is given by

$$I_E = \frac{I_B \beta_I}{m} = \frac{(V_{CC} - 2V_D)\beta_I}{mR_B}$$

This means that R_C must be chosen small enough that

$$V_{CC} - nI_E R_C \geqq 2V_D$$

in order to ensure that the emitters of Q_2 are in fact reverse-biased. The logic 1 voltage thus depends upon R_C and n and is given by

$$V_{CC} > v_1 > 2V_D$$

It is important to note that for both cases 1 and 2 the collector junction of Q_2 is forward-biased. This means that storage time of Q_2 is not a consideration. It also means that during transient conditions when one or more inputs change from v_1 to v_0, reverse base current can flow from Q_3 through Q_2 to the drive gate. This will materially reduce the storage time of Q_3.

Effects of the Parasitic *P-N-P*

We have already seen how to include the effects of the parasitic *p-n-p* in the analysis of the inverter. In the TTL gate, the input transistor has a parasitic *p-n-p* which must also be taken into account; this is shown in Fig. 12-14. In the DTL circuit, Q_p had negligible effect because its emitter junction was shorted. This is not the case in the TTL circuit; note that the emitter of Q_p is now forward-biased, and its collector is reverse-biased, and it is thus in the forward active region. Some of the current I_B is now diverted through Q_p to the substrate, thus reducing I_{B3}.

We can find I_{B3} by using the four-layer model for Q_2. First we consider all emitters to have the same effect as a single emitter and we assume it to be reverse-biased. Then in the four-layer model

$$I_{s1}(e^{q\phi_F/mkT} - 1) \approx -I_{s1}$$

Next we see that the substrate junction voltage is $\phi_s = -V_{B3} \approx -V_D$, and therefore

$$I_{s3}(e^{q\phi_s/mkT} - 1) \approx -I_{s3}$$

Using these results in the four-layer-model equations, we get

$$I_2 + I_3 = I_{B3} = \alpha_N I_{s1} + (1 - \alpha_S)\lambda_C - I_{s3}(1 - \alpha_{SI})$$

$$I_1 + I_2 = I_B = -I_{s1}(1 - \alpha_N) + (1 - \alpha_I)\lambda_C + \alpha_{SI} I_{s3}$$

where $\lambda_C = I_{s2}(e^{q\phi_C/mkT} - 1)$

$\lambda_E = I_{s1}(e^{q\phi_E/mkT} - 1)$

Next we assume that I_{B3} and I_B are much larger than I_{s1} and I_{s3}, and terms involving I_{s1} and I_{s3} are neglected. Solving for I_{B3}, we obtain

$$I_{B3} \approx I_B(\beta_I + 1)(1 - \alpha_S)$$

Thus we see that I_{B3} is reduced by a factor $1 - \alpha_S$ from the value it would have in the discrete circuit. Since typical integrated devices have very low β_I but significant α_S, I_{B3} will generally be less than I_B.

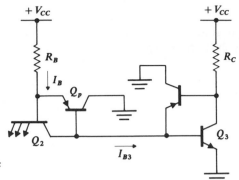

FIGURE 12-14
Parasitic *p-n-p* transistors in the basic
TTL gate.

DC Fan-out of TTL

With the information now at hand, we can calculate the dc fan-out of the basic TTL gate by considering the two conditions of the inverter transistor Q_3. When Q_3 is saturated, its collector current is

$$I_{C3} = [V_{CC} - V_{CE(\text{sat})}]G_C + nI_L$$

The load gate currents I_L will be largest when each of the n-load gates has only the emitter connected to Q_3 in the low state. Then

$$I_L \approx [V_{CC} - V_{CE(\text{sat})} - V_D]G_B$$

The bias current I_B is

$$I_B = (V_{CC} - 2V_D)G_B$$

From our preceding analysis, the base current for the inverter is

$$I_{B3} = I_B(\beta_I + 1)(1 - \alpha_S)$$

Now for the inverter to saturate with $v_{C3} = V_{CE(\text{sat})}$, it is required that

$$I_{C3} = \beta'_N I_{B3}$$

where β'_N is the forced β obtained from (12-1). Combining these results, we find

$$n = \frac{\beta'_N(\beta_I + 1)(1 - \alpha_S)(V_{CC} - 2V_D) - [V_{CC} - V_{CE(\text{sat})}](G_C/G_B)}{V_{CC} - V_{CE(\text{sat})} - V_D} \qquad (12\text{-}10)$$

When Q_3 is cut off, the criterion to be imposed is that all of the n emitters connected to Q_3 must be reverse-biased. Now the worst condition will be when each of the n-load gates has all of its input emitters reverse-biased. For this case, the current in each emitter is

$$I_L = \frac{(V_{CC} - 2V_D)G_B\beta_I}{m}$$

It is therefore required that

$$(V_{CC} - 2V_D)G_C = \frac{n}{m} \beta_I G_B (V_{CC} - 2V_D)$$

or

$$\frac{G_C}{G_B} = \frac{n}{m} \beta_I$$

Inserting this result in (12-10), we obtain

$$n = \frac{\beta'_N (1 + \beta_I)(1 - \alpha_S)(V_{CC} - 2V_D)}{[V_{CC} - V_{CE(\text{sat})} - V_D] + [(\beta_I/m)(V_{CC} - V_{CE(\text{sat})})]} \qquad (12\text{-}11)$$

EXAMPLE 12I *Calculation of n.* To obtain some idea of the dc fan-out, we consider a device with $\alpha_N = 0.98$, $\alpha_I = 0.1$, $\alpha_S = 0.8$, $r_{C(\text{sat})} = 100\ \Omega$. We assume that $V_{CE(\text{sat})} = 0.3$ is necessary to cut off Q_3, and that $m = 4$, $V_{CC} = 7V_D$. With these values, we find

$$n = 2.1$$

This low value of *n* arises because of the relatively low value used for $V_{CE(\text{sat})}$, which led to $\beta'_N = 11$ as we saw in Sec. 12-1. Note that *n* depends directly on β'_N, so we expect that a significant increase in *n* would result if we could slightly increase $V_{CE(\text{sat})}$. Since small changes of $V_{CE(\text{sat})}$ cause large changes of β'_N, we need to investigate the detailed behavior of v_{B3} when Q_3 is cut off. This can easily be done with the four-layer model. In particular, we need to know the collector-emitter voltage of Q_2 when Q_3 is cut off, since

$$v_{B3} = V_{CE(\text{sat})} + \phi_{CE2}$$

The conditions for the analysis are:

1   The current $I_{B3} \approx 0$.
2   The substrate is reverse-biased.
3 I_B flows into the base of Q_2.

The four-layer-model equations now become

$$I_1 + I_2 - I_B - (1 - \alpha_N)\lambda_{E2} + (1 - \alpha_I)\lambda_{C2} + \alpha_{SI} I_{s3}$$

$$I_2 + I_3 = 0 = -\alpha_N \lambda_{E2} + (1 - \alpha_S)\lambda_{C2} - I_{s3}(1 - \alpha_{SI})$$

We next assume that both junctions are sufficiently forward-biased that $\lambda \approx I_s\, e^{q\phi/mkT}$ and we neglect the terms involving I_{s3}. Then we find

$$\phi_{CE2} \approx \frac{mkT}{q} \ln \left[\frac{I_{s2}}{I_{s1}} \frac{1 - \alpha_S}{\alpha_N} \right] \qquad (12\text{-}12)$$

Inserting $I_{s2}/I_{s1} = 4$ and $\alpha_S = 0.8$ in (12-12), we obtain

$$\phi_{CE2} = -5.7\ \text{mV}$$

The negative sign indicates that the collector junction is forward-biased slightly more than the emitter junction. For the values we have used, then, we can safely assume that

$$V_{B3} \approx V_{CE(\text{sat})} \qquad\qquad ////$$

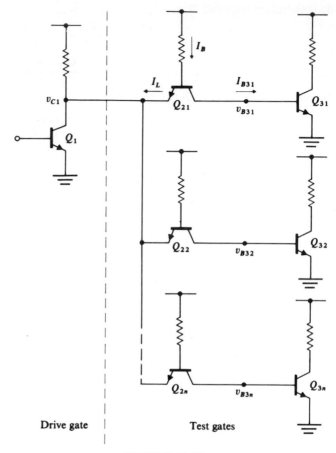

FIGURE 12-15
Load-distribution problems in a TTL gate.

Example 12I indicates the importance of making transistors with low $r_{C(sat)}$. It will be recalled that for $V_{CE(sat)} = 0.3$, in our example $I_C r_{C(sat)}$ was 0.2, making $\phi_{CE(sat)} = 0.1$. It was this low value of $\phi_{CE(sat)}$ which led to the low value for β'_N and the resulting low n. We found in Sec. 12-1 that if $r_{C(sat)}$ is reduced to 75 Ω, $\beta'_N = 31$ is obtained for $V_{C3(sat)} = 0.3$, leading to $n = 6$.

Load Distribution Problems

Unfortunately, the output of the saturated inverter in the TTL gate is not completely independent of the load conditions on the drive gate. To see this we refer to Fig. 12-15, which shows a number of gates connected to the output of a single drive gate. We are interested in the case for which Q_1 is cut off. For this case,

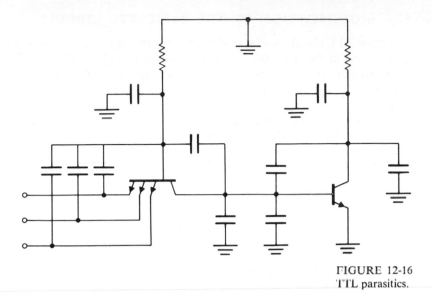

FIGURE 12-16
TTL parasitics.

all the Q_{3n} should be saturated and the Q_{2n} should be in the inverse active region. Suppose also that the emitter-base voltage of the Q_{2n} is close to zero. Note that in a logic system the gates designated "test gates" in Fig. 12-15 will probably be on different chips, and the Q_{3n} may not be identical. Suppose, for example, that v_{B31} is larger than all the other v_{B3}. Then Q_{21} is the closest of the Q_{2n} to saturation. As n is increased, Q_{21} will saturate first. Since its emitter is then forward-biased, some of the current I_B will be diverted into the emitter, and I_{B31} will be reduced. It is easily seen that the largest base current will flow to the inverter with lowest v_B. This is known as *current hogging*. When the current I_{B31} is reduced, for example, $V_{CE(sat)}$ of Q_{31} will increase. Since the design of the TTL gate depends critically on $V_{CE(sat)}$, current hogging can be an important problem. What is needed to alleviate the problem is either a smaller R_C, or a nonlinear resistor to prevent v_{C1} from being dependent upon n. Note that this problem did not arise with DTL circuits because there is no current flow in the input diodes when the inputs are in the high state. Thus the drive gate is effectively decoupled from the test gates in DTL.

TTL Layout

The basic TTL circuit requires only three isolation regions: one for resistors, one for Q_2, and one for Q_3. Parasitics are shown in Fig. 12-16; once the layout is completed and the dimensions are known, the values of the capacitances can be calculated.

12-6   MODIFICATION OF THE BASIC TTL CIRCUIT

The basic TTL circuit can be modified to incorporate an active pull-up transistor; this is done in the MC 500 TTL circuit shown in Fig. 12-17. In this circuit Q_4 serves to drive the active pull-up Q_5. When Q_3 is on, Q_4 is saturated and v_{C4} is between V_D and $2V_D$. Diode D ensures that for these conditions no emitter current flows in the active pull-up Q_5. When Q_3 is cut off, Q_4 is also off, and R_1 provides base current to turn on Q_5. For this case Q_5 can now supply a large emitter current with very little change of the output voltage. As long as V_{CC} is chosen large enough, the current-hogging problem will be negligible.

The approximate logic voltages are easily determined. When Q_3 is saturated, the output is in the logic 0 state, and

$$v_0 = V_{CE(\text{sat})}$$

When Q_3 is cut off, the output rises until Q_5 saturates. The output is then in the logic 1 state; if β_N is large the voltage v_1 is

$$v_1 \approx V_{CC} - 2V_D$$

and v_1 is approximately independent of load current.

Note that the use of Q_4 has introduced an additional offset voltage, and when Q_3 is on,

$$v_{B2} = 3V_D$$

For Q_2 to be in the inverse active region for this case, it is necessary that

$$V_{CC} > v_{B2} + 2V_D = 5V_D$$

since the inputs are driven by identical gates whose inverters are cut off. The minimum logic swing is then

$$v_l = v_1 - v_0 = 3V_D - V_{CE(\text{sat})}$$

Use of Q_4 makes the circuit less dependent upon $V_{CE(\text{sat})}$ because of the additional offset introduced by Q_4. However, note that with Q_4 in the circuit, there is no longer a possibility for the drive gates to cause reverse base drive from Q_3 through Q_2. In order to keep the storage time low, R_2 is inserted to provide some reverse base current.

To see what $V_{CE(\text{sat})}$ is required for this circuit, we note that when Q_3 is off, $v_{B4} \approx V_{CE(\text{sat})}$. The collector current of Q_4 is then

$$I_{C4} \approx I_s\, e^{q\phi_{E4}/mkT}$$

and ϕ_{E4} is given by

$$\phi_{E4} = V_{CE(\text{sat})} - I_{C4} R_2$$

If the collector current is not to exceed some value $I_{C4(\text{max})}$, the following relationship must obtain:

$$V_{CE(\text{sat})} \lesseqgtr R_2 I_{C4(\text{max})} + \frac{mkT}{q} \ln \frac{I_{C4(\text{max})}}{I_s}$$

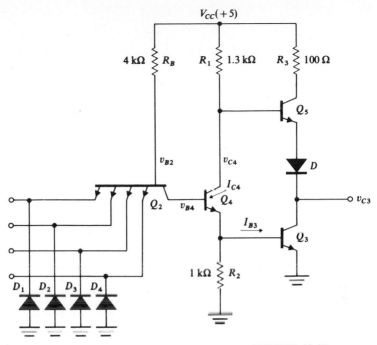

FIGURE 12-17
The MC 500 TTL circuit.

EXAMPLE 12J *Calculation of $V_{CE(\text{sat})}$.*   For the circuit of Fig. 12-17, suppose that we do not allow the drop across R_1 due to I_{C4} to exceed 0.13 V when Q_3 is cut off. Then

$$I_{C4(\text{max})} = 0.1 \text{ mA}$$

Let $I_s = 10^{-10}$, and $R_2 = 1 \text{ k}\Omega$.   Then

$$V_{CE(\text{sat})} \leq 0.1 + 0.026 \ln 10^6$$

$$= 0.460 \text{ V}$$

While $V_{CE(\text{sat})}$ in this circuit is not critical as far as the cutoff of Q_3 is concerned, it must be low to keep Q_4 cut off.   ////

Exercise 12-11   Calculate v_{C4} when Q_3 is cut off if $V_{CE(\text{sat})}$ of the drive gates is V_D.

Exercise 12-12   Calculate the approximate value of I_{B3} when Q_3 is on if $\beta_I = 0.1$.

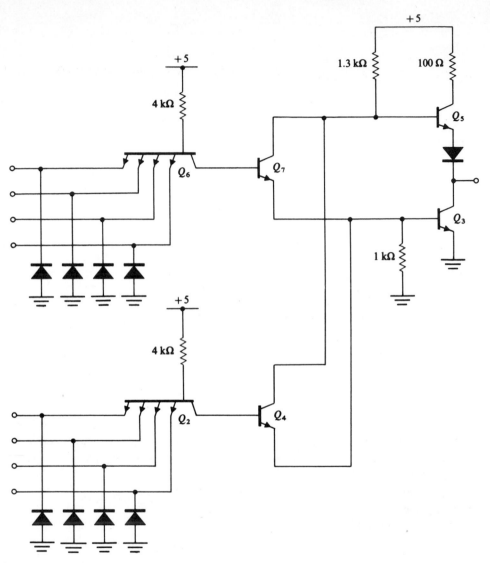

FIGURE 12-18
The MC 505 AND-OR-INVERT circuit.

Exercise 12-13   Show that when Q_3 is saturated the collector current of Q_5 is negligible.

Exercise 12-14   A capacitive load $C = 1000$ pF is connected to the output of the MC 500 circuit.   When one input of the circuit changes from v_1 to v_0, find dv_{C3}/dt if $\beta_N = 50$ and storage time and parasitics are negligible.

In the TTL circuit, large transient currents flow in the ground lead and the inputs when changes of state occur.   Any parasitic inductance in the leads can give rise to overshoot, which could result in false output from the gate.   To prevent negative overshoot at the inputs, diodes D_1, D_2, D_3, and D_4 are added as clamps.

The MC 500 has a fan-out capability $n = 15$, average propagation delay of 10 ns, and power dissipation of 15 mW.

Other Logic Operations

While all logic functions can be implemented with only the NAND operation, it is often convenient to have other operations available in order to reduce the number of gates required and the power dissipation.   The TTL circuit is easily modified to perform the AND-OR-INVERT operation; this is accomplished by the MC 505 circuit shown in Fig. 12-18.   In this circuit the AND operation is performed by the emitters of Q_2 and Q_6.   The OR operation is performed by Q_4 and Q_7, since either or both of these turning on will cause Q_3 to saturate.   At the same time, Q_6 and Q_7 provide offset voltage in the base circuit of Q_3, and also control the active pull-up Q_5.

Increasing the Speed of the TTL Gate

The speed of the TTL circuit can be increased by providing increased reverse base drive for the saturating inverter and much larger active pull-up current capability. This is accomplished in the MC 2103 circuit by reducing the value of R_2 and adding an additional emitter follower to drive the active pull-up transistor, as shown in Fig. 12-19.   Here Q_6 functions as the additional emitter follower.   The value of R_2 has been reduced by 40 percent, as has R_1, so that the requirement on $V_{CE(sat)}$ is still the same, but the reverse base drive for Q_3 is increased.   Note that it is not necessary that Q_6 be completely cut off when Q_3 is saturated, nor is a diode required in the collector circuit of Q_3 in order to ensure cutoff of Q_5.   With Q_4 saturated, R_4 will cause Q_6 to be slightly on, and the base-emitter voltage of Q_5 will be

$$V_{BE5} = V_D - [V_D + V_{CE(sat)}] = -V_{CE(sat)}$$

and Q_5 is cut off.

The MC 2103 has a propagation delay of 6.0 ns and a power dissipation of 22 mW.

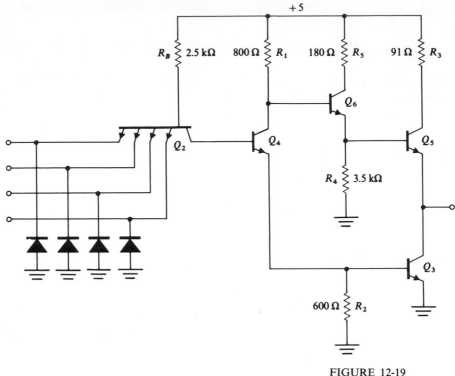

FIGURE 12-19
The MC 2103 TTL circuit.

12-7 RESISTOR-TRANSISTOR LOGIC (RTL)

One of the simplest forms of integrated logic circuits is resistor-transistor logic (RTL), the circuit for which is shown in Fig. 12-20. This circuit performs the NOR operation: if any of the m inputs is in the high state, the transistor connected to that input saturates and the output is low; while if all inputs are low, all m gate transistors are cut off and the output is high. To see how the fan-out number and the resistances R_B and R_C are related, we consider two cases.

Case 1 *All Q_m are cut off.* This can only occur if the Q_D connected to each input is saturated with output voltage $V_{CE(sat)}$. Note that there are no offset diodes in this circuit, so $V_{CE(sat)}$ must be chosen small enough to ensure cutoff of the Q_m; this will require some particular forced β to be used. Let this be β'_N.

The output of the test gate with all transistors cut off is easily seen to be

$$v_{C1} = V_D + \frac{(V_{CC} - V_D)(R_B/n)}{R_C + R_B/n}$$

$$= v_1$$

Note that the logic 1 voltage depends on n.

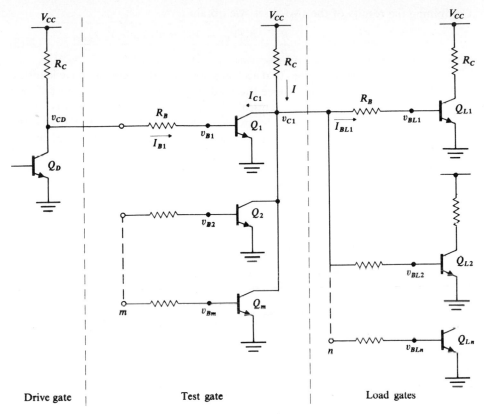

FIGURE 12-20
Resistor-transistor logic circuit.

Case 2   *Q_D is cut off and Q_1 is saturated.*   Now

$$v_{C1} = V_{CE(\text{sat})} = v_0$$

If we assume that Q_D is connected to a total of n gates, then for this case $v_{CD} = v_1$. The base current I_{B1} is given by

$$I_{B1} = \frac{(1/n)(V_{CC} - V_D)}{R_C + R_B/n}$$

The collector current of Q_1 for this case is

$$I_{C1} = \frac{V_{CC} - V_{CE(\text{sat})}}{R_C}$$

If the transistor is to have $v_{C1} = V_{CE(\text{sat})}$, the collector and base currents must be related by

$$I_{C1} = \beta'_N I_{B1}$$

Combining the results of the two cases, we obtain

$$\frac{R_B}{R_C} = \frac{\beta'_N(V_{CC} - V_D) - n[V_{CC} - V_{CE(\text{sat})}]}{V_{CC} - V_{CE(\text{sat})}} \qquad (12\text{-}13)$$

The above analysis assumes that all transistors are identical and have identical base voltages when saturated. Since the load transistors Q_L may be on different chips, they will not necessarily be identical; thus a current-hogging situation similar to that encountered in TTL exists. Suppose that the maximum base-voltage variation among transistors is $\pm \Delta V_B$; the worst situation will be when one of the Q_L has largest base voltage and all others have smallest base voltage. Let $v_{BL1} = V_{BL} + \Delta V_B$ and $v_{BL2}, v_{BL3}, \ldots, v_{BLn} = V_{BL} - \Delta V_B$. Then Q_{L1} will have the smallest base current. The decrease of I_{BL1} is easily found to be

$$\Delta I_{BL1} = \frac{-\Delta V_B[R_B + 2R_C(n-1)]}{R_B(R_B + nR_C)} \qquad (12\text{-}14)$$

Now to be sure that Q_{L1} saturates, it is necessary that

$$(I_{BL1} + \Delta I_{BL1})\beta'_N = I_{CL1}$$

or
$$\beta'_N\left(\frac{V_{CC} - V_D}{nR_C + R_B} - \Delta I_{BL1}\right) = \frac{V_{CC} - V_{CE(\text{sat})}}{R_C} \qquad (12\text{-}15)$$

Inserting (12-14) in (12-15) and rearranging, we can solve for R_B/R_C; the result is

$$\frac{R_B}{R_C} = \left\{ \left(\frac{\beta'_N(V_{CC} - V_D + \Delta V_B)}{2[V_{CC} - V_{CE(\text{sat})}]} - \frac{n}{2}\right) \right.$$
$$\left. \times \left\{ 1 + \sqrt{1 + \frac{4\Delta V_B(2n-2)\beta'_N}{\left[V_{CC} - V_{CE(\text{sat})}\right]\left[\frac{\beta'_N(V_{CC} - V_D - \Delta V_B)}{V_{CC} - V_{CE(\text{sat})}} - \frac{n}{2}\right]^2}} \right\} \qquad (12\text{-}16)$$

Exercise 12-15   Derive Eqs. (12-14) and (12-16). If $R_B = 0 = \Delta V_B$, what is the maximum fan-out that can be obtained?

It should be noted that any reverse base drive for the saturating inverter must flow through R_B. As R_B is increased, the effects of ΔV_B during saturation become less important, but the reverse base current during turnoff is also decreased.

RTL Layout

The layout of the RTL circuit is very simple, owing to the particular configuration of the transistors. Since all collectors are connected together, all gate transistors can be placed in a single isolation region; both resistors can be put in a second region, making a total of two isolation regions. Typical values for R_C and R_B are 640 and 450 Ω, respectively. With such small-value resistors, little space will be consumed by the circuit.

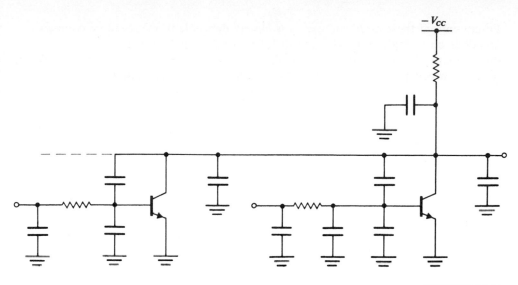

FIGURE 12-21
RTL parasitics.

Parasitic capacitances for the RTL circuit are shown in Fig. 12-21. With the values given above for R_C and R_B, and with $V_{CC} = 3.0$, the circuit has a propagation delay of 12 ns and a power dissipation of 19 mW.

Exercise 12-16 By making use of both base and epitaxial resistors, show how the RTL gate could be laid out in a single isolation region.

12-8 MERGED TRANSISTOR LOGIC (MTL)[3]

The logic circuits we have thus far described are fabricated having various combinations of number of inputs and number of gates per chip. For example, the MC 408 is a quad two-input TTL and NAND gate. To implement a particular logic function, one selects the necessary integrated circuits and interconnects them on a printed circuit board. If the logic system is quite complex, many packages will be required. The obvious disadvantages of this procedure for large systems are:

1 The total cost of the packages may be high.
2 The cost of assembling the system on the printed circuit board may be high.
3 The interconnections on the printed circuit board contribute parasitics which degrade performance.
4 There will probably be a significant number of unused inputs, since standard packages are being used for a special-purpose function. This means that one is paying for considerable capability that is unused.

To circumvent these disadvantages, it is clearly desirable to realize more complex functions on a single chip. Instead of fabricating individual gates and flip-flops, the next level of sophistication is to implement system building blocks such as shift registers, decoders, etc., on a chip. But such large-scale integration (LSI) requires

1 Increase of the density of elements on a chip
2 Increase of chip size
3 Simplification of the processing to achieve the increased yield required by increased chip size
4 Simplification of the basic circuits to reduce the complexity of the metallization interconnections on the chip

It will quickly be realized that a major contributor to both the size of a circuit and the complexity of the processing is the isolation problem. If a circuit could be designed which required no isolation regions, the area consumed by isolation diffusions (a significant amount) would be available for other devices. Furthermore, not only would the isolation diffusion be unnecessary, but with no isolation junction there would be no need for an epitaxial layer or for buried layers. The parasitic substrate capacitance would not exist, so circuit performance would be improved. A circuit which required no isolation regions would of necessity be a rather simple circuit, and therefore the space required and the complexity of the interconnections would be low.

Merged transistor logic (MTL), also known as Integrated Injection Logic (I^2L), possesses all of the above properties. No isolation is used, so the fabrication requires only four masks. The circuit is quite similar to RTL, except that no base resistors are used, and the gate transistors are operated in the inverted connection. The basic cross section and circuit diagram for a two-input MTL NOR gate are shown in Fig. 12-22.

Since all emitters are connected together in the circuit, a common *n* region can be used for them. An n^+ diffusion is used for the emitters, and an *n*-type substrate is used. The normal *p* diffusion is performed for base regions, and an *n* diffusion is used for the collectors; thus the collectors are at the surface, in contrast to the ordinary four-layer *n-p-n*. To avoid the need for resistors, lateral *p-n-p* transistors are used as current sources to supply base current. As Fig. 12-22*a* shows, the lateral *p-n-p* and the *n-p-n* transistors are merged, since they share common *p* regions; the base regions p_2 and p_3 of the *n-p-n* transistors are also the collectors of the *p-n-p* transistor whose emitter is p_1. (Although the regions are separately numbered, it is clear that all *n* regions except n_1 are fabricated with a single diffusion and all *p* regions are fabricated with a single diffusion.) It is important to note that since the collectors are at the surface, multiple collectors can be made for the *n-p-n* transistors; this is not possible with the four-layer *n-p-n* as it is conventionally employed. This property makes possible the implementation of the NOT operation as well as the NOR, as is shown in Fig. 12-22*b*. Emitter current I_0 for all the *p-n-p* transistors on the chip can be obtained by connecting all the *p-n-p* emitters together and supplying them through a single resistor connected to a positive supply voltage.

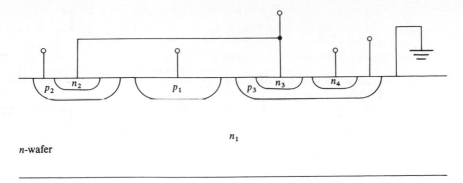

$$n_1$$

n-wafer

(*a*)

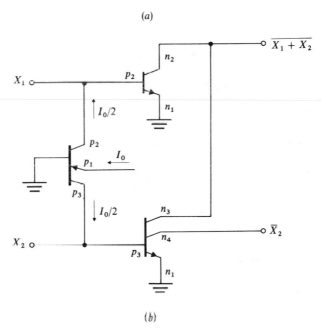

(*b*)

FIGURE 12-22
(*a*) Cross section of an MTL gate; (*b*) circuit diagram.

A two-input MTL gate with drive and load gates is shown in Fig. 12-23. The logic voltages v_1 and v_0 can easily be determined as follows.

Case 1 Q_{D1} *is cut off.* If Q_{D1} is cut off, I flows into the base of Q_1 and Q_1 saturates. Note that in this condition the collector current I_{C1} is the sum of all the *p-n-p* collector currents in the load gate transistors Q_L:

$$I_{C1} = nI$$

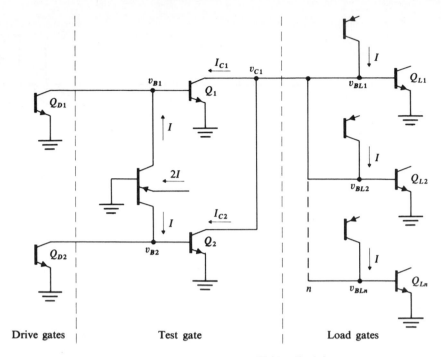

FIGURE 12-23
MTL gate with drive and load gates.

If both Q_{D1} and Q_{D2} are cut off, Q_1 and Q_2 both saturate and share this current equally.   The output voltage is

$$v_{C1} = V_{CE(sat)} = v_0$$

It is clear that $V_{CE(sat)}$ must be sufficiently low that all the Q_L are cut off; thus we expect that overdrive for Q_1 and Q_2 will be necessary.

Case 2   *Both Q_{D1} and Q_{D2} are saturated.*   Now Q_1 and Q_2 are cut off and

$$v_{C1} \approx V_D = v_1$$

while the logic swing is

$$v_l = V_D - V_{CE(sat)}$$

A current-hogging problem exists in this case.   Note that no base resistors are present to isolate the Q_L from one another.   However, since all Q_L are on the same chip the only variation of the V_{BL} arises from the fact that the Q_L may each have different fan-out numbers and therefore different collector currents.

   To see how serious the current-hogging problem is, we use the four-layer model, omitting the layer corresponding to the substrate; the resulting equations are the Ebers-Moll equations for the transistors. It is to be emphasized, however, that $I_1 = -I_C$ and $I_2 = I_E$ for the transistor used in the inverted connection, as in the MTL circuit. We solve for ϕ_C to obtain

$$\phi_C = \frac{mkT}{q} \ln \frac{I_B + I_C(1 - \alpha_N)}{(1 - \alpha_N \alpha_I)I_{s2}}$$

where ϕ_C is the base-to-ground voltage. Since α_N is nearly unity, ϕ_C is a weak function of I_C. This indicates that current hogging is not a serious problem.

The forced β necessary to reduce the saturation voltage to $\phi_{EC(sat)}$ is

$$\beta'_N = \frac{\alpha_I(I_{s2}/I_{s1})e^{q\phi_{EC(sat)}/mkT} - 1}{(1 - \alpha_N) + (1 - \alpha_I)(I_{s2}/I_{s1})e^{q\phi_{EC(sat)}/mkT}}$$

Since there are no resistors involved in the circuit, the fan-out is easily calculated. The base current of Q_1, for example, is

$$I_{B1} = I$$

and the collector current is

$$I_{C1} = nI$$

To ensure saturation it is necessary that

$$I_{C1} = \beta'_N I_{B1}$$

from which we obtain

$$n = \beta'_N$$

Since β'_N is always somewhat less than β_I, this appears to severely limit the fan-out capability of the circuit. However, mitigating circumstances exist. Since no epitaxial layer is involved, and since breakdown voltage is not a consideration, the n-type substrate can be much more heavily doped than in the epitaxial case. This will have two effects. First, the injection efficiency from the substrate into the base will increase, making α_I larger. Second, the bulk resistance in the substrate will be very low owing to the heavy doping and the thickness of the substrate compared with the usual epitaxial-layer thickness. Therefore $r_{C(sat)}$ will be much lower than for the conventional epitaxial transistor; this means that for a given $V_{EC(sat)}$ a larger $\phi_{EC(sat)}$, and consequently a larger β'_N, can be used.

Parasitics for the MTL gate are very low, involving only the collector-base and emitter-base junction capacitances of Q_1 and Q_2. Typical propagation delays are 12 ns for a gate with a power dissipation of 100 μW. A single gate circuit occupies only 16 mils^2 with 0.30-mil lines, 0.15-mil spacing, and 0.20 $\times$ 0.20 mil contact windows.

Although the logic swing for the gate of Fig. 12-23 is small, MTL circuits can be made compatible with TTL circuits by connecting the output of the gate to a resistor and 5-V power supply.

12-9   TERMINAL CHARACTERIZATION OF LOGIC CIRCUITS[4]

Our discussion thus far has dealt with the details of operation of several logic circuits. For purposes of measurement and for system design and specification, it is convenient to develop a method of characterization which involves only the terminal behavior. This is done in terms of the input-output voltage transfer characteristic. To obtain some feeling for the transfer characteristic, we consider the simple RTL gate with a single transistor, as shown in Fig. 12-24a. The collector current as a function of v_{in} is plotted in Fig. 12-24b with the load line corresponding to R_C. Figure 12-24c shows the RTL inverter driving a single load gate; this load gate now causes the load line for the inverter to be nonlinear, as shown in Fig. 12-24d. When v_{in} is high, v_{out} is low and Q_2 is cut off; no current flows in its base circuit and the slope of the load line is determined by R_C. As v_{out} rises, Q_2 turns on and base current flows to it; now the slope of the load line is determined by R_B and R_C in parallel. Figure 12-24d can now be used to construct a plot of v_{out} versus v_{in}; this is the transfer characteristic shown in Fig. 12-24e.

Properties of Transfer Characteristics

It will be noted that the circuit of Fig. 12-24c is an inverting circuit; this is manifested in the transfer characteristic by the negative slope. Note also that the slope of the transfer characteristic is dv_{out}/dv_{in} which is the gain of the circuit. One can see by inspection of Fig. 12-24e that this transfer characteristic has two points at which the gain is -1.

In our discussion of transfer characteristics, we make the following assumptions:

1 The logic circuit output is approximately a dependent voltage source; that is, the output voltage is independent of the *output current*.

2 The transfer characteristic is obtained using one input, with all other inputs connected to v_0 for a NOR gate, or to v_1 for a NAND gate.

The first approximation is valid for circuits having an active pull-up, but poor for such circuits as RTL.

The operating points of a logic circuit can be determined from the transfer characteristic by making use of the constraint that the voltages v_1 and v_0 must be the same for both input and output; that is, input and output levels must be compatible. Consider a general logic circuit which may be either inverting or non-inverting, having a transfer characteristic $v_{out} = f(v_{in})$. Now form the inverse function $v_{in} = g(v_{out})$. (To do this graphically, one merely interchanges the axes on the transfer-function plot.) It is assumed that $f(v_{in})$ and $g(v_{out})$ are single-valued.

Next consider the functions $f(v)$ and $g(v)$; since $g(v)$ is obtained from $f(v)$ by interchanging horizontal and vertical axes of the transfer characteristic, we see that $g(v)$ can be written as

$$g(v) = f[f(v)]$$

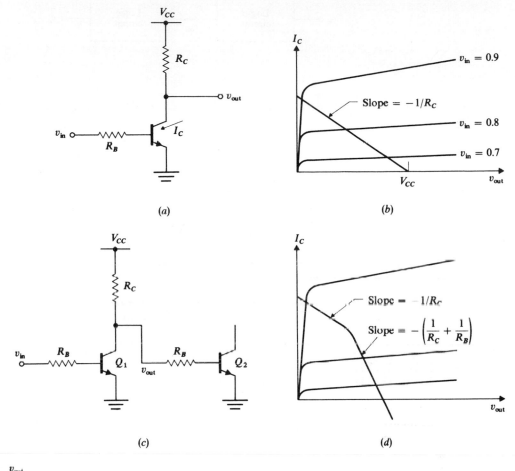

(a)

(b)

(c)

(d)

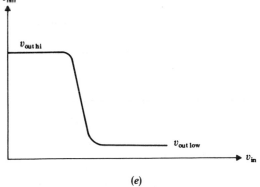

(e)

FIGURE 12-24
(a) RTL inverter; (b) v_{out} versus I_C; (c) RTL inverter driving one load gate; (d) v_{out} versus I_C; (e) transfer characteristic v_{out} versus v_{in}.

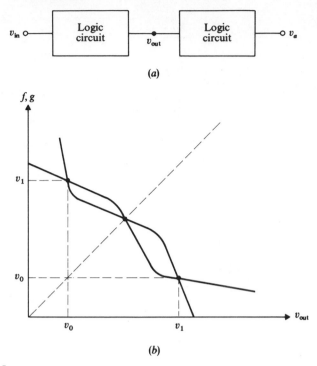

FIGURE 12-25
(a) Two identical logic circuits connected in cascade; (b) $f(v_{out})$ and $g(v_{out})$ for the connection of (a), showing the operating points.

Note that $g(v)$ could also be obtained by rotating $f(v)$ about the 45° line, which is also the unity-gain locus.

Next we let two identical circuits be connected as shown in Fig. 12-25a; the circuit used can be either inverting or noninverting. Since we have constrained input and output levels to be compatible, the following condition must obtain:

When v_{in} is either v_1 or v_0, v_a must equal v_{in}.

This restriction has important consequences as far as $f(v)$ and $g(v)$ are concerned. Note that

$$v_{out} = f(v_{in})$$

and since both circuits are identical

$$v_a = f(v_{out})$$

It is also true that

$$v_{in} = g(v_{out})$$

and since it is required for compatibility that $v_a = v_{in}$, we have

$$g(v_{out}) = f(v_{out})$$

Now let us plot $f(v_{out})$ and $g(v_{out})$ on the same set of axes as shown in Fig. 12-25b. Then the possible operating points are the values of v_{out} at which the two curves intersect; these are at v_0, v_1, and a third point which occurs on the 45° line because of the quadrantal symmetry existing between f and g. This third point is not a permitted operating point because it requires v_{in} to differ from both v_1 and v_0; that is, it is not one of the two logic voltages.

Note that in the above procedure for obtaining the operating points, nothing was postulated about any part of f being approximately horizontal. Most practical transfer characteristics, however, have two regions which are nearly horizontal, as is the case in Fig. 12-24e. For such a case, it is not necessary to replot the entire transfer characteristic to obtain the operating points; inspection shows that they will lie at the intersections of $f(v_{in})$ with the lines $v_{in} = v_{out(low)}$ and $v_{in} = v_{out(hi)}$.

Threshold Point

Next we replot $v_{out} = f(v_{in})$ and draw a line between the operating points $v_{out} = v_1$ and $v_{out} = v_0$ as shown in Fig. 12-26a; this line will have a slope of -1 for an inverting gate and $+1$ for a noninverting gate. This fact is a result of the way $g(v)$ was obtained from $f(v)$. Consider the inverting case. The slope of the line is

$$s = \frac{f(v_1) - f(v_0)}{v_1 - v_0}$$

But since v_1 and v_0 are the operating points,

$$f(v_1) = v_0 \qquad \text{and} \qquad f(v_0) = v_1$$

Thus
$$s = \frac{v_0 - v_1}{v_1 - v_0} = -1$$

The threshold point is defined as the third intersection of this line with $f(v_{in})$, and is denoted v_t, as shown in Fig. 12-26a.

Transition Width

As was previously noted, there are two points at which the slope of the transfer characteristic is -1; these are denoted $v_{\mu1}$ and $v_{\mu0}$. The transition width v_w of the transfer characteristic is defined as the change in *input* voltage required to change the output from $v_{\mu1}$ to $v_{\mu0}$:

$$v_w \triangleq |v_{in}(v_{\mu1}) - v_{in}(v_{\mu0})|$$

Since $v_{in} = g(v_{out})$ this can also be written as

$$v_w = |g(v_{\mu1}) - g(v_{\mu0})|$$

For transfer characteristics with nearly horizontal sections, v_w is a measure of the large-signal voltage gain of the circuit, since it essentially represents the minimum input voltage change required to cause the output to change from one logic state to the other.

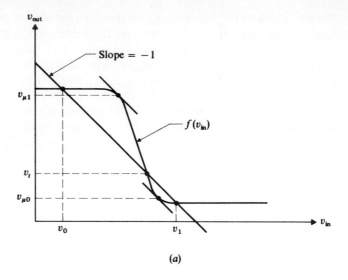

(a)

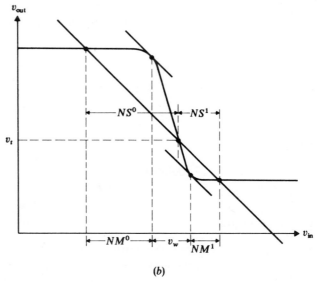

(b)

FIGURE 12-26
(a) Definition of various points of the transfer characteristic; (b) noise performance.

Noise Margin

The noise margin NM of the circuit is defined as the *input* voltage change required to cause the output to change from an operating point to the nearest unity-gain point.   Note that the noise margins for the input in the 0 and 1 states may be different, depending upon the shape of the transfer characteristic.   The notation used is

$$NM^0 = \text{noise margin with the } input \text{ at } v_0$$
$$NM^1 = \text{noise margin with the } input \text{ at } v_1$$

From Fig. 12-26b these are seen to be

$$NM^0 = |v_{in}(v_{\mu 1}) - v_{in}(v_1)| = |g(v_{\mu 1}) - g(v_1)|$$
$$NM^1 = |v_{in}(v_0) - v_{in}(v_{\mu 0})| = |g(v_0) - g(v_{\mu 0})|$$

The noise margin is a measure of how much the input can deviate from the operating point before further input changes are amplified by the circuit.

Noise Sensitivity

Noise sensitivity NS is the *input* voltage change required to cause the output voltage to change from an operating point to the threshold point.   Again, noise sensitivity for $v_{in} = v_1$ may differ from that for $v_{in} = v_0$.   The notation used is

$$NS^0 = \text{noise sensitivity with } v_{in} = v_0$$
$$NS^1 = \text{noise sensitivity with } v_{in} = v_1$$

From Fig. 12-26b the noise sensitivities are seen to be

$$NS^0 = |v_{in}(v_t) - v_{in}(v_1)| = |g(v_t) - g(v_1)|$$
$$NS^1 = |v_{in}(v_0) - v_{in}(v_t)| = |g(v_0) - g(v_t)|$$

If $v_t = (v_1 + v_0)/2$, $NS^0 = NS^1$ For this case the noise sensitivity is a measure of how much change of input voltage can occur before it is no longer possible to correctly determine the output logic state.

Noise Immunity

The noise immunity NI is defined as the ratio of noise sensitivity to logic swing:

$$NI^0 = \frac{NS^0}{v_l}$$

$$NI^1 = \frac{NS^1}{v_l}$$

Since noise in logic circuits is usually proportional to the logic swing, noise immunity is often a better criterion for evaluating noise performance than is noise sensitivity.

12-10 COMPARISON OF SATURATING LOGIC CIRCUITS

The noise performance described in the previous section is based on the transfer characteristic. Unfortunately, it is difficult to calculate the unity-gain points and the threshold point analytically, and for a given circuit, recourse must generally be made to measurement. However, comparison of the speed-power properties of saturating logic circuits can be made by using the data available in the manufacturers' specifications.

As we saw earlier in this chapter, the propagation delay of a circuit can often be reduced by adding active pull-up transistors, by providing more reverse base drive, etc. All of these measures, however, increase the power dissipation of the circuit. Thus a useful figure of merit for a logic circuit is the product of power and average propagation delay. Table 12-3 gives the fan-out, power, propagation delay, and power-delay product for the various circuits discussed in this chapter.

Table 12-3 SPEED AND POWER CHARACTERISTICS FOR VARIOUS SATURATING LOGIC CIRCUITS

Circuit	n	$P_{D(max)}$, mW	$\bar{t}_D$, ns	$P_{D(max)}\bar{t}_D$, pJ
MC 932, DTL	25	43	35	1500
MC 500, TTL	15	15	10	150
MC 2103, TTL	11	22	6	132
MC 505, AND-OR-INVERT	15	20	12	240
MC 903, RTL	5	19	12	233
MTL, NOR	15	0.1	12	1.2

PROBLEMS

12-1 In the calculation of t'', do not assume that when $i_B = 0$ the collector voltage is V_T. Do a two-part analysis in which $v_C(t'')$ is calculated and used as the initial condition for the second part. Calculate the total t''.

12-2 What logic operations do TTL and RTL perform if negative logic is used rather than positive logic? What does the MC 505 perform for negative logic?

12-3 Calculate the substrate current of the multiple-emitter transistor of Fig. 12-14.

— *12-4* A possible DTL NOR circuit is shown in Fig. P12-4. Find v_1 and v_0, and discuss the operation of the circuit.

⌐*12-5* The circuit for the MC 660 is shown in Fig. P12-5. Find v_1 and v_0.

12-6 Lay out an RTL gate with $R_B = 640\ \Omega$ and $R_C = 450\ \Omega$. Use 0.5-mil geometry, estimate the parasitic capacitances, and estimate the times t_D, t', t_s, and t''.

12-7 Suggest a circuit for performing AND-OR-INVERT in a DTL-type structure. What v_1 and v_0 would your circuit have?

— *12-8* A noninverting logic circuit is to be used in a logic system. If it has transfer characteristic A in Fig. P12-8, find the operating points. Repeat for transfer characteristic B.

12-9 A logic circuit has the piecewise-linear transfer characteristic shown in Fig. P12-9. Define and calculate the noise margins.

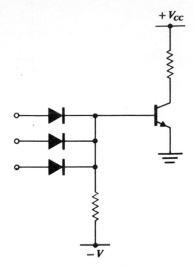

FIGURE P12-4

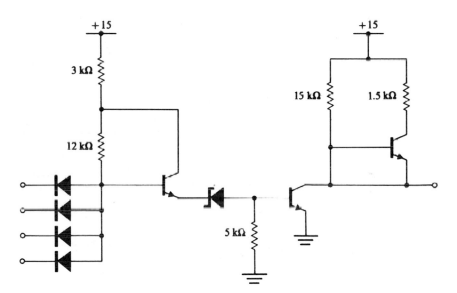

FIGURE P12-5

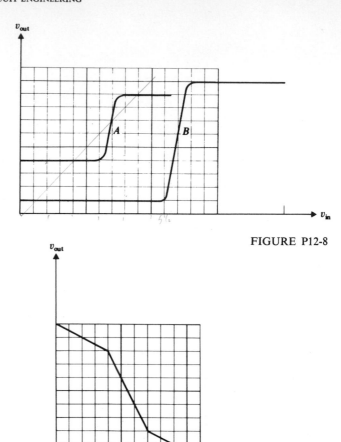

FIGURE P12-8

FIGURE P12-9

12-10 Using the transistor model of Fig. 12-3b, calculate and sketch the transfer characteristic of an RTL gate with $R_B = 640\ \Omega$ and $R_C = 450\ \Omega$.

12-11 A two-input MTL gate is fabricated using 0.5-mil geometry. Estimate the parasitic capacitances and the times t_D and t'.

REFERENCES

1 PHISTER, M.: "Logical Design of Digital Computers," John Wiley & Sons, Inc., New York, 1961.

2 MEYER, C. S., D. K. LYNN, and D. J. HAMILTON: "Analysis and Design of Integrated Circuits," chaps. 6 to 11, McGraw-Hill Book Company, New York, 1968, gives a detailed analysis of bipolar logic circuits.

3 BERGER, H. H., and S. K. WIEDMANN: Merged-transistor Logic (MTL)—A Low-cost Bipolar Logic Concept, *IEEE J. Solid-State Circuits*, vol. SC-7, pp. 340–346, 1972

4 MEYER et al.: op. cit., chap. 6.

13

NONSATURATING LOGIC CIRCUITS

In logic circuits employing a saturating inverter, storage time accounts for a considerable portion of the propagation delay. This storage time results from the excess minority carrier charge that is stored in the transistor during saturated operation. As forward base drive is increased, for a fixed level of collector current the forward bias of the collector junction increases. This causes the minority carrier charge in both the base and collector regions to increase substantially above the charge which would be required to maintain the same collector current were the transistor in the active region.

In Fig. 13-1 are sketched the approximate minority carrier charge distributions for an idealized one-dimensional transistor. If the transistor is just-saturated, that is, $I_B = I_C/\beta_N$, a minority carrier charge Q is stored in the base as shown in Fig. 13-1a. Now if the base current is increased, the collector junction becomes forward-biased. Little change occurs in the collector current, but a large amount of excess charge is stored in both base and collector regions as shown in Fig. 13-1b. In an integrated transistor, the collector region is much larger in extent than the base region, and the impurity level in the collector is relatively low. Consequently the total excess charge stored in the collector region may considerably exceed that in the base region.

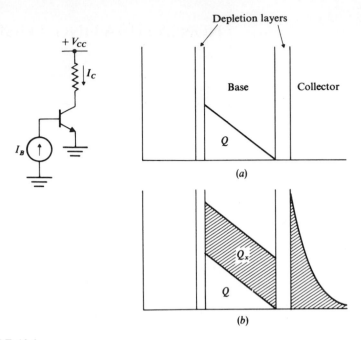

FIGURE 13-1
(a) Stored minority carrier charge for just-saturated operation $(I_B = I_C/\beta_N)$;
(b) excess charge (shaded areas) for $I_B > I_C/\beta_N$.

To turn off the transistor, it is necessary to remove all of the excess charge before the transistor can enter the forward active region of operation. If the base current is made zero, removal of the charge will occur by recombination; if reverse base current is applied, charge can be removed by the reverse base current. The amount of excess charge stored is a nonlinear function of the forward bias of the collector junction, and does not become significant until the collector junction voltage is several hundred millivolts. Forward voltages of this magnitude occur, as we have seen, when large enough base current is used that the forced β is significantly less than β_N.

If a saturating inverter is to be used, one of the following measures must be taken if storage time is to be reduced.

1 The circuit must be designed for $\beta'_N \approx \beta_N$. This means that $V_{CE(\text{sat})}$ will be large. Moreover, it will not be possible to maintain $\beta'_N \approx \beta_N$ if variations of β_N occur.

2 Reverse base drive comparable in magnitude to the forward base current must be applied in order to remove the excess charge rapidly.

3 The circuit configuration must be modified to prevent saturation of the transistor.

Option 2 is undesirable since it is usually inconvenient to apply reverse base drive in a logic circuit. We now consider modifications of the basic inverter circuit to prevent saturation.

13-1 THE FEEDBACK CLAMP CIRCUIT

The simplest modification of the circuit is to employ diodes as shown in Fig. 13-2a to prevent forward bias of the collector junction.[1] If both D_1 and D_2 have a forward voltage V_D, the collector-base voltage will be zero, and the transistor is just-saturated. For this condition,

$$I_C = \beta_N I_B$$

Applying Kirchhoff's current law to collector and base terminals, we can also write

$$I_C = I_L + I_D$$

and

$$I_B = I_{in} - I_D$$

Combining these relations, we obtain

$$I_C = (I_L + I_{in})\alpha_N$$

and

$$I_B = \frac{I_L + I_{in}}{\beta_N + 1}$$

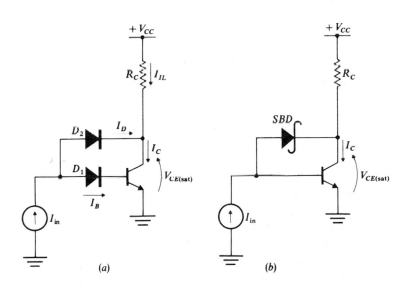

FIGURE 13-2
(a) Feedback clamp circuit to prevent saturation; (b) Schottky-barrier diode clamp.

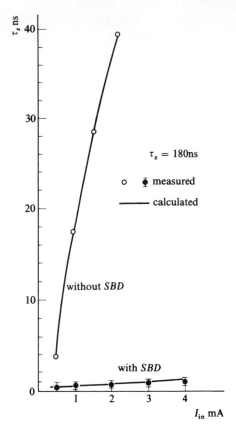

FIGURE 13-3
Typical performance for an inverter with
SBD clamp (after Tarui et al.[2]).

Thus we see that as the input current I_{in} is increased, most of I_{in} flows through D_2 to the collector rather than through D_1 to the base. D_1 serves only to provide an offset voltage.

Unfortunately, D_2 will have excess minority carrier charge storage, and will cause a storage time problem even though the transistor does not saturate. Some improvement over the saturating inverter can be obtained by choosing a type a diode configuration.

A Schottky-barrier diode (SBD) can be used to advantage in the feedback clamp circuit.[2,3] Since minority carriers are not stored in the SBD, it has negligible storage time; furthermore the forward voltage of the SBD is typically 300 to 400 mV less than that of a silicon p-n junction diode. Therefore the offset diode D_1 can be omitted. The collector junction is now forward-biased, but the forward bias is sufficiently small that negligible excess charge is stored in the transistor. Elimination of the offset diode has the further advantage that $V_{CE(sat)}$ is now considerably less than V_D. Typical performance for an inverter with SBD clamp is shown in Fig. 13-3.

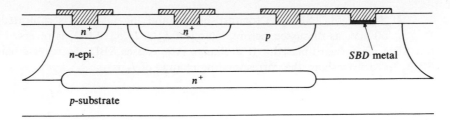

FIGURE 13-4
Cross section of the inverter with SBD clamp.

With the omission of the offset diode, the inverter with SBD clamp is easily fabricated, as is shown in the cross section of Fig. 13-4. Two collector contact windows are used, but the n^+ contact diffusion is employed at only one. At the other window, the metal for the SBD is first deposited to form the SBD in the collector region. Metallization is then applied to connect the SBD to the base contact; thus the increase of surface area resulting from the SBD is very small.

Feedback clamp circuits can be used not only to prevent storage time problems arising from saturation, but also to reduce delays caused by parasitic capacitances at the base of the transistor.[4] If the transistor operation must change from saturation to cutoff, the change of base voltage is several hundred millivolts, and turn-on or turnoff delays result partly from the parasitic capacitance having to be charged and discharged by this amount. If two SBDs are employed, as shown in Fig. 13-5, well-determined collector voltages can be maintained without the transistor being cut off.

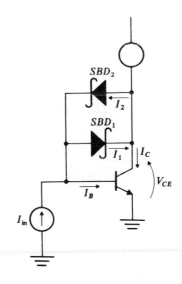

FIGURE 13-5
Shunt-feedback Schottky-clamped inverter.

If $I_{in} > I_L/\beta_N$, SBD$_1$ is forward-biased and the inverter is saturated with $V_{CE} \approx 400$ mV, as discussed before. Now if $I_{in} < I_L/\beta_N$, SBD$_1$ is reverse-biased, but SBD$_2$ is forward-biased. If a change ΔI_{in} causes SBD$_1$ to reverse-bias and SBD$_2$ to forward-bias, the corresponding change of I_B is

$$\Delta I_B = \frac{\Delta I_{in}}{\beta_N + 1}$$

This change of I_B will be accompanied by a change of V_{BE} of only a few millivolts.

At the output, V_{CE} is determined by the forward voltage of the SBDs and the emitter junction voltage. Let V_{ds} be the forward SBD voltage; the emitter junction voltage is $V_{BE} \approx V_D$. For $I_{in} > I_L/\beta_N$, we obtain

$$V_{CE} \approx V_D - V_{ds} \approx 400 \text{ mV}$$

while for $I_{in} < I_L/\beta_N$

$$V_{CE} \approx V_D + V_{ds} \approx 1100 \text{ mV}$$

The voltage swing at the base of the transistor is typically less than 50 mV; thus delays caused by parasitics at this point are minimized.

13-2 USE OF A TRANSISTOR TO CONTROL SATURATION[5]

While the use of SBDs provides a very effective means of controlling saturation of the inverter, it requires extra processing steps. If SBDs are being used elsewhere in the circuit, then the extra processing steps are already being employed. In lieu of SBDs it is possible to use a standard n-p-n bipolar transistor to control saturation of the inverter, thus making the nonsaturating circuit completely compatible with the standard process. This is done as shown in Fig. 13-6.

In this circuit, R_0 is several hundred ohms and v_A is about 100 mV more positive than v_B. During the turn-on transient of Q_1, the emitter of Q_0 is reverse-biased, and the only effect of Q_0 results from the small parasitic capacitance it introduces in the base circuit of Q_1. As long as v_{CE} is larger than about 200 mV, little current flows in Q_0. Now as v_{CE} begins to fall below 200 mV, Q_0 begins to turn on, and it applies negative feedback to the base of Q_1, thus preventing further saturation. If we define m to be the ratio of injected collector current of Q_1 without Q_0 to injected collector current with Q_0, it can be shown that the storage time is given approximately by

$$t_s \approx \left[\frac{1}{\omega_N} + \frac{1}{\omega_I m(1 - \alpha_I)} \right] \frac{I_1}{I_2}$$

where ω_N is the alpha cutoff frequency in the normal mode of operation and ω_I is the alpha cutoff frequency in the inverted mode. Typically m is of the order of

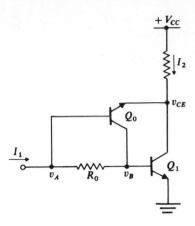

FIGURE 13-6
Use of a bipolar transistor to control
saturation of the inverter.

50, and the storage time of the inverter is reduced by a factor of 20 from that
without Q_0.

When this method of saturation control is used with TTL, the circuit can be
modified slightly, as shown in Fig. 13-7a.   Here the base voltage for Q_0 is obtained
by the voltage divider R_1 and R_2.   Since Q_0 and Q_2 have common-collector con-

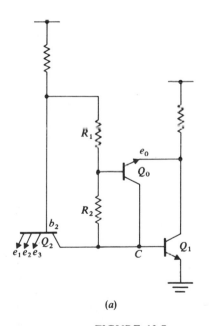

(a)

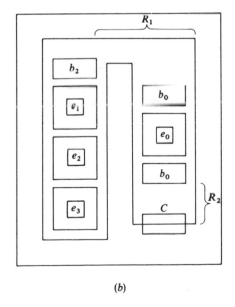

(b)

FIGURE 13-7
(a) Saturation control of a TTL circuit; (b) layout of Q_0, Q_2, R_1, R_2 (after
Wiedmann[5]).

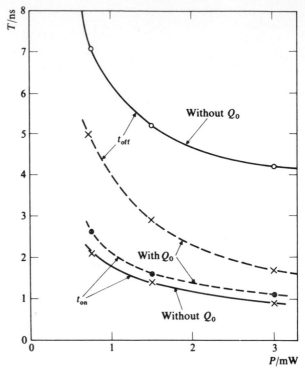

FIGURE 13-8
Turn-on and turnoff delays for the circuit of Fig. 13-7 (after Wiedmann[5]).

nections, both can be placed in a single isolation region.   Resistors R_1 and R_2 can be obtained by extending the base diffusion of Q_2, as shown in Fig. 13-7b.   Total turn-on and turnoff delays for the TTL circuit are shown in Fig. 13-8.   Note that t_{off} in Fig. 13-8 includes both storage time and turnoff transient.   Since a reduction of t_s by a factor of about 20 has been accomplished, most of t_{off} is due to turnoff transient.   This could be reduced by use of an active pull-up transistor.

13-3   EMITTER-COUPLED LOGIC (ECL)[6]

If saturation of an inverter is to be prevented without the use of clamp circuits, some means must be made available to accurately control the collector-base voltage. Clearly a common-emitter configuration in which base current is the independent variable cannot be used, because variations of β_N, which can be quite large, will cause variations of the collector current, and hence the collector voltage.   Although the circuit may be designed not to saturate for one value of β_N, the variations nor-

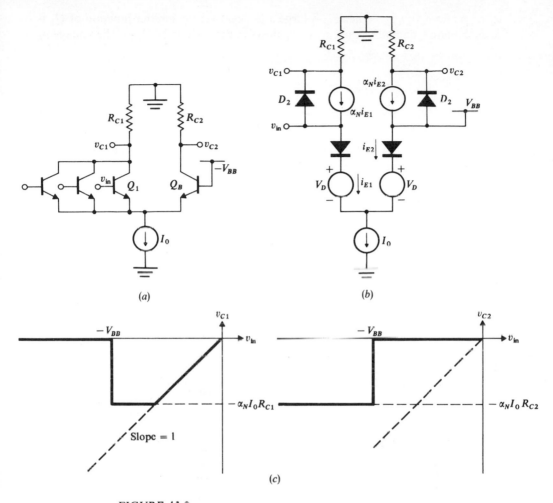

FIGURE 13-9
(*a*) Differential amplifier with multiple inputs; (*b*) equivalent circuit with first-order transistor models; (*c*) transfer characteristics.

mally encountered in β_N are sure to produce saturation. What is required is some way to control the emitter current. The collector current is $\alpha_N i_E$, and variations of α_N will be at most 2 percent.

Consider the differential amplifier of Fig. 13-9*a* with one input connected to a supply voltage $-V_{BB}$ and with multiple transistors, each having an input, forming the other half of the differential amplifier. We analyze first the case with only a single transistor on the left side, and we use a first-order model to represent the transistors, as shown in Fig. 13-9*b*. Let us first plot the transfer characteristics

$v_{C1}(v_{in})$ and $v_{C2}(v_{in})$. When v_{in} is large and negative, the emitter junction of Q_1 is reverse-biased, and all of the current I_0 flows in the emitter of Q_B. The voltage v_E is

$$v_E = -V_{BB} - V_D$$

while

$$v_{C1} = 0$$

and

$$v_{C2} = -\alpha_N I_0 R_{C2}$$

Now when $v_{in} = -V_{BB}$ the base voltages of Q_1 and Q_B are equal, and the two transistors each have emitter current $I_0/2$. For this case

$$v_{C1} = -\frac{\alpha_N I_0 R_{C1}}{2}$$

$$v_{C2} = -\frac{\alpha_N I_0 R_{C2}}{2}$$

As v_{in} becomes slightly more positive than $-V_{BB}$, the emitter diode of Q_B becomes reverse-biased, and all of I_0 flows in the emitter of Q_1. Now we have

$$v_{C1} = -\alpha_N I_0 R_{C1}$$
$$v_{C2} = 0$$

As v_{in} is made more positive, no change in v_{C2} occurs since Q_B is cut off. No change in v_{C1} occurs until $v_{in} = v_{C1}$, at which point the collector junction of Q_1 becomes forward-biased, that is, Q_1 saturates. Note that saturation occurs not because the collector voltage decreases but because the base voltage increases. Saturation clearly occurs when

$$v_{in} = v_{C1} = -\alpha_N I_0 R_{C1}$$

Further increases of v_{in} are accompanied by equal increases of v_{C1}, as is shown in Fig. 13-9c.

An important aspect of the transfer characteristics of Fig. 13-9c is that as long as

$$v_{in} < -\alpha_N I_0 R_{C1}$$

both v_{C1} and v_{C2} depend on α_N, not β_N. We have therefore achieved the objective of making the -collector voltage relatively independent of transistor parameter variations. Note also that the circuit has both an inverting and a noninverting transfer characteristic available.

Now if we consider the multiple input transistors, we see that if all but one of the inputs is low, the transistors with low inputs will be cut off and the circuit behaves as described above. If more than one input is high and all high inputs are at the same voltage, the emitters of those transistors equally share the current I_0. But since all collectors are connected together, the current in R_{C1} is still $\alpha_N I_0$; the transfer characteristics do not change for multiple input transistors.

Next we investigate the possibility of using the circuit to perform logic operations. We impose the following restrictions:

1 For high-speed operation, the circuit is to be constrained so that the transistors never have a collector-base voltage less than zero; that is, they never saturate.

2 For logic system use, the input and output levels must be compatible; that is, v_1 must have the same value at input and output, as must v_0.

Let $v_{C1}(v_{in}) = f_1(v_{in})$ and $v_{C2}(v_{in}) = f_2(v_{in})$. From the analysis of Chap. 12, we know that if input and output are to be compatible, the operating points must be those values of v_{in} which produce

$$\cdot \; f_1(v_{in}) = g_1(v_{C1})$$

and

$$f_2(v_{in}) = g_2(v_{C2})$$

This means that the operating points are the intersections of $f(v_{in})$ with $f(v_{in})$ rotated about the unity-gain line. Inspection of Fig. 13-9c shows that for v_{C1} the intersections occur only along that part of the transfer characteristic which is coincident with the unity-gain line, that is, with Q_1 saturated. Similarly, for v_{C2} there is only a single intersection, which occurs for $v_{in} - 0$. We therefore conclude that in its present form the circuit cannot meet the compatibility requirements necessary for performing logic.

In order to make the circuit suitable for compatible logic operation, it is necessary that the transfer characteristics be translated vertically in the v_C versus v_{in} plane. This is done by adding an offset voltage to the outputs, as shown in Fig. 13-10a. Here the emitter followers perform two functions: first, they provide an offset of V_D with their emitter junctions; second, they provide current gain so that the voltages v_{C1} and v_{C2} are isolated from any loads connected to the outputs.

If any input or combination of inputs is high, $v_{(out)1}$ will be low. If all inputs are low, $v_{(out)1}$ will be high. Therefore if compatibility of input and output levels obtains, $v_{(out)1}$ represents the NOR operation. Similar reasoning shows that $v_{(out)2}$ represents the OR operation. It is noteworthy that both OR and NOR operations are obtained from this circuit without unduly increasing its complexity.

A surprising amount of information can be obtained about the operation and design of the circuit by using the first-order transistor models employed in Fig. 13-10b. We first assume that the emitter junctions of the emitter followers are always forward-biased, and the offset voltage thus provided is V_D. We invoke the following requirements for the design.

1 When any input is v_0, the transistor connected to that input is cut off.
2 When any input is v_1, the transistor connected to that input is at the edge of the saturation region; that is, the collector-base voltage is zero.
3 Input and output levels must be compatible.

We now calculate v_1 and v_0 subject to the above constraints.

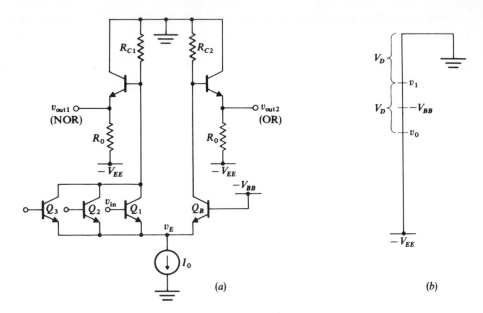

FIGURE 13-10
(a) ECL NOR-OR circuit with emitter followers providing offset; (b) relative voltages for the circuit.

Case 1 *All inputs are v_0.* If all inputs are v_0, all input transistors are cut off. Then

$$v_{C1} = 0$$

and

$$v_{(out)1} = -V_D$$

Now if the circuit is to perform the NOR operation, when all inputs are v_0 the output must be v_1. Therefore

$$v_1 = -V_D \qquad (13\text{-}1)$$

Case 2 *Any input or combination of inputs is v_1.* If $v_{in} = v_1$ and the transistors are at the edge of saturation, then

$$v_{C1} = v_1$$

and

$$v_{(out)1} = v_1 - V_D$$

Since the NOR operation is being performed, when $v_{in} = v_1$, $v_{(out)1}$ must be v_0. Thus we have

$$v_0 = v_1 - V_D \qquad (13\text{-}2)$$

Combining (13-1) and (13-2), we find

$$v_0 = -2V_D$$

The logic swing is

$$v_l = |v_1 - v_0| = V_D$$

We can easily see what relationship must exist among the supply voltage V_{BB}, the bias current, and the resistors R_{C1} and R_{C2}. Note that the logic swing must be given by

$$v_l = \alpha_N I_0 R_{C1}$$

If $v_{(out)2}$ is to be compatible with v_{in}, the logic swing must also be

$$v_l = \alpha_N I_0 R_{C2}$$

and thus

$$R_{C1} = R_{C2}$$

This is only true when a bias current source is used, as in Fig. 13-9a.

When all inputs are v_0, in order for all input transistors to be cut off,

$$v_E < v_{in} - V_D = v_1 - V_D = -2V_D$$

For this case, Q_B is on, and

$$-V_{BB} = v_E + V_D$$

or

$$-V_{BB} > -2V_D$$

If any input is v_1, v_E is

$$v_E = v_{in} - V_D = v_1 - V_D$$

For Q_B to be cut off,

$$-V_{BB} < v_E + V_D$$

or

$$-V_{BB} < -V_D$$

Combining the results for the two cases, we have

$$-2V_D < -V_{BB} < -V_D \qquad (13.3)$$

A convenient choice is

$$-V_{BB} = \frac{v_1 + v_0}{2} = -\frac{3}{2}V_D$$

The relative voltages in the circuit are shown in Fig. 13-10b.

It should be noted that the configuration of the ECL circuit is such that the current in the ground lead is constant, regardless of the state of the inputs and outputs. This means that noise generated by the change of state of the inputs and outputs will be much less severe in ECL than in, for example, TTL.

Use of an Emitter Bias Resistor

The current source I_0 can be obtained by using a bias current-source circuit like that described in Chap. 9, or simply by using a resistor R_E connected between v_E and $-V_{EE}$. This has several effects on the behavior of the circuit:

1   The transfer characteristic v_{C1} versus v_{in} is slightly altered.
2   Because the current in R_E changes as the input changes from v_0 to v_1, in order for the NOR and OR outputs to be compatible it is necessary that R_{C1} and R_{C2} have unequal values.
3   The power supply and ground currents will no longer be constant as v_{in} changes from v_0 to v_1.

To establish the bias current I_0, we select R_E so that the current in R_E is I_0 when Q_B is on and all other transistors are off.   Then

$$R_E = \frac{V_{EE} - V_{BB} - V_D}{I_0}$$

The logic swing at the OR output is

$$v_{1OR} = \alpha_N I_0 R_{C2} = \alpha_N(V_{EE} - V_{BB} - V_D)\frac{R_{C2}}{R_E}$$

Note that v_{1OR} depends on a ratio of resistors rather than absolute resistor values. Now when Q_1 is on and Q_B is off, if $v_{in} = v_1$ the current in R_E is

$$I_E = \frac{v_1 - V_D + V_{EE}}{R_E}$$

$$= \frac{V_{EE} - 2V_D}{R_E}$$

and the logic swing at the NOR output is

$$v_{INOR} = \alpha_N(V_{EE} - 2V_D)\frac{R_{C1}}{R_E}$$

For compatibility $v_{INOR} = v_{1OR}$; this will also ensure compatibility of both v_1 and v_0 on NOR and OR sides.   Combining the above results we find

$$\frac{R_{C2}}{R_{C1}} = \frac{V_{EE} - 2V_D}{V_{EE} - V_{BB} - V_D}$$

If $-V_{BB} = (v_1 + v_0)/2 = -\frac{3}{2}V_D$, we obtain

$$\frac{R_{C2}}{R_{C1}} = \frac{V_{EE} - 2V_D}{V_{EE} - \frac{5}{2}V_D} \tag{13-4}$$

When v_{in} is just slightly more positive than $-V_{BB}$, Q_B is cut off and the collector current of Q_1 is

$$I_{C1} = \alpha_N I_0 = \frac{\alpha_N(V_{EE} - V_D - V_{BB})}{R_E}$$

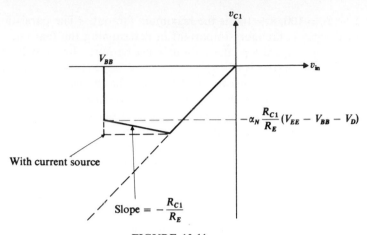

FIGURE 13-11
Transfer characteristics with I_0 replaced by R_E.

As v_{in} increases so does v_E, and the current in R_E increases correspondingly; thus the transfer characteristic has slope $-(R_{C1}/R_E)$, as shown in Fig. 13-11. Saturation of Q_1 occurs when $v_{in} = v_1$.

The change ΔI of the power supply current is, if we ignore R_0,

$$\Delta I = \frac{(v_1 - V_D + V_{EE}) - (-V_{BB} - V_D + V_{EE})}{R_E}$$

$$= \frac{V_D}{2R_E}$$

for the case for which

$$-V_{BB} - \frac{v_1 + v_0}{2}$$

ECL Design Considerations

In the design of the ECL circuit, dc fan-out and fan-in are not determining factors. This is easily seen by considering the circuit of Fig. 13-10a. The base current of Q_1 is $I_0(1 - \alpha_N)$; if n ECL circuits are connected to the output of a test gate, the load current is $nI_0(1 - \alpha_N)$. But change of current in R_{C1} or R_{C2} caused by this load is only

$$\Delta I_R = nI_0(1 - \alpha_N)(1 - \alpha_N)$$

If $\Delta I_R \leq 0.01\alpha_N I_0$, negligible change of v_{C1} or v_{C2} will result from this load; thus n can be as large as

$$n = \frac{0.01\beta_N}{1 - \alpha_N}$$

For $\beta_N = 100$, $n = 100$ is the maximum fan-out. The parasitic capacitance of the load gates is far more important in determining the fan-out. In order to obtain high-speed operation, the fan-out must be much lower than the dc limit; typical maximum fan-out is $n = 15$.

Our simple transistor model used in the analysis of the circuit does not include anything which limits the fan-in. As was the case for fan-out, parasitic capacitances become the determining factor; a typical value is $m = 12$.

Our first-order analysis has indicated that the logic swing is determined by a ratio of resistors, and that the logic voltages are determined by power supply voltages and diode offset voltages. Clearly we do not have sufficient information to design the circuit on the basis of logic voltages, fan-out, and fan-in; either speed considerations or power-dissipation limitations or both must be invoked as well. Speed considerations are dealt with later in this chapter by means of transient analysis; power dissipation can easily be calculated.

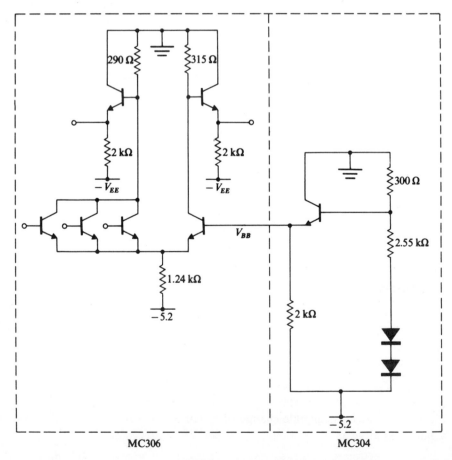

MC306   MC304

FIGURE 13-12
MC 306 three-input ECL gate with MC 304 bias driver.

In practical circuits the logic swing is not made exactly V_D. Note that the value of v_0 is not critical as long as $v_0 \leq (v_1 - V_D)$; this will always ensure that the NOR-side transistors are cut off when the input is v_0. It is common practice to use v_0 slightly less than $v_1 - V_D$ in order to allow for variations of v_1 and v_0 with temperature. When this is done, R_{C2}/R_{C1} will differ slightly from that given in (13-4). The MC 306 three-input ECL gate is shown in Fig. 13-12, together with the MC 304 bias driver. Logic and supply voltages are

$$v_1 = -0.750$$
$$v_0 = -1.55$$
$$v_l = 0.800$$
$$V_{BB} = -1.15$$
$$V_{EE} = -5.20$$

The bias driver uses two diodes in a voltage divider to produce a temperature coefficient for V_{BB} that maintains equal noise margins on OR and NOR outputs as temperature varies.

13-4 NONLINEAR ANALYSIS OF THE ECL CIRCUIT

While the analysis with the first-order transistor model yields considerable information regarding logic levels, it does not suffice for the determination of noise margin, transition width, etc., because these depend on the detailed nature of the transfer characteristic. For this a nonlinear analysis is necessary. Fortunately, since the transistors do not saturate, the substrate p-n-p transistors are always cut off, and contribute only parasitic capacitance. If we assume I_{s3} to be negligible, the four-layer-model equations become the Ebers Moll equations for the three-layer device:

$$\begin{bmatrix} I_1 \\ I_2 \end{bmatrix} - \begin{bmatrix} 1 & -\alpha_I \\ -\alpha_N & 1 \end{bmatrix} \begin{bmatrix} I_{s1}(e^{q\phi_E/mkT} - 1) \\ I_{s2}(e^{q\psi_C/mkT} - 1) \end{bmatrix}$$

For the transistor with reverse-biased collector, these become

$$I_E = I_{s1}(e^{q\phi_E/mkT} - 1) + \alpha_I I_{s2}$$
$$-I_C = -\alpha_N I_{s1}(e^{q\phi_E/mkT} - 1) - I_{s2}$$

If we are only interested in the transfer characteristic in regions where $I_C \gg I_{s2}$ and $e^{q\phi_E/mkT} \gg 1$, we can make the approximation

$$I_C \approx \alpha_N I_{s1} e^{q\phi_E/mkT} \approx \alpha_N I_E$$

Now for simplicity, consider the case where a current source is used to obtain I_0, and where only a single input transistor is on, as shown in Fig. 13-13a. We assume that the emitter followers have zero base current, and contribute only an offset voltage V_D. From the circuit of Fig. 13-13, we note that

$$\phi_{E1} = v_{in} - v_E$$
$$\phi_{E2} = -V_{BB} - v_E$$

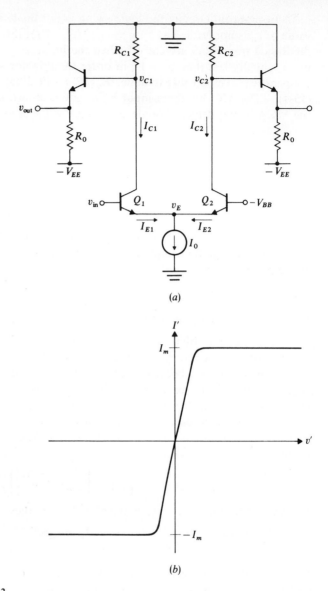

(a)

(b)

FIGURE 13-13
(a) Circuit used in nonlinear analysis; (b) transfer characteristics I' versus v'.

Then the emitter currents are

$$I_{E1} = I_{s1}e^{q(v_{in}-v_E)/mkT}$$
$$I_{E2} = I_{s1}e^{q(-V_{BB}-v_E)/mkT}$$

and these are related to the bias current by

$$I_0 = I_{E1} + I_{E2}$$

Taking the ratio of emitter currents, we obtain

$$\frac{I_{E1}}{I_{E2}} = e^{q(v_{in}+V_{BB})/mkT}$$

from which

$$I_0 = I_{E1}[1 + e^{-q(v_{in}+V_{BB})/mkT}]$$
$$= I_{E2}[1 + e^{q(v_{in}+V_{BB})/mkT}]$$

The collector currents are now easily obtained:

$$I_{C1} = \alpha_N I_{E1} = \frac{\alpha_N I_0}{1 + e^{-q(v_{in}+V_{BB})/mkT}} \qquad (13\text{-}5)$$

$$I_{C2} = \alpha_N I_{E2} = \frac{\alpha_N I_0}{1 + e^{q(v_{in}+V_{BB})/mkT}} \qquad (13\text{-}6)$$

In (13-5) and (13-6) note that:

1 When $v_{in} = -V_{BB}$, $I_{C1} = I_{C2} = \alpha_N I_0/2$.
2 Odd-function symmetry exists about $v_{in} = -V_{BB}$.

To take advantage of the symmetry, we define new variables

$$I_m \triangleq \frac{\alpha_N I_0}{2}$$
$$I' \triangleq I_{C1} - I_m$$
$$v' \triangleq v_{in} + V_{BB}$$

In terms of these variables, the transfer function is now as shown in Fig. 13-13b; it can easily be expressed analytically as

$$I' = I_{C1} - I_m = \frac{2I_m}{1 + e^{-q(v')/mkT}} - I_m = I_m \frac{1 - e^{(-qv')/mkT}}{1 + e^{(-qv')/mkT}} = I_m \tanh \frac{qv'}{2mkT} \qquad (13\text{-}7)$$

While the transfer characteristic we seek is actually v_{out} versus v_{in}, it can be obtained from (13-7) by recalling that

$$v_{out} = -I_{C1}R_{C1} - v' \qquad (13\text{-}8)$$

and

$$v' = v_{in} + V_{BB}$$

It is not necessary to carry this out, however, since we can obtain most of the information we need from (13-7).

Unity-gain, Points and Transition Width

To calculate the unity-gain points, we note that

$$\frac{\partial v_{out}}{\partial v_{in}} = \frac{-R_{C1}\partial I_{C1}}{\partial v_{in}} = -R_{C1}\frac{\partial I_{C1}}{\partial I'}\frac{\partial I'}{\partial v'}\frac{\partial v'}{\partial v_{in}} = -R_{C1}\frac{\partial I'}{\partial v'}$$

Using the expression for I', we obtain

$$\frac{\partial v_{out}}{\partial v_{in}} = -R_{C1}I_m\frac{q}{2mkT}\operatorname{sech}^2\frac{qv'}{2mkT}$$

Now the unity-gain points are those values of v' for which $\partial v_{out}/\partial v_{in} = -1$. Let these be designated v'_μ. Then we have

$$-1 = -\frac{R_{C1}I_m(q/2mkT)}{\cosh^2(qv'_\mu/2mkT)}$$

Solving for v'_μ, we obtain the two unity-gain points $v'_{\mu 1}$ and $v'_{\mu 0}$:

$$v'_{\mu 1} = -v'_{\mu 0} = \frac{2mkT}{q}\cosh^{-1}\sqrt{\frac{q\alpha_N I_0 R_{C1}}{4mkT}} \tag{13-9}$$

The transition width v'_w is

$$v'_w = |v'_{\mu 1} - v'_{\mu 0}| = 2v'_{\mu 1} = \frac{4mkT}{q}\cosh^{-1}\sqrt{\frac{q\alpha_N I_0 R_{C1}}{4mkT}} \tag{13-10}$$

This can be expressed in terms of the logic swing v_l by noting that $v_l = \alpha_N I_0 R_{C1}$; inserting this in (13-10) we obtain

$$\cosh\frac{qv'_w}{4mkT} = \sqrt{\frac{qv_l}{4mkT}}$$

Now we make the approximation that

$$\cosh\frac{qv'_w}{4mkT} \approx \frac{1}{2}e^{qvw'/4mkT}$$

Then

$$v'_w \approx \frac{2mkT}{q}\ln\frac{v_l q}{mkT} \tag{13-11}$$

This relationship illustrates the important fact that, because v_l depends essentially only on junction offset voltage and not on resistors or voltages in the circuit, the transition width is independent of circuit parameters. At 300°K, the transition width is

$$v'_w = 178 \text{ mV}$$

This means that a change of v_{in} from $v_{in} = -V_{BB} - 89.5$ mV to $v_{in} = -V_{BB} + 89.5$ mV suffices to cause the output to change from one unity-gain point to the other. (Recall that the unity-gain points are *not* the operating points.)

Noise Immunity

Because of the symmetry of the transfer characteristic, it can be shown that the threshold point occurs at $v' = 0$. The symmetry also means that $NI^0 = NI^1 \triangle NI$, and that

$$NI = \frac{v'_w}{2} = 0.311$$

Noise Margin

Again because of the symmetry of the transfer characteristic, $NM^0 = NM^1 \triangle NM$ and

$$NM = \frac{v_l}{2} - v'_{\mu 1} = \frac{v_l}{2} - \frac{v'_w}{2} = \frac{V_D}{2} - \frac{v_w}{2} = 0.311$$

Effects of Fan-In on the Transfer Characteristic

The preceding analysis assumed that only one input transistor is involved in the determination of the transfer characteristic. If m_1 input transistors have identical base voltages, they will each share the current on the input side of the gate, and each will have a lower base-emitter voltage than would a single transistor. This fact will slightly alter the transfer characteristic, since now the effective I_{E1} is

$$\frac{I_{E1}}{m_1} = I_{s1} e^{q(v_{in} - v_E)/mkT}$$

which leads to

$$\frac{I_{E1}}{I_{E2}} = m_1 e^{q(v_{in} + V_{BB})/mkT}$$

As a result, the input voltage which produces the threshold voltage is now less than V_{BB}, and is given by

$$v_{in} = V_{BB} - \frac{mkT}{q} \ln m_1 \qquad (13\text{-}12)$$

Exercise 13-1 Show that the input threshold voltage is given by (13-12).

13-5 FIRST-ORDER TRANSIENT ANALYSIS OF ECL

ECL Layout

Referring to Fig. 13-12, we can easily determine the number of isolation regions required for the ECL circuit. All resistors can be placed in a single isolation region,

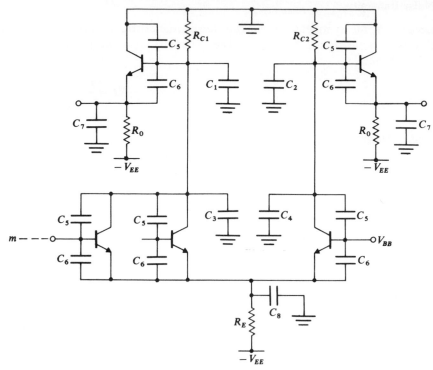

FIGURE 13-14
ECL parasitics.

and since the emitter followers also have their collectors connected to the most positive supply voltage, these transistors can also share the resistor isolation region. All input transistors have collectors connected together and can share a single isolation region.   A separate region is required for the OR-side transistor.   The isolation regions required are

Resistors and emitter followers: 1
Input transistors: 1
OR-side transistor: 1
    Total: 3

If the bias driver is to be fabricated on the same chip, its resistors and emitter follower can share the isolation region with those of the ECL circuit, but two separate regions are required for the diodes.

The parasitic capacitances accompanying the circuit are shown in Fig. 13-14. Here all resistors are represented by pi-section lumped models, with half the total

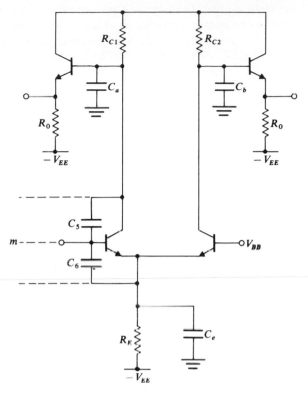

FIGURE 13-15
Equivalent circuit for ECL parasitics.

capacitance attached to each end. In the analysis of transient behavior, we linearize all capacitances by using appropriate average values. Combining capacitances where appropriate, we obtain the equivalent circuit of Fig. 13-15. The capacitances in the equivalent circuit are

$$C_a = C_3 + C_1 + C_5 + \frac{C_7}{\beta_N}$$

$$C_b = C_4 + C_2 + 2C_5 + \frac{C_7}{\beta_N}$$

$$C_e = C_8 + C_6$$

In the first-order analysis of ECL, it was noted that R_E, R_{C1}, and R_{C2} must be chosen to give the speed-power compromise desired by the designer. The speed of operation depends on the resistor values used, and the parasitic capacitances in the circuit. Because the parasitic capacitances of the resistors depend upon the resistor area, the design of the circuit for a particular power-delay-time product requires an iterative procedure which is best implemented by computer. For our purposes, we therefore select reasonable values for the resistors, calculate the capacitances, and use a first-order analysis to estimate the transient behavior. The following values are chosen:

$$R_{C1} = 250 \ \Omega$$
$$R_{C2} = 280 \ \Omega$$
$$R_E = 1 \ \text{k}\Omega$$
$$R_0 = 1.5 \ \text{k}\Omega$$
$$V_{EE} = 5.2 \ \text{V}$$
$$V_{BB} = 1.15 \ \text{V}$$

A base diffusion sheet resistance of 200 Ω per square is assumed. Resistors R_{C1}, R_{C2}, and R_E are 1 mil wide while R_0 is 0.5 mil wide. The capacitance values are

$$C_8 \approx 0.8 \ \text{pF}$$
$$C_1 \approx 0.3 \ \text{pF}$$
$$C_7 \approx 0.7 \ \text{pF}$$
$$C_5 \approx 2.0 \ \text{pF}$$
$$C_6 \approx 1.0 \ \text{pF}$$

For an epitaxial-layer thickness of 12 μm, the substrate capacitance C_1 is approximately 4.4 pF. However, since no transistors saturate, the parasitic substrate p-n-p transistors are always cut off. Therefore it is not necessary to use a thick epitaxial layer to reduce the current gain of the substrate p-n-p transistors. If a 5-μm epitaxial layer is used, C_1 is reduced to approximately 3.2 pF.

For our analysis, we consider only a three-input ECL gate; note that since Q_B is in an isolation region by itself, its substrate capacitance will be slightly larger than $C_3/3$:

$$C_4 \approx 1.3 \ \text{pF}$$

In order to obtain a tractable analysis, we make the following approximations:

1 Parasitic effects dominate transient behavior.
2 Emitter-follower transistors are never cut off.
3 The input transistors do not enter the forward active region of operation until the emitter-base voltage is approximately V_D.
4 When the transistor is in the forward active region, the emitter-base voltage is approximately constant and equal to V_D.

The circuit to be analyzed for turn-on time is shown in Fig. 13-16. We denote by n_D the fan-out of the drive gate, by m_1 the number of input transistors in the test

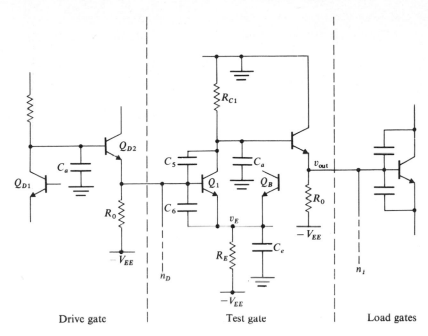

FIGURE 13-16
Equivalent circuit for transient analysis.

gate, and by n_T the fan-out of the test gate. From the above approximations it is clear that the turn-on of Q_1 will be made up of two distinct parts:

1 A delay as Q_1 changes its operation from cutoff to the forward active region

2 A turn-on transient as the collector voltage of Q_1 changes while Q_1 is in the forward active region

Turn-on Delay t_{D1}

The equivalent circuit which applies during the turn-on delay is shown in Fig. 13-17a. Since Q_B is in the forward active region during the turn-on delay, little change of the emitter voltage occurs. Therefore C_E can be neglected and C_6 can be assumed to be connected to ground. There will also be little change of v_{C1} until Q_1 enters the forward active region of operation; therefore C_5 can be assumed to be in parallel with C_6. Finally, with the emitter followers always in the active region, R_0 will have negligible effect in comparison with R_{C1}. By making use of these approximations we have reduced the equivalent circuit to a first-order network. Focusing attention on the base of Q_1, we see that the time constant is

$$\tau_1 - \frac{R_{C1}}{\beta_N} C_a'$$

where
$$C_a' = \beta_N C_a + (n_D + m_1)(C_5 + C_6)$$

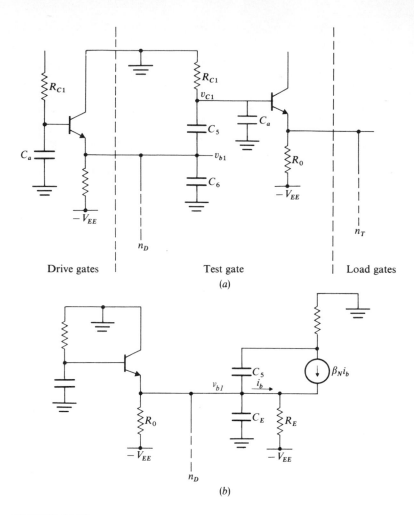

FIGURE 13-17
(*a*) Equivalent circuit for turn-on delay; (*b*) equivalent circuit for turn-on transient.

The initial and final values of v_{b1} are known from the dc analysis we have already performed. Since v_{b1} is connected to the output of the drive gate, the initial value is v_0, and the final value is v_1. It will be recalled that the circuit was designed so that V_{BB} is the average of v_1 and v_0; when $v_{b1} = V_{BB}$ both Q_1 and Q_B have equal emitter-base voltages and the value is approximately V_D. We may therefore consider Q_1 to enter the active region when $v_{b1} \approx V_{BB}$. The approximate expression for $v_{b1}(t)$ during the turn-on delay is

$$v_{b1}(t) = v_1 + (v_0 - v_1)e^{-t/\tau_1}$$

The turn-on delay is found by setting $v_{b1}(t_{D1}) = V_{BB}$:

$$t_{D1} = \tau_1 \ln 2$$

For the capacitance values previously calculated and for $m = 3$, $n_D = 15$, $\beta_N = 50$, we obtain

$$t_{D1} \approx 1.2 \text{ ns}$$

Turn-on Transient t'

For tractability of analysis, we make the approximation that once Q_1 enters the forward active region of operation, Q_B quickly cuts off and has no further influence on the behavior of the circuit. During the ensuing transient, the equivalent circuit is as shown in Fig. 13-17b; here $C_E = C_8 + mC_6$. Approximating the circuit of Fig. 13-17b by a first-order network and focusing attention on v_{b1}, we find the time constant to be

$$\tau' \approx \frac{R_{C1} C'}{\beta_N}$$

where

$$C' \approx \beta_N C_a + \frac{n_D C_E}{\beta_N} + n_D C_5 \left(1 + \frac{R_{C1}}{R_E}\right)$$

For the values previously used, we obtain

$$\tau' \approx 1.7 \text{ ns}$$

The transient time of the output voltage of the drive gate is

$$t' \approx 2.2\tau' = 3.7 \text{ ns}$$

Fall Time of the Output

It will be noted that the transient time of the output of the drive gate is also the rise time of the emitter voltage of Q_1 and Q_B. The fall time of the collector voltage v_{C1}, and hence the output voltage v_{out}, depends upon both the rise time of the emitter voltage and the collector circuit of Q_1. The circuit can no longer be reduced to a first-order equivalent circuit. For simplicity, however, we treat the collector circuit separately, and then combine the results of emitter and collector circuits to obtain an estimate of the fall time. The collector of Q_1 has in parallel with R_{C1} an equivalent capacitance C'' consisting of all parasitic capacitances of the load gates reflected through the emitter follower. Thus

$$C'' \approx C_a + \frac{n_T(C_5 + C_6)}{\beta_N}$$

We combine the effects of the collector circuit and the rise of the emitter voltage by using Elmore's method, which yields the fall time t_{f1}:

$$t_{f1}^2 \approx (2.2\tau')^2 + (2.2\tau'')^2$$

where

$$\tau'' = R_{C1} C'' \approx 1.05 \text{ ns}$$

For the values previously calculated, we find

$$t_{f1} \approx 4.4 \text{ ns}$$

The turn-on time for the collector voltage of Q_1 is

$$t_{on} = t_{D1} + t_{f1} = 5.6 \text{ ns}$$

Turnoff Transient t_{off}

The turnoff time for the collector voltage of Q_1 is the same as the rise time for the output voltage of Q_{D2}, since all gates are assumed to be identical. Thus we have

$$t_{off} = t_{D1} + t' = 4.9 \text{ ns}$$

If we define the average propagation delay t_D by

$$t_D = \frac{t_{on} + t_{off}}{2}$$

we obtain for the ECL circuit we have analyzed

$$t_D \approx 5.2 \text{ ns}$$

The geometry we have used in laying out the circuit is rather large, and it would be possible to reduce the size of the circuit by using $\frac{1}{2}$-mil-wide lines for R_E, R_{C1}, and R_{C2}, and 0.25-mil lines for R_0. The size of the transistors could also be reduced, and the corresponding decrease of all parasitic capacitances would lead to a reduction of the average propagation delay. It should be borne in mind, however, that a great many simplifying approximations were made in the transient analysis, which is at best crude. It is interesting to note, however, that the results of the analysis agree reasonably well with the values specified for the MC 1007 ECL circuit. This circuit uses $R_{C1} = 290 \ \Omega$, $R_E = 1.18 \text{ k}\Omega$, $V_{EE} = 5.2$, $V_{BB} = 1.175$, and has an average propagation delay of 9 ns, with a fan-out of 15. With a fan-out of 1, the average propagation delay of the MC 1007 circuit is 4 ns.

PROBLEMS

13-1 In the circuits of Fig. P13-1, the current I_1 is made zero at $t = 0$. Calculate the time elapsed until the collector voltage begins to change.

13-2 Construct the transfer characteristic for the circuit of Fig. 13-10a; use this to find the operating points v_1 and v_0.

13-3 The emitter-follower outputs of the ECL gates are replaced by Darlington pairs. Find v_1 and v_0.

13-4 Design an ECL circuit using a resistor R_E instead of a current source, and impose the restriction that the collector junction of the input transistors is never reverse-biased by less than 100 mV. Calculate v_1 and v_0.

13-5 Design an ECL circuit with a logic swing greater than 6 V, without adding extra processing steps. (*Hint:* Use the breakdown of an emitter junction.)

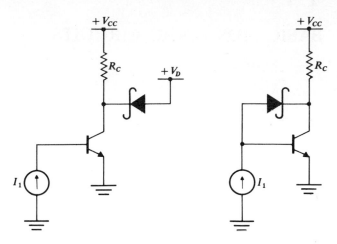

13-6 Plot the transfer characteristic obtained from a nonlinear analysis of ECL; use it to graphically determine the operating points.

13-7 Calculate and plot the input resistance of the ECL gate for $v_0 \leq v_{in} \leq v_1$.

13-8 Use electrothermal models to calculate the temperature coefficient of V_{BB} in the bias driver.

REFERENCES

1 BAKER, R. H.: Maximum Efficiency Switching Circuits, *Tech. Rept.* TR-110, M.I.T. Lincoln Laboratories, 1956.

2 TARUI, Y., Y. HAYASHI, H. TESHIMA, and T. SEKIGAWA: Transistor Schottky-barrier-diode Integrated Logic Circuit, *IEEE J. Solid-State Circuits*, vol. SC-4, pp. 3–12, 1969.

3 KURZ, B., and M. BARRON: Improved Schottky Clamped (T^2L) Circuits, *IEEE J. Solid-State Circuits*, vol. SC-7, pp. 175–179, 1972.

4 KAN, D.: Shunt-feedback Schottky-clamped Logic Gates, *IEEE J. Solid-State Circuits*, vol. SC-7, pp. 405–411, 1972.

5 WIEDMANN, S. K.: A Novel Saturation Control in TTL Circuits, *IEEE J. Solid-State Circuits*, vol. SC-7, pp. 243–251, 1972.

6 MEYER, C. S., D. K. LYNN, and D. J. HAMILTON: "Analysis and Design of Integrated Circuits," chap. 7, McGraw-Hill Book Company, New York, 1968, gives a detailed analysis of ECL.

14

BASIC MOS LOGIC CIRCUITS

The MOSFET plays an important role in integrated logic circuits, particularly in large-scale integration (LSI). This is because the MOSFET can be made with very small geometry, and therefore large device densities can be realized. Processing technology has advanced to a degree that makes possible high-yield production of LSI chips containing thousands of devices. The combination of high-density circuits and high-yield production has been responsible for the emergence of such diverse items as large-scale random-access memories, READ-only memories, sophisticated but low-cost electronic calculators, and electronic wristwatches.

The inverter circuit is always a basic part of logic circuits, as we saw in Chap. 12. In this chapter we consider inverter circuits of several types, and we investigate the use of the inverter in static and dynamic logic circuits.

14-1 MOS INVERTERS[1,2]

Before we begin a detailed consideration of the inverter, we first review the static characteristics of the MOSFET, and we focus attention on enhancement-mode devices. It will be recalled from Chap 6 that p-channel enhancement-mode devices

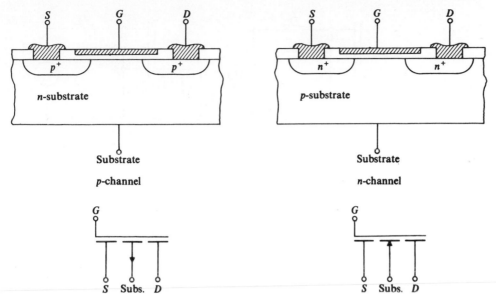

FIGURE 14-1
p-channel and *n*-channel enhancement-mode MOSFETs.

are made by starting with *n*-type material, while *n*-channel devices are made by starting with *p*-type material, as shown in Fig. 14-1. If we define polarities as shown in Fig. 14-2*a*, the idealized *IV* characteristics for the MOSFET will be as shown in Fig. 14-2*b*. It is important to note that *all voltages are measured relative to the source;* in the curves of Fig. 14-2*b*, V_T is the threshold voltage derived in Chap. 6. For the polarities shown in Fig. 14-2*a*, V_T will be positive for both *n*-channel and *p*-channel enhancement-mode devices. The important aspects of device behavior can be summarized as follows:

1 No inversion channel forms unless $V_G > V_T$; therefore I_D is zero for all $V_G \leq V_T$.
2 For $V_G > V_T$, an inversion channel extends between source and drain for all V_D such that $V_G - V_D > V_T$. For these voltages the device behaves, between source and drain terminals, as a nonlinear resistor, and I_D is given by

$$I_D = \mu \frac{Z}{L} C_0 \left[(V_G - V_T)V_D - \frac{V_D{}^2}{2} \right] \qquad (14\text{-}1)$$

where μ = majority carrier mobility
$$C_0 = \frac{K_0 \varepsilon_0}{d_0}$$
d_0 = gate oxide thickness
Z = channel width
L = channel length
This region of operation is called the *nonsaturated* or *triode* region.

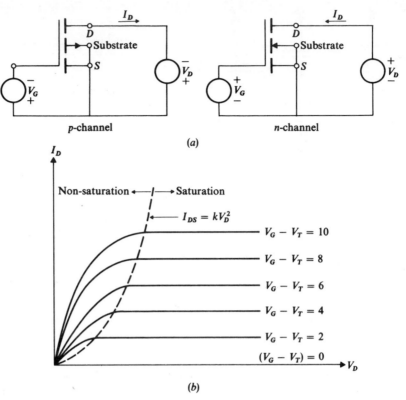

FIGURE 14-2
(*a*) Definition of polarities for MOSFETs; (*b*) IV characteristics.

3 If $V_G > V_T$, but V_D is large enough that $V_G - V_D \leq V_T$, an inversion channel will form, but it will not extend all the way to the drain diffusion. The channel current is now independent of V_D, and is given by

$$I_D = \mu \frac{Z}{L} \frac{C_0}{2} (V_G - V_T)^2 \qquad (14\text{-}2)$$

For this condition, called *saturation*, one may think of the inversion channel as being *pinched-off* at the drain end. For any given V_G, the drain voltage V_{DS} at which saturation occurs is that for which

$$V_G - V_{DS} = V_T$$

or

$$V_{DS} = V_G - V_T \qquad (14\text{-}3)$$

and the drain current I_{DS} at the edge of saturation is

$$I_{DS} = \mu \frac{Z}{L} \frac{C_0}{2} (V_G - V_T)^2 \qquad (14\text{-}4)$$

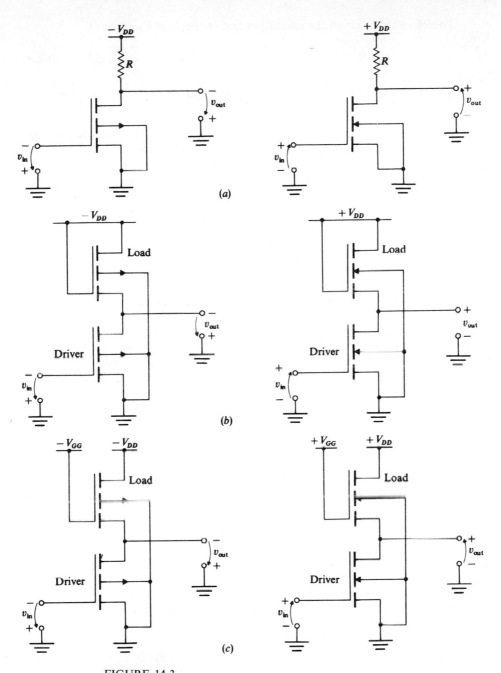

FIGURE 14-3
(a) Diffused resistor load; (b) saturated MOSFET load; (c) nonsaturated MOSFET load.

Therefore the locus of the saturation line in the $I_D - V_D$ plane is

$$I_{DS} = \mu \frac{Z}{L} \frac{C_0}{2} V_D{}^2 \qquad (14\text{-}5)$$

4 There is negligible dc gate current for both saturated and nonsaturated operation.

An inverter circuit can be formed by using a MOSFET with a diffused-resistor load, as shown in Fig. 14-3a. However, it will be recalled that to keep the area of the MOSFET small, Z and L will be small, with typical Z/L ratios being about 10. Typical values of drain current will be tens of microamperes. For values of V_{DD} of 5 to 15 V, it can be seen that large-value resistors would be required; this fact nullifies the advantages gained from the small geometry and high density of the active devices. Therefore, diffused resistors are not used, but rather an active device is used as a load. The load device can be operated in the saturated mode, as shown in Fig. 14-3b, or in the nonsaturated mode, as shown in Fig. 14-3c; in the latter case it is necessary that

$$V_{GG} - V_T \geqq V_{DD}$$

to ensure nonsaturated operation. The circuit of Fig. 14-3b has the advantage that only a single power supply is required; however, it should be noted that in the circuit of Fig. 14-3c no dc current flows in the V_{GG} supply. As we shall later see, for a given V_{DD} the logic swing is larger for the circuit of Fig. 14-3c.

14-2   THE INVERTER WITH SATURATED LOAD

When the MOSFET is operated with gate connected to drain,

$$V_G = V_D$$

and we see that there can be no inversion at the drain end of the channel. Moreover, no drain current can flow unless $V_D > V_T$. Stated another way, for very small values of I_D there will be an offset of V_T between source and drain. For all $V_D > V_T$, drain current will flow, and the device will be in the saturated mode.

The drain current for $V_G = V_D$ is sketched superimposed on the $I_D V_D$ characteristics in Fig. 14-4a. When the device is used as a load, the gate-source voltage is $V_G = V_{DD} - v_{out}$, and the $I_D v_{out}$ characteristic is as shown in Fig. 14-4b. It is convenient to define a *conduction parameter* k_L for the load device as

$$k_L \triangleq \frac{\mu_L(Z_L/L_L)C_{0L}}{2}$$

The drain current can be written in terms of this parameter as

$$I_D = k_L(V_G - V_T)^2$$

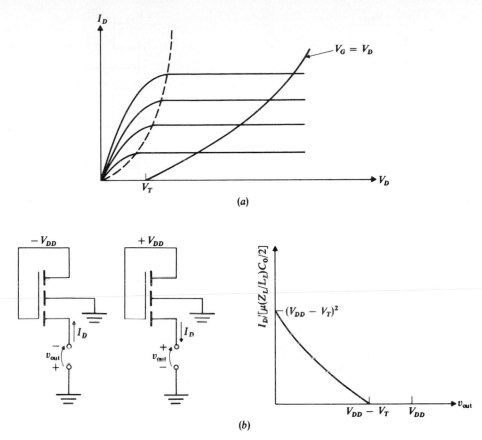

FIGURE 14-4
(a) $I_D V_D$ characteristic for the case $V_G = V_D$; (b) $I_D v_{out}$ characteristic for the load device.

for a saturated device; for the circuit of Fig. 14-4b the drain current becomes

$$I_D = k_L(V_{DD} - v_{out} - V_T)^2$$

When the driver transistor is included in the inverter circuit, the characteristic of Fig. 14-4b becomes a load line for the driver; the result is shown in Fig. 14-5. Since the source-substrate voltage is different for driver and load devices, the threshold voltages will be slightly different, as was discussed in Sec. 6-4. However, for simplicity we assume equal threshold voltages for load and driver transistors. Since both devices are made with the same oxide thickness and with the same substrate material, they will have equal mobility and equal C_0; therefore the conduction parameters for load and driver devices differ only if the Z/L ratios differ. In Fig. 14-5, the drain current is not normalized to the conduction factor.

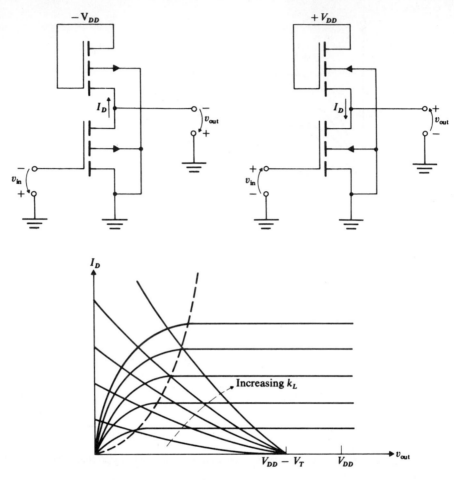

FIGURE 14-5
Load lines of load transistor on $I_D v_{out}$ characteristic of the driver transistor.

It can be seen qualitatively that as k_L increases, the inverter performance is degraded because the output voltage when the driver is on is increased, thus reducing the logic swing. We shall later see that the ratio of conduction factors of the driver and load devices is important in determining the minimum output voltage; this ratio is

$$\beta_R \triangleq \frac{k_D}{k_L}$$

For typical inverters, β_R is of the order of 25. The surface geometry for the inverter is shown in Fig. 14-6; it will be noted that to achieve high β_R, the channel of

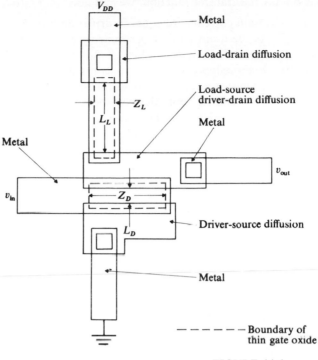

FIGURE 14-6
Layout of the inverter.

the load resistor must be long and narrow and that of the driver short and wide. It is interesting to calculate the β_R which would be obtained for two devices whose channels use the same amount of surface area. If we assume that the minimum line width is used for Z_L and L_D, so that $L_D = Z_L$, β_R is given by

$$\beta_R = \frac{Z_D L_L}{Z_L L_D} = \frac{Z_D L_L}{L_D{}^2}$$

But for both areas to be equal, $Z_D L_D = Z_L L_L$, from which we find

$$\beta_R = \left(\frac{Z_D}{L_D}\right)^2$$

Inverter Transfer Function

Since the inverter is to be used in logic circuits, we are interested in its transfer function

$$v_{out}(v_{in})$$

To obtain the transfer function, we first note qualitatively several important facts:

1   Drain currents are equal in driver and load.

2   As the drain current becomes very small, it can be seen from Fig. 14-5 that regardless of the value of β_R, $v_{out} \rightarrow V_{DD} - V_T$. That v_{out} does not reach the supply voltage is a consequence of the offset occurring in the $I_D V_D$ characteristic of the load transistor, as previously discussed.

3   For small v_{in}, both devices operate in the saturation region.

4   For good inverter performance, when v_{in} is large the driver should operate in the nonsaturated region.

The transfer function can be obtained quantitatively by equating drain currents and solving for $v_{out}(v_{in})$.   For the load transistor,

$$V_G = V_{DD} - v_{out} = V_D$$

and
$$I_D = k_L(V_{DD} - v_{out} - V_T)^2 \qquad (14\text{-}6)$$

while for the driver, when it is saturated,

$$V_G = v_{in}$$
$$V_D = v_{out}$$

and
$$I_D = k_D(v_{in} - V_T)^2 \qquad (14\text{-}7)$$

Combining (14-6) and (14-7), we obtain

$$v_{out} = -\left(\frac{k_D}{k_L}\right)^{1/2}(v_{in} - V_T) + (V_{DD} - V_T) \qquad (14\text{-}8)$$

Equation (14-8) applies only for $v_{in} \geq V_T$, since that restriction was implicit in (14-7).   As long as both devices are saturated, (14-8) shows that the transfer function is linear; its slope is

$$\frac{dv_{out}}{dv_{in}} = -\left(\frac{k_D}{k_L}\right)^{1/2} = -\beta_R^{1/2}$$

It is interesting to note that for equal-area devices, as previously discussed

$$\frac{dv_{out}}{dv_{in}} = \frac{-Z_D}{L_D}$$

As v_{in} is further increased, the load transistor remains saturated but the driver enters the nonsaturated region; this occurs when

$$V_D = V_G - V_T$$

that is, when

$$v_{out} = v_{in} - V_T$$

Inserting this in (14-8) and solving, we find

$$v_{in} = V_T + \frac{V_{DD} - V_T}{1 + (\beta_R)^{1/2}} \qquad (14\text{-}9)$$

and
$$v_{out} = \frac{V_{DD} - V_T}{1 + (\beta_R)^{1/2}} \qquad (14\text{-}10)$$

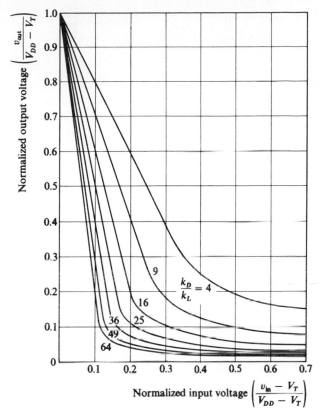

FIGURE 14-7
Inverter transfer function (after Penney et al.[1]).

For all v_{in} larger than that given by (14-9) the driver is nonsaturated and its drain current is given by

$$I_D = k_D[2(V_G - V_T)V_D - V_D{}^2]$$
$$= k_D[2(v_{in} - V_T)v_{out} - v_{out}{}^2] \qquad (14\text{-}11)$$

Combining (14-11) and (14-6), we obtain

$$v_{out}{}^2(1 + \beta_R) - v_{out}[2V_H + 2\beta_R(v_{in} - V_T)] + V_H{}^2 = 0 \qquad (14\text{-}12)$$

where $V_H \triangleq V_{DD} - V_T$.
Equation (14-12) can be solved for $v_{out}(v_{in})$; the transfer function is plotted in Fig. 14-7.

Exercise 14-1 Derive Eq. (14-9).

Exercise 14-2 Solve Eq. (14-12) and obtain $v_{out}(v_{in})$ for the case where the driver is not saturated.

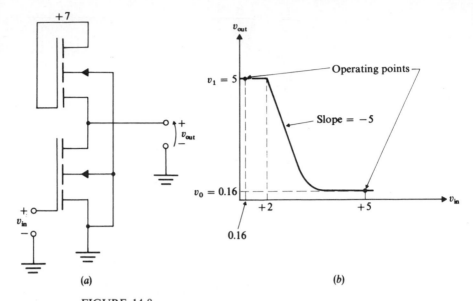

FIGURE 14-8
(*a*) Inverter circuit for Example 14A; (*b*) transfer function and operating points.

Operating Points

The operating points can be determined by assuming the test inverter to be driven by an identical inverter. When the input signal is small, the drive transistor of the test inverter is cut off, and the output voltage is

$$v_{\text{out}} = V_{DD} - V_T \qquad (14\text{-}13)$$

This will be the case for any input voltage as long as the restriction $v_{\text{in}} < V_T$ holds.

When the input voltage of the test inverter is $V_{DD} - V_T$, the drive transistor is assumed to be nonsaturated; the output voltage is then obtained by substituting $v_{\text{in}} = V_{DD} - V_T$ in (14-12) and solving for v_{out}. The result is the second operating point. An alternate method for finding operating points graphically is to plot $v_{\text{out}}(v_{\text{in}})$ and $v_{\text{in}}(v_{\text{out}})$ on the same set of axes as discussed in Chap. 12; the intersections are then the operating points.

EXAMPLE 14A *Calculation of Operating Points.* Suppose an inverter is fabricated with *n*-channel enhancement-mode devices having $V_T = 2$ V, and $\beta_R = 25$. The supply voltage is $V_{DD} = 7$ V. The circuit is shown in Fig. 14-8*a*. If it is assumed that positive logic is used, the logic 1 voltage is

$$v_1 = V_H = V_{DD} - V_T = 5 \text{ V}$$

To find the zero-state voltage v_0, we let $v_{\text{in}} = 5$ in (14-12):

$$26 v_{\text{out}}^2 - 160 v_{\text{out}} + 25 = 0$$

Solving for v_{out} we obtain

$$v_{out} = 0.16 \text{ V} \triangleq v_0$$

The logic swing is

$$v_l = |v_1 - v_0| = 4.84 \text{ V}$$

The operating points are shown in Fig. 14-8b; the slope of the transfer characteristic in the linear region is $-(\beta_R)^{1/2} = -5$.   ////

Unity-gain Points and Noise Margin

The idealized transfer characteristic has no unity-gain point when both devices are saturated. When the driver is nonsaturated, the unity-gain point can be found by solving (14-12) for v_{in}, taking dv_{in}/dv_{out}, setting this result equal to -1, and solving for v_{out} and v_{in}. The noise margin NM was defined in Chap. 12 as the input voltage change required to cause the output voltage to change from an operating point to the nearest unity-gain point. Since the noise margins may differ for the two operating points, the notation used is

$$NM^0 = \text{noise margin with input at } v_0$$

$$NM^1 = \text{noise margin with input at } v_1$$

Because the inverter has no unity-gain point when v_{in} is low, we define the noise margin for this case as the input voltage difference between the operating point and the point at which the output voltage begins to change.

EXAMPLE 14B   *Calculation of Noise Margins.*   We consider again the inverter with the values of Example 14A. For that case, (14-12) becomes

$$v_{in} = \frac{v_{out}^2 + 3.46v_{out} + 0.96}{1.93v_{out}} \qquad (14\text{-}14)$$

from which we find

$$\frac{dv_{in}}{dv_{out}} = 0.521 - \frac{0.50}{v_{out}^2}$$

Setting $dv_{in}/dv_{out} = -1$ and solving for v_{out}, we obtain

$$v_{out} = 1.04 \text{ V}$$

Inserting this result in (14-14), we get

$$v_{in} = 2.83 \text{ V}$$

We next calculate

$$NM^1 = 5 - 2.83 = 2.17 \text{ V}$$

From Fig. 14-8b we see that

$$NM^0 = 2 - 0.16 = 1.84 \text{ V} \qquad ////$$

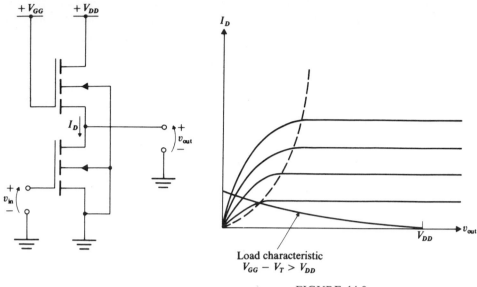

Load characteristic
$V_{GG} - V_T > V_{DD}$

FIGURE 14-9
The inverter with nonsaturated load.

14-3 THE INVERTER WITH NONSATURATED LOAD

If two supply voltages V_{GG} and V_{DD} are used as shown in Fig. 14-9 the load transistor can be made to operate always in the nonsaturation region, provided that

$$V_{GG} - V_{DD} > V_T$$

This configuration has the advantage that for $v_{in} = 0$, $v_{out} = V_{DD}$ since there is no offset in the load transistor characteristic; this means that for a given V_{DD}, the logic swing will be larger for the nonsaturated load inverter than for the saturated load inverter. Because the use of a second supply voltage introduces another variable, the analysis becomes rather cumbersome. For simplicity we consider the specific case for which

$$V_{GG} - 2V_T = V_{DD}$$

The current in the load is

$$
\begin{aligned}
I_D &= k_L[2(V_{GG} - v_{out} - V_T)(V_{DD} - v_{out}) - (V_{DD} - v_{out})^2] \\
&= k_L[2(V_{DD} - v_{out} + V_T)(V_{DD} - v_{out}) - (V_{DD} - v_{out})^2] \quad (14\text{-}15)
\end{aligned}
$$

For small $v_{in} - V_T$ the driver is saturated and its current is

$$I_D = k_D(v_{in} - V_T)^2 \quad (14\text{-}16)$$

We now make a change of variables

$$v'_{out} \triangleq V_{DD} - v_{out}$$
$$v'_{in} \triangleq v_{in} - V_T$$

Substituting these in (14-15) and (14-16) and combining the results, we obtain

$$v_{out}'^2 + 2V_T v_{out}' - \beta_R v_{in}'^2 = 0$$

from which we find

$$v_{out}' = -V_T + (V_T^2 + \beta_R v_{in}'^2)^{1/2} \qquad (14\text{-}17)$$

The unity-gain point occurs where $dv_{out}'/dv_{in}' = 1$; this occurs for

$$v_{in}' = V_T[\beta_R(\beta_R - 1)]^{-1/2} \qquad (14\text{-}18)$$

The input voltage for this case is

$$v_{in} = V_T + V_T[\beta_R(\beta_R - 1)]^{-1/2} \approx V_T \qquad (14\text{-}19)$$

As $v_{in} \to V_{DD}$, the driver becomes nonsaturated and the driver current is

$$I_D = k_D[2(v_{in} - V_T)(v_{out}) - v_{out}^2] \qquad (14\text{-}20)$$

Combining (14-20) and (14-15), we get

$$v_{out}'^2 + 2V_T v_{out}' = \beta_R[2v_{in}'(V_{DD} - v_{out}') - (V_{DD} - v_{out}')^2] \qquad (14\text{-}21)$$

Equation (14-21) can be solved for $v_{out}(v_{in})$ and from the solution the second unity-gain point can be determined. The operating points and noise margins can be determined by procedures similar to those used for the saturated load inverter.

The shape of the transfer characteristic for the unsaturated load will be similar to that for the saturated load, except that the logic swing of the former will be approximately V_T larger than that of the latter.

14-4 THE INVERTER WITH DEPLETION-MODE LOAD

The larger logic swing of the nonsaturated load inverter can be obtained without the use of an extra supply voltage if a depletion-mode MOSFET is used as a load. The depletion-mode device has an inversion channel with $V_G = 0$, and its *IV* characteristics are as shown in Fig. 14-10. Here the pinchoff voltage V_T is the gate-source voltage required to pinch off the inversion channel; its sign will be opposite that of the threshold voltage. The inversion channel will pinch off at the drain end when $V_D = V_{DS}$, where

$$V_{DS} = V_G - V_T$$

The drain current I_{DS} for this drain voltage is

$$I_{DS} = k[2(V_{DS} - V_T)V_{DS} - V_{DS}^2]$$

Thus the locus of the saturation drain current in the $I_D V_D$ plane is

$$I_{DS} = kV_D^2$$

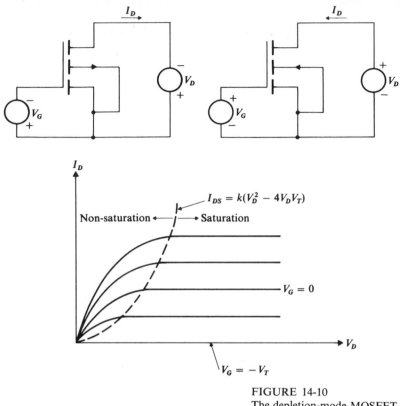

FIGURE 14-10
The depletion-mode MOSFET.

It will be noted that if gate and drain are connected together, the inversion channel cannot pinch off at the drain; for this connection the device is always operating in the nonsaturated mode. Its current is

$$I_D = k[2(V_D - V_T)V_D - V_D{}^2]$$
$$= k(V_D{}^2 - 2V_D V_T) \tag{14-22}$$

The depletion-mode device can be used as a load; its IV characteristic is as shown in Fig. 14-11. It is important to note that *extra processing steps are required if depletion devices are to be used.* For example, if both enhancement and depletion p-channel devices are to be used in an inverter, the net impurity concentration at the surface of the n material will have to be lower for the depletion device than for the enhancement device, unless different gate metals are being used. This is usually accomplished by using ion implantation for the channel of one type of device. Ion implantation enables the close control of the impurity concentration; such control is essential if reasonable tolerances are to be maintained

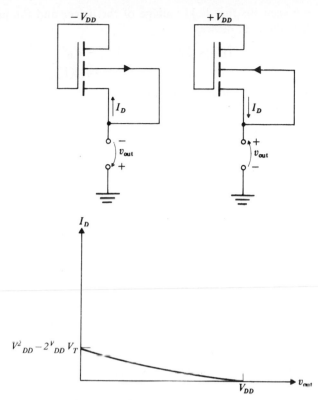

FIGURE 14-11
IV characteristic of the depletion device as a load.

on threshold and pinchoff voltages. Since the two types of devices have different impurity concentrations in their channels, the pinchoff voltage and threshold voltage will not necessarily have the same magnitude.

EXAMPLE 14C *Calculation of the Depletion-load IV Characteristic.* Consider an n-channel depletion-mode device with a pinchoff voltage $V_T = -3.0$ V. The $I_D v_{out}$ characteristic for this device with gate and drain connected is

$$I_D = k[v_{out}^2 - v_{out}(2V_{DD} - 2V_T) + V_{DD}(V_{DD} - 2V_T)]$$
$$= k[v_{out}^2 - 2v_{out}(V_{DD} + 3) + V_{DD}(V_{DD} + 6)] \qquad ////$$

Transfer Characteristic

The transfer characteristic for the inverter with depletion load is calculated in the same manner as for the other inverters, except that it is now necessary to distinguish between the threshold voltage of the driver and the pinchoff voltage of the load. For the load device,

$$I_L = k_L[2(V_G - V_{TL})V_D - V_D{}^2]$$
$$= k_L[2(V_{DD} - v_{out} - V_{TL})(V_{DD} - v_{out}) - (V_{DD} - v_{out})^2] \qquad (14\text{-}23)$$

while for the driver in saturation

$$I_D = k_D(v_{in} - V_{TD})^2 \qquad (14\text{-}24)$$

Changing variables to

$$v'_{out} = V_{DD} - v_{out}$$

and

$$v''_{in} = v_{in} - V_{TD}$$

we combine (14-23) and (14-24) and solve to obtain

$$-v'_{out} = -V_{TL} + (V_{TL}{}^2 + \beta_R v''_{in}{}^2)^{1/2} \qquad (14\text{-}25)$$

When the driver becomes nonsaturated,

$$I_D = k_D[2(v_{in} - V_{TD})(v_{out}) - v_{out}{}^2] \qquad (14\text{-}26)$$

The transfer function in this region is again found by combining (14-26) and (14-23). As can be seen from the above results, its shape will be similar to that of the nonsaturated load inverter.

14-5 THE COMPLEMENTARY INVERTER (CMOS)

In the inverter circuits we have thus far discussed, both transistors were either p-channel or n-channel devices. Early MOS logic circuits were of the p-channel enhancement-mode type, because p-channel devices were most easily fabricated. As the processing technology developed, such processes as ion implantation and silicon-gate technology made possible high-yield production of n-channel devices, and of both enhancement- and depletion-mode devices on the same chip. Further advances have made possible the fabrication of low-threshold n-channel and p-channel enhancement-mode devices on the same chip. With such complementary devices available, it is possible to design logic circuits in which the dc power dissipation is zero; power is dissipated only when switching occurs. Complementary MOS logic circuits are designated CMOS or COS/MOS circuits.

In order to fabricate both n- and p-channel devices on the same chip, some means must be provided for isolation; this is done by using an n-type substrate and providing a p-type diffusion for the n-channel devices, as shown in Fig. 14-12a. The p-channel device is to be used as a load, with n substrate and *source* connected to V_{DD}; in this configuration it has the IV characteristic shown in Fig. 14-12b. The inverter is formed by connecting the p-channel load device and the n-channel driver as shown in Fig. 14-13a. To verify inverter operation, we see that with $v_{in} = 0$, the n-channel device is cut off, and $I_D = 0$. But the gate voltage for the p-channel load is $V_G = V_{DD}$; if $V_{DD} > V_{TL}$, the p-channel load will have an inversion channel. With $I_D = 0$, the output will be $v_{out} = +V_{DD}$.

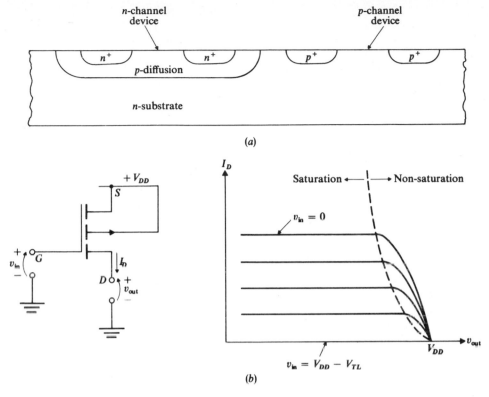

FIGURE 14-12
(*a*) Cross section of the CMOS chip; (*b*) *IV* characteristic of the *p*-channel device used as a load.

If $v_{in} = +V_{DD}$, the load device is cut off, and $I_n = 0$, but if $V_{DD} > V_{TD}$ the driver has an inversion channel. With $I_D = 0$, the output will be $v_{out} = 0$. The input and output voltages are seen to be compatible, and the circuit performs logic inversion. For positive logic, $v_1 = V_{DD}$ and $v_0 = 0$. Since $I_D = 0$ both for $v_{out} = v_1$ and $v_{out} = v_0$, no dc power is dissipated in either state.

Transfer Characteristic

To calculate the transfer characteristic, we assume for simplicity that

$$V_{TL} = V_{TD} = V_T$$

For small v_{in}, the driver is cut off and the load is nonsaturated. As v_{in} increases, no I_D flows until $v_{in} = V_T$, so there will be no change of v_{out}, and $v_{out} = V_{DD}$. For v_{in} slightly larger than V_T, I_D begins to flow, the load is nonsaturated, and the driver is saturated. The output voltage begins to decrease. As v_{in} continues to increase, I_D increases because V_{Gn} for the driver is increasing; however, V_{Gp} for the

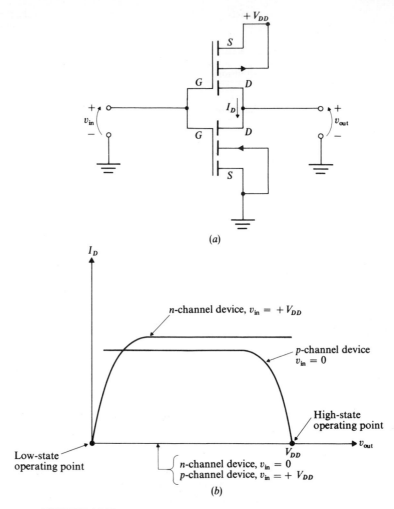

FIGURE 14-13
(*a*) The CMOS inverter; (*b*) *n*-channel and *p*-channel *IV* characteristics.

load is decreasing and the load saturates. At this point both devices are saturated; if the devices were ideal, dv_{out}/dv_{in} would be infinite.

The salient features of the transfer characteristic are shown in Fig. 14-14; the output voltages at which saturation of either device occurs are easily determined as follows. The load device will become saturated when

$$V_{Dp} = V_{Gp} - V_T$$

that is, for

$$V_{DD} - v_{out} = V_{DD} - v_{in} - V_T$$

which yields

$$v_{out} - v_{in} = V_T \qquad (14\text{-}27)$$

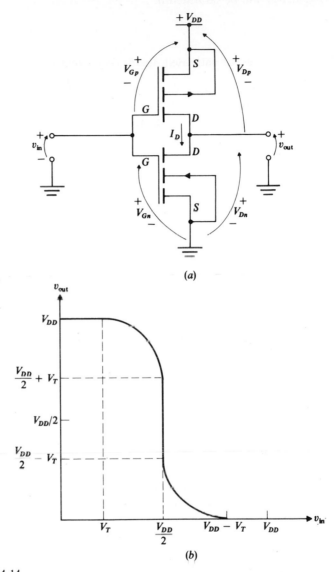

FIGURE 14-14
(a) Definition of polarities for the CMOS inverter; (b) transfer characteristic
for $V_{TL} = V_{TD}$, $k_L = k_D$.

The driver will be saturated if

$$V_{Dn} = V_{Gn} - V_T$$
$$v_{out} = v_{in} - V_T \qquad (14\text{-}28)$$

or

If both devices are saturated, the drain current is

$$I_D = k_L(V_{Gp} - V_T)^2$$
$$= k_L(V_{DD} - v_{in} - V_T)^2 \qquad (14\text{-}29)$$

But the drain current must also be

$$I_D = k_D(V_{Gn} - V_T)^2$$
$$= k_D(v_{in} - V_T)^2 \qquad (14\text{-}30)$$

If we assume for simplicity that the devices are truly complementary, then $k_L = k_D$; combining (14-29) and (14-30) we obtain

$$v_{in} = \frac{V_{DD}}{2} \qquad (14\text{-}31)$$

If $k_L \neq k_D$, the result is

$$v_{in} = \frac{V_{DD}}{1 + \sqrt{\beta_R}} - \frac{V_T(1 - \sqrt{\beta_R})}{1 + \sqrt{\beta_R}} \qquad (14\text{-}32)$$

We can now substitute (14-31) in (14-27) and (14-28) to obtain the output voltages at which each device saturates; the results are

$$v_{out} = \frac{V_{DD}}{2} + V_T$$

when the load saturates, and

$$v_{out} = \frac{V_{DD}}{2} - V_T$$

when the driver saturates.

When the input voltage reaches $V_{DD} - V_T$, $V_{Gp} < V_T$ and no inversion channel exists in the load. The drain current is zero and no further change of the output voltage can occur.

Exercise 14-3 Derive Eq. (14-32).

Exercise 14-4 Find the two values of v_{out} for which load and driver saturate if $k_L \neq k_D$.

The transfer characteristic is symmetrical about $v_{in} = V_{DD}/2$ for $k_L = k_D \triangleq k$, so it is necessary only to consider the case for which the load is nonsaturated and the driver is saturated. For the driver,

$$I_D = k(V_{Gn} - V_T)^2$$
$$= k(v_{in} - V_T)^2 \qquad (14\text{-}33)$$

and for the load

$$I_D = k[2(V_{Gp} - V_T)(V_{Dp}) - (V_{Dp})^2]$$
$$= k[2(V_{DD} - v_{in} - V_T)(V_{DD} - v_{out}) - (V_{DD} - v_{out})^2] \qquad (14\text{-}34)$$

Combining (14-33) and (14-34) and solving, we obtain

$$v_{out} = v_{in} + V_T + [(v_{in} + V_T)^2 + V_{DD}^2 - 2v_{in} V_{DD} - 2V_{DD} V_T - (v_{in} - V_T)^2]^{1/2}$$
$$(14\text{-}35)$$

The unity-gain point can be found from (14-35) in the usual manner by setting $dv_{out}/dv_{in} = -1$ and solving for v_{out} and v_{in}:

$$v_{in} = \frac{V_T}{4} + \frac{3V_{DD}}{8}$$

EXAMPLE 14D *Calculation of Noise Margin.* Consider a CMOS inverter with $V_{DD} = 5$, $V_T = 1.0$, and $k_L - k_D$. From (14-35), the transfer characteristic is given by

$$v_{out} = 1 + v_{in} + (15 - 6v_{in})^{1/2}$$

The unity-gain point occurs at

$$v_{in} = \frac{15 - \frac{9}{4}}{6} = 2.13$$

The noise margin is therefore seen to be

$$NM^0 \approx 2.13 \text{ V}$$

By symmetry, $NM^1 \approx 2.13$ V. ////

14-6 POWER DISSIPATION

We have already seen that the dc power dissipation of the CMOS inverter is zero. For the other inverter configurations, however, maximum I_D flows when the driver transistor has maximum V_G. For this condition, the driver drain voltage is nearly zero, but the load has maximum V_D, and therefore maximum power dissipation. The power dissipation for this case is easily calculated.

Inverter with Saturated Load

For the inverter with saturated load, the conditions are:

1 Load transistor saturated
2 $v_{out} \approx 0$
3 $v_{in} = V_{DD} - V_T$

For the load transistor, $V_G = V_D \approx V_{DD}$. The drain current is

$$I_D \approx k_L(V_{DD} - V_T)^2$$

and the power is

$$P = V_{DD}I_D = k_L V_{DD}(V_{DD} - V_T)^2$$

Inverter with Nonsaturated Load

For this inverter, the conditions are:

1 Load transistor nonsaturated
2 $v_{out} \approx 0$
3 $v_{in} = V_{DD}$

For the load transistor, $V_G = V_{GG}$ and $V_D = V_{DD}$. The drain current is

$$I_D \approx k_L[2(V_{GG} - V_T)(V_{DD}) - V_{DD}^2]$$

and the power is

$$P = V_{DD}I_D = k_L V_{DD}^2[2(V_{GG} - V_T) - V_{DD}]$$

Inverter with Depletion Load

For this case, the conditions are:

1 Load transistor nonsaturated
2 $v_{out} \approx 0$
3 $v_{in} \approx V_{DD}$

The load transistor has $V_G = V_{DD}$ and $V_D = V_{DD}$, and the drain current is

$$I_D = k_L[2(V_{DD} - V_T)V_{DD} - V_{DD}^2]$$

and the power is

$$P = I_D V_{DD} = k_L(V_{DD}^3 - 2V_T V_{DD}^2)$$

14-7 INVERTER TRANSIENT RESPONSE

In MOS logic circuits, inverters will be driving other MOS transistors, and the output current will be almost entirely capacitive, as shown in Fig. 14-15. If it is assumed that the input driving waveform is a step function, the turn-on and turnoff times can be calculated by solving the nonlinear differential equations which apply during each interval. Detailed solutions for the transient times are given in the literature. For our purposes, however, the detailed solution is too cumbersome. We wish to obtain only an estimate of the times; we do not require the details of the waveforms. To obtain an estimate we make many approximations at the outset, and we use an "average current" method. The method is illustrated by two examples.

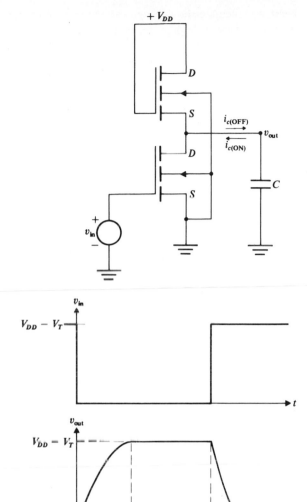

FIGURE 14-15
Saturated-load inverter driving a capacitance.

The Inverter with Saturated Load

To estimate t_{OFF} and t_{ON} for the circuit of Fig. 14-15, we make the following approximations:

> *1* When the output is in the high state, the voltages are $v_{\text{out}} = V_{DD} - V_T = V_H$ and $v_{\text{in}} \approx 0$.
> *2* When the output is in the low state, the voltages are $v_{\text{out}} \approx 0$ and $v_{\text{in}} \approx V_H$.

Next we calculate the average capacitor currents which flow as the output changes from low state to high state, and vice versa. These currents are

$$I_{c(OFF)} \triangleq \frac{\int_0^{V_H} i_{c(OFF)}(v_{out}) \, dv_{out}}{\int_0^{V_H} dv_{out}} \qquad \text{for } v_{in} = 0 \qquad (14\text{-}36)$$

and

$$I_{c(ON)} \triangleq \frac{\int_0^{V_H} i_{c(ON)}(v_{out}) \, dv_{out}}{\int_0^{V_H} dv_{out}} \qquad \text{for } v_{in} = V_H \qquad (14\text{-}37)$$

We then use these average currents to estimate t_{OFF} and t_{ON} as follows:

$$t_{OFF} \approx \frac{CV_H}{I_{c(OFF)}}$$

$$t_{ON} \approx \frac{CV_H}{I_{c(ON)}}$$

Consider first the case for which the input changes from high state to low state, causing the output to change from 0 to V_H. For $t > 0$, the driver will be cut off, and the current $i_{c(OFF)}$ consists entirely of the load transistor drain current. Since the load transistor is always saturated, the current $I_{c(OFF)}$ is easily found to be

$$I_{c(OFF)} = \frac{1}{V_H} \int_0^{V_H} k_L(V_{DD} - v_{out} - V_T)^2 \, dv_{out}$$

$$= \frac{1}{V_H} \int_0^{V_H} k_L(V_H - v_{out})^2 \, dv_{out}$$

$$= \frac{k_L}{3} V_H^2 \qquad (14\text{-}38)$$

The estimate of t_{OFF} is then

$$t_{OFF} \approx \frac{CV_H}{I_{c(OFF)}} = \frac{3C}{k_L V_H} \qquad (14\text{-}39)$$

Next consider the case for which the input changes from 0 to V_H. The driver is now turned on; it will be saturated until v_{out} drops to $V_H - V_T$, after which it will remain nonsaturated. The average driver drain current is

$$\langle I_D \rangle = \frac{\int_{V_H - V_T}^{V_H} k_D(V_H - V_T)^2 \, dv_{out} + \int_{V_H - V_T}^{0} k_D[2(V_H - V_T)v_{out} - v_{out}^2] \, dv_{out}}{\int_{V_H}^{0} dv_{out}}$$

$$= k_D(V_H - V_T)^2 \left(\frac{2}{3} + \frac{V_T}{3V_H} \right) \qquad (14\text{-}40)$$

The average capacitor current during turn-on is

$$I_{c(ON)} = \langle I_D \rangle - \langle I_L \rangle$$

where $\langle I_L \rangle$ is the average load transistor drain current; note that $\langle I_L \rangle$ is given by (14-38). Combining (14-38) and (14-40), we obtain

$$t_{ON} = \frac{3CV_H}{k_D} \left[\left(2 + \frac{V_T}{V_H} \right)(V_H - V_T)^2 - \frac{V_H^2}{\beta_R} \right]^{-1} \qquad (14\text{-}41)$$

Reasoning intuitively, we see that because k_L is usually much less than k_D, much less average capacitor current is available during t_{OFF} than during t_{ON}; therefore we expect the former to be larger.

EXAMPLE 14E *Calculation of t_{ON} and t_{OFF}*. Consider an inverter with $V_{DD} = 7$, $V_T = 2$, $\beta_R = 25$. For this case $V_H = 5$ and $V_H - V_T = 3$. The times are

$$t_{ON} \approx 0.728 \frac{C}{k_D}$$

$$t_{OFF} \approx 0.60 \frac{C}{k_L}$$

The ratio t_{OFF}/t_{ON} is

$$\frac{t_{OFF}}{t_{ON}} = 0.824 \beta_R \qquad \text{////}$$

The CMOS Inverter

The CMOS inverter driving a capacitor is shown in Fig. 14-16. If both transistors have identical V_T and k_n, t_{ON} and t_{OFF} will be identical since only one device is conducting current during each interval. If we consider t_{ON}, the p-channel transistor is cut off, and the n-channel transistor is saturated until v_{out} drops to $V_{DD} - V_T$, after which it remains nonsaturated. The average capacitor current is

$$I_{c(ON)} = \frac{\int_{V_{DD}}^{V_{DD}-V_T} k_D(V_{DD} - V_T)^2 \, dv_{out} + \int_{V_{DD}-V_T}^{0} k_D[2(V_{DD} - V_T)v_{out} - v_{out}^2] \, dv_{out}}{\int_{V_{DD}}^{0} dv_{out}}$$

$$= \frac{k_D}{V_{DD}} (V_{DD} - V_T)^2 \left(\frac{2V_{DD}}{3} + \frac{V_T}{3} \right)$$

The turn-on time is then estimated by

$$t_{ON} = \frac{3CV_{DD}^2}{k_D(V_{DD} - V_T)^2(2V_{DD} + V_T)} \qquad (14\text{-}42)$$

EXAMPLE 14F *Calculation of t_{ON}*. If we consider the devices to have $V_T = 2$ as in Example 14E, and if we choose $V_{DD} = 5$ so that the CMOS inverter will have approximately the same logic swing as that of Example 14E, we find

$$t_{ON} = 0.695 \frac{C}{k_D}$$

If the supply voltage is increased to $V_{DD} = 10$, we obtain

$$t_{ON} = 0.169 \frac{C}{k_D} \qquad \text{////}$$

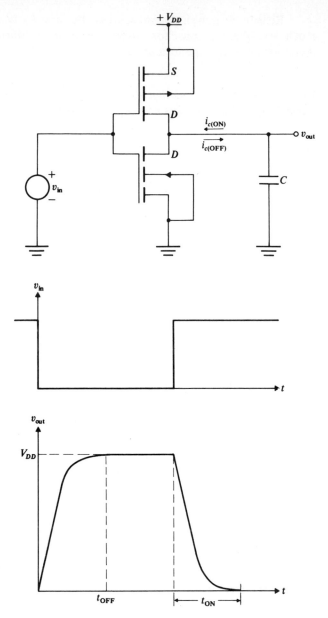

FIGURE 14-16
CMOS inverter driving a capacitance.

EXAMPLE 14G *Calculation of* t_{ON} *Using the Exact Solution.* When the transistor is in the nonsaturated region, the nonlinear differential equation to be solved is

$$\frac{C \, dv_{out}}{dt} = -k[2(V_{DD} - V_T)v_{out} - v_{out}{}^2]$$

Rearranging terms and using a partial fraction expansion, we can rewrite this as

$$2(V_{DD} - V_T)\frac{k}{C} \, dt = \frac{-dv_{out}}{v_{out}} + \frac{dv_{out}}{v_{out} - 2(V_{DD} - V_T)}$$

Integrating and solving for the time t_1 elapsed as v_{out} changes from $V_{DD} - V_T$ to $0.1V_{DD}$, we find

$$t_1 = \frac{C}{2k(V_{DD} - V_T)} \ln \frac{1.9V_{DD} - V_T}{0.1V_{DD}}$$

The total 0 to 90 percent turn-on time is t_1 plus the time elapsed as v_{out} changes from V_{DD} to $V_{DD} - V_T$, that is, the time t_0 during which the transistor is in the saturated region. This latter time is

$$t_0 = \frac{CV_T}{k(V_{DD} - V_T)^2}$$

and the total turn-on time is

$$t_{ON} = t_0 + t_1 = \frac{C}{k(V_{DD} - V_T)^2}\left(V_T + \frac{V_{DD} - V_T}{2} \ln \frac{1.9V_{DD} - 2V_T}{0.1V_{DD}}\right)$$

For $V_{DD} = 5$ and $V_T = 2$, we obtain

$$t_{ON} = 0.622 \frac{C}{k}$$

while for $V_{DD} = 10$ and $V_T = 2$, we find

$$t_{ON} = 0.201 \frac{C}{k}$$

These results, when compared with those of Example 14F, indicate that the average current method gives a reasonable estimate for t_{ON}. ////

CMOS Inverter Power

Although the dc power dissipation of the CMOS inverter is zero, the transient power is not. For a capacitive output load, the energy stored in the capacitor during each half cycle of the input is $CV_{DD}{}^2/2$. The energy stored in the capacitor during one complete cycle divided by the period of the input is the average power dissipated in the inverter:

$$P = CV_{DD}{}^2 f$$

where f is the input frequency.

14-8   LOGIC CIRCUITS[4]

Protection

MOS inverters and logic circuits are highly susceptible to damage from static charge.   Because the device area is so small, the gate capacitance is very small, with the result that a small static charge produces a large voltage.   The breakdown voltage of the thin gate oxide is of the order of 40 V.   If sufficient static charge is present to develop a voltage in excess of the breakdown voltage, the thin oxide can rupture and cause a gate-channel short, destroying the usefulness of the device.   Mere handling of the circuits during packing and unpacking for shipment can generate sufficient static charge to destroy the devices, as can voltage transients occurring during operation.   In order to avoid this problem, some form of protective circuit is included on the chip.   An example of this is the use of a protection diode on all gates which are connected to signal input pads on the chip, as shown in Fig. 14-17a.   For CMOS circuits, protection diodes can easily be fabricated by using the p diffusion for n-channel devices, the n-channel source diffusion, and the p-channel source diffusion, as shown in Fig. 14-17b.   If the gate voltage rises due to some undesired transient, the diode becomes forward-biased when the gate voltage reaches V_{DD}, and prevents further increase of the gate voltage.   The diode is designed to have a breakdown voltage of about 30 V; this limits negative gate voltages to a maximum magnitude of $V_{DD} - 30$.

Static Logic Circuits

Static logic circuits of NAND and NOR type can easily be implemented by modifying the basic inverter to allow more than a single input.   We illustrate the method with the n-channel inverter with saturated load; extension of the methods to p-channel inverters and other types of load is obvious.

If additional driver transistors are added in series, as shown in Fig. 14-18a, no drain current can flow unless all drivers have their gates at the high-state voltage, which is $V_{DD} - V_T$ for the saturated load inverter.   If any gate is at the low-state voltage, that device will be cut off, preventing the flow of drain current; the output will therefore be high.   When all gates are high, all driver transistors are on; drain current flows, causing the output to be low.   We see that the circuit performs the NAND operation for positive logic and the NOR operation for negative logic.

In our earlier discussion of the inverter using two devices of the same channel type, we saw that it was necessary to maintain a β_R of about 25 in order to have a reasonable transfer characteristic with an acceptable low-state output voltage. In the circuit of Fig. 14-18a, it is clear that if the same driver transistors are used as in the inverter, the low-state output voltage with n series driver transistors will be approximately n times that of the inverter.   If the inverter driver has a conduction factor k_D, it will be necessary for each series driver of the logic circuit to have a conduction factor nk_D in order to keep the low-state voltage of the circuit the same as that for the inverter, that is, to maintain the same effective β_R for the logic

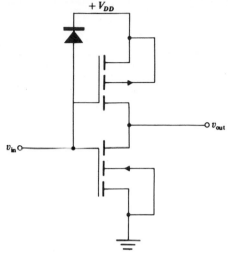

(a)

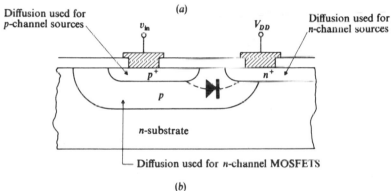

(b)

FIGURE 14-17
(a) CMOS inverter with input protection diode; (b) fabrication of the protection diode.

circuit as for the inverter. Since there are n series drivers each with conduction factor nk_D, the channel area of the drivers is n^2 times the channel area of the inverter driver.

If drivers are added in parallel with the inverter driver, each with its own input, as shown in Fig. 14-18b, any one driver in the on condition will cause the output to be low. Only if all inputs are low will the output be high. The circuit therefore performs the NOR operation for positive logic and the NAND operation for negative logic. Note that for this circuit each driver transistor need only have the same k_D as the original inverter driver for the circuit to have the same effective β_R as the inverter.

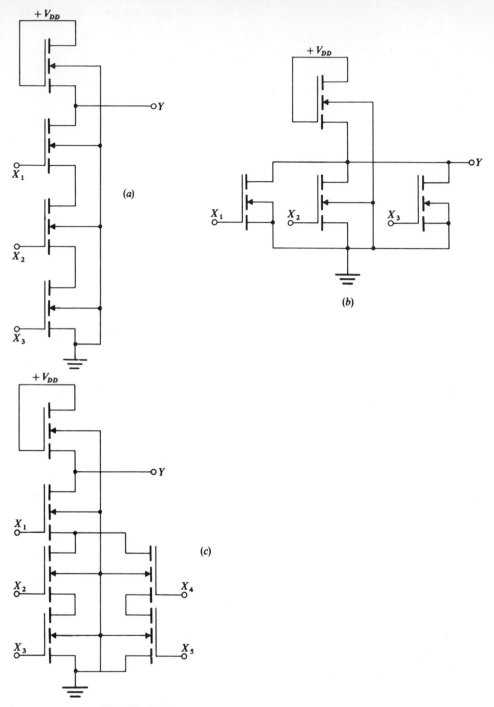

FIGURE 14-18
(a) Use of series driver transistors to perform NAND for positive logic and NOR for negative logic. (b) Use of parallel driver transistors to perform NOR for positive logic and NAND for negative logic. (c) Combination of series and parallel driver transistors.

The basic inverter circuit can be modified to perform more complex functions by using both series and parallel drivers; an example is shown in Fig. 14-18c.

Exercise 14-5   Suppose that a fabrication process has minimum line width W.   Compare the total channel area of the circuit of Fig. 14-18a with that of Fig. 14-18b if both circuits are to have an effective $\beta_R = 25$.

Exercise 14-6   What logic operation is performed by the circuit of Fig. 14-18c for positive logic?   What is performed for negative logic?

Exercise 14-7   For the circumstances of Exercise 14-5, what is the total channel area of the circuit of Fig. 14-18c?

The basic CMOS inverter can also be modified to perform logic operations.[5] Since both driver and "load" transistors must be supplied with gate signals, it is necessary that additional transistors always be added in pairs.   In the circuit of Fig. 14-19a, if X_1 is high and X_2 is low, Q_1 is on, and the gate voltage of Q_4 is of the proper value to turn Q_4 on.   However, Q_3 is off; since Q_3 and Q_4 are in series, no drain current flows in either.   Therefore the output is low.   The output can be high only if both X_1 and X_2 are low; then Q_3 and Q_4 are both on.   Although Q_3 and Q_4 are in series, no dc drain current flows in them when they are on, so it is not necessary to increase their conduction factors, as was required in all-p-channel or all-n-channel circuits.   All devices can have the same k_p as those of the basic inverter.   The MC 14001AL quad two-input NOR circuit is shown in Fig. 14-19b.

The NAND operation is performed by the circuit of Fig. 14-20a; again it is necessary that each input connect to both n- and p-channel devices.   If X_1 is high and X_2 is low, Q_3 is off but Q_4 is on and the output is high. Q_1 also has sufficient gate voltage to turn it on, but it is prevented from turning on since it is n series with Q_2, which is off.   The output can be low only if both Q_1 and Q_2 are on; for this condition Q_3 and Q_4 are both off.   The circuit therefore performs the NAND operation.   A quad two-input NAND circuit, the MC 14011AL, is shown in Fig. 14-20b.

More complex operations can be performed by combinations of transistors; from the above it can be seen that when n-channel devices are used in series, each must have its gate connected to the gate of a parallel-connected p-channel device, and vice versa.   An example of a more complex circuit is the MC 14507AL EXCLUSIVE-OR circuit shown in Fig. 14-21.

The AND-OR-INVERT operation is performed by the circuit of Fig. 14-22. This circuit is formed by connecting an inverter with two complementary pair transistors, all of which are contained in the MC 14007AL.

Exercise 14-8   Verify that the circuit of Fig. 14-21 performs the operation $Y = A\bar{B} + B\bar{A}$.

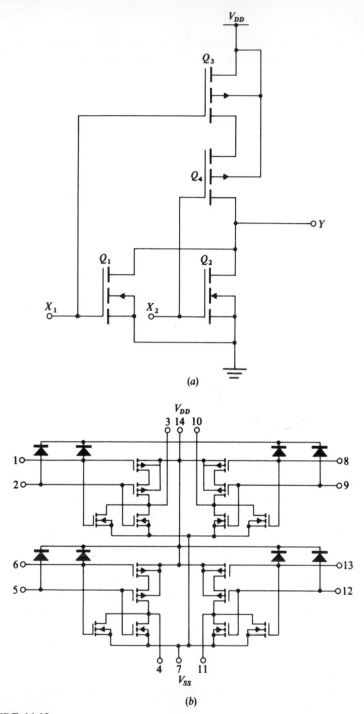

(a)

(b)

FIGURE 14-19

(a) CMOS circuit performs NOR for positive logic; (b) the MC 14001AL quad two-input NOR circuit.

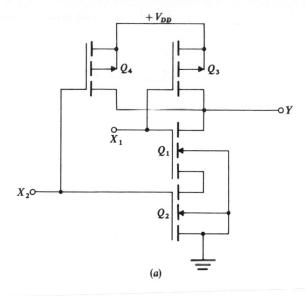

(a)

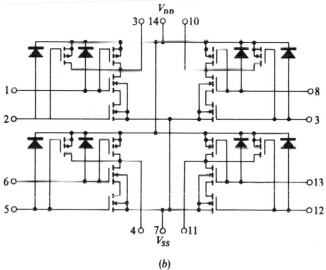

(b)

FIGURE 14-20
(a) CMOS NAND circuit for positive logic; (b) the MC 14011AL quad two-input NAND circuit.

(1/4 of device shown, but all pin numbers indicated)

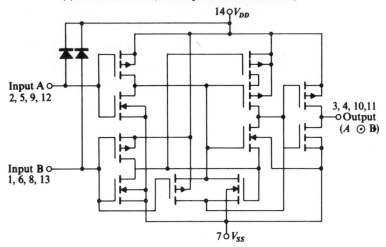

FIGURE 14-21
The MC 14507AL EXCLUSIVE-OR circuit.

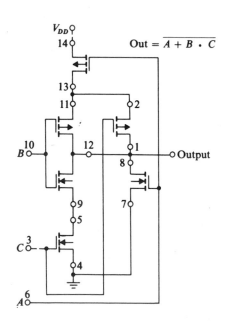

FIGURE 14-22
AND-OR-INVERT using the MC
14007AL

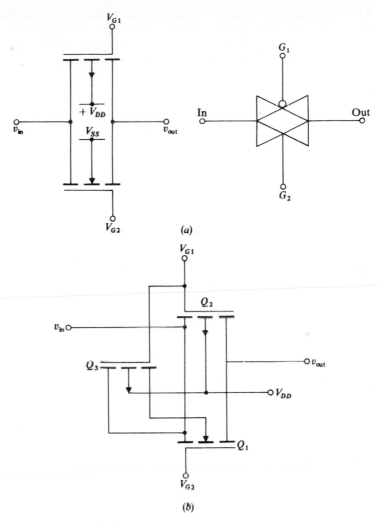

FIGURE 14-23
(a) Basic transmission circuit; (b) modified circuit for low resistance.

Transmission Circuit

With CMOS circuits it is possible to fabricate a transmission circuit, useful in both digital and analog applications. The basic transmission circuit is shown in Fig. 14-23a, together with its symbol. This circuit performs a function similar to that of the well-known diode bridge. If $V_{G1} = V_{SS}$ and $V_{G2} = V_{DD}$, both transistors are on. The input voltage v_{in} (which must be between V_{SS} and V_{DD}) is then connected to the output through the parallel on resistance of the channels of the two transistors. As v_{in} approaches V_{DD}, the n-channel device cuts off but the p-channel device remains nonsaturated; as v_{in} approaches V_{SS}, the p-channel device cuts off but the n-channel device remains nonsaturated. Therefore there is always a nonsaturated transistor between input and output. Moreover, no offset voltage exists when the output current is zero.

If $V_{G1} = V_{DD}$ and $V_{G2} = V_{SS}$, both transistors are off, and the resistance between v_{in} and v_{out} is of the order of 10^9 Ω.

A modified transmission circuit with low on-resistance is shown in Fig. 14-23b. When V_{G1} is low and V_{G2} is high, the circuit is enabled; as v_{in} increases, Q_1 is driven toward the saturation mode, with a corresponding decrease of its source-drain conductance. But this effect is offset by coupling the input through Q_3 to the substrate of Q_1, forcing the substrate to act as a second gate. The increase of v_{in} at the substrate of Q_1 causes an increase of the channel conductance of Q_1. When V_{G1} is high and V_{G2} is low, Q_3 is cut off, as are Q_1 and Q_2. It is possible to make use of the n-channel device substrate in a CMOS circuit because this substrate is a p diffusion into the n-type wafer. Therefore each n-channel substrate is isolated from every other device. The substrate of the p-channel device cannot be used in this manner because the n wafer is the substrate for all p-channel devices.

Dynamic Logic

We have seen that static logic circuits which use devices having a single channel type have two disadvantages:

1 The dc power dissipation is not zero.
2 Load and driver devices must have different conduction factors.

Both disadvantages can be overcome by using a *ratioless dynamic circuit* in which timing signals function as power supplies, and equal conduction factors are used. A 1-bit delay circuit of this type is shown in Fig. 14-24a; although p-channel devices are used in this example, the same methods apply to n-channel devices. The timing diagram is shown in Fig. 14-24b. If v_{in} goes from high to low state at $t = 0$, when the ϕ_1 clock is low, Q_1, Q_2, and Q_3 turn on and drive C_1 low. This causes C_2 to be driven low also. Q_4, Q_5, and Q_6 are off because the ϕ_2 clock is off. The ϕ_1 clock turns off before the ϕ_2 clock turns on, so the information regarding the state of v_{in} is stored on C_1 and C_2. When the ϕ_2 clock appears, Q_4, Q_5, and Q_6 turn on; this drives C_2 and C_3 high and also drives v_{out} high. Thus v_{out} is the complement of v_{in}, delayed by 1-bit time. When the ϕ_2 clock turns off, the output remains high since Q_5 and Q_6 are cut off and C_4 retains its charge.

The use of two clock phases permits the input voltage to change without destroying the information at the output, since input information changes with the ϕ_1 clock but the change is not coupled to the output circuit until the ϕ_2 clock appears. All capacitors in the circuit are not lumped capacitors but rather are the capacitances of the MOS transistors.

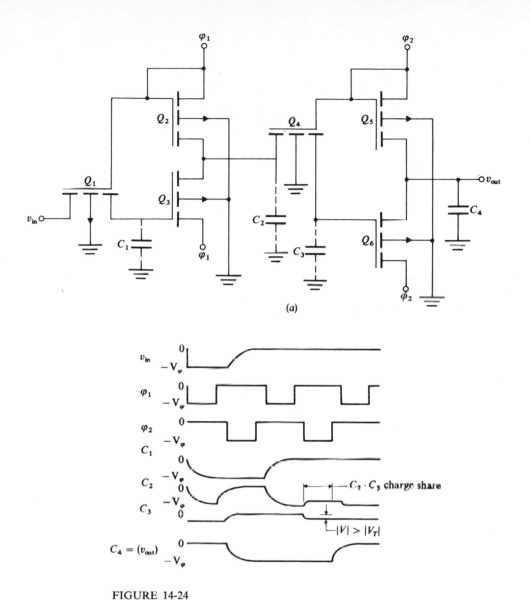

FIGURE 14-24
(*a*) One-bit-time-delay ratioless two-phase circuit; (*b*) timing diagram (after Penney et al.[1]).

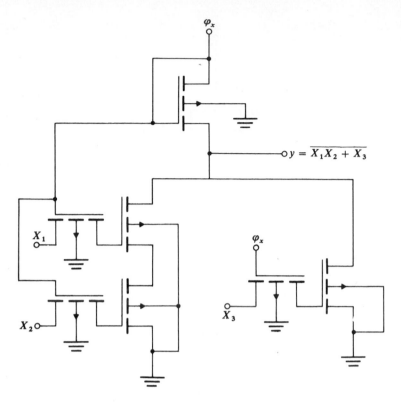

FIGURE 14-25
Two-phase ratioless AND-OR-INVERT circuit.

More complex logic operations can be performed by adding more transistors to the ratioless configuration; for example, an AND-OR-INVERT circuit for negative logic is shown in Fig. 14-25. Here the input information is sampled by the ϕ_x clock; the output is then available at the end of the ϕ_x clock period. It would be transferred to the next circuit by a clock of different phase.

PROBLEMS

14-1 Show that the inverter of Fig. 14-3*b* with $\beta_R = 1$ is of little value. Do this by calculating the operating points and the logic swing.

14-2 If the inverter of Example 14A is operated with $V_{DD} = 12$, calculate the operating points and the noise margin.

14-3 Calculate the operating points, unity-gain points, and noise margins for an *n*-channel inverter with nonsaturated load if $V_{DD} = 5$, $V_T = 2$, $V_{GG} = 9$, $\beta_R = 25$.

14-4 Calculate the operating points, unity-gain points, and noise margins for an *n*-channel depletion-load inverter with $V_{TD} = 2.0$, $V_{TL} = -2.0$, $V_{DD} = 5$, $\beta_R = 25$.

14-5 A CMOS inverter has complementary transistors with $V_T = 1.0$. If $\beta_R = 1$ and $V_{DD} = 15$, calculate the transfer characteristic, the unity-gain points, and the noise margin.

14-6 A CMOS inverter has $\beta_R = 5$, $V_{TL} = 2.0$ V, and $V_{TD} = 1.0$ V. Calculate the transfer characteristic, the unity-gain points, and the noise margins.

14-7 Estimate t_{ON} and t_{OFF} for the nonsaturated load inverter and the depletion-load inverter. Assume $V_{DD} = 5$, $V_T = 2$ for both cases. Repeat for $V_{DD} = 10$.

14-8 Lay out the circuits of Fig. 14-18 for $\beta_R = 25$, and minimum line width and spacing of 0.2 mil. Assume 0.2-mil registration clearances.

14-9 For the circuit of Fig. 14-18a, suppose that the series driver transistors all have the same k_D as the original inverter driver. Assume that the bottom two transistors have high inputs. Find the transfer function $v_{out}(v_{in})$ where v_{in} is the gate voltage of the top transistor.

REFERENCES

1 PENNEY, W. M., ET AL.: "MOS Integrated Circuits," Van Nostrand Reinhold Company, New York, 1972.

2 CARR, W. N., and J. P. MIZE: "MOS/LSI Design and Application," McGraw-Hill Book Company, New York, 1972.

3 MASUHARA, T., M. NAGATA, and H. HASHIMOTO: A High-performance *n*-Channel MOS LSI using Depletion-type Load Elements, *IEEE Trans. Solid-State Circuits*, vol. SC-7, pp. 224–231, 1972.

4 COOK, P. W., D. L. CRITCHLOW, and L. M. TERMAN: Comparison of MOSFET Logic Circuits, *IEEE J. Solid-State Circuits*, vol. SC-8, pp. 348–356, 1973.

5 "McMOS Integrated Circuits Data Book," Motorola, Inc., 1973.

15

APPLICATIONS OF DIGITAL INTEGRATED
CIRCUITS

In the preceding three chapters, we saw how circuits can be designed to perform various logic functions. In this chapter we consider briefly and qualitatively some representative applications of these circuits. We begin with ways of forming more complex logic functions by the wired-function method, and we then consider adders, flip-flops, shift registers, and memories.

15-1 WIRED LOGIC FUNCTIONS

In Chaps. 12 through 14 we dealt with the design of individual logic circuits such as NAND, NOR, AND-OR-INVERT, etc., and we described some commercially available circuit packages. Some of these packages contained several circuits on one chip, such as a quad two-input NOR, etc. It seldom occurs that the user of integrated circuits is also the designer of the circuits themselves; the user is therefore forced to accept available packages and build his system from them. The effectiveness of certain logic circuits can sometimes be increased by wiring together

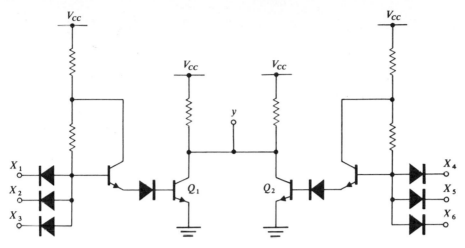

FIGURE 15-1
Simple DTL circuits with outputs wired together.

their outputs. Consider, for example, the DTL circuits shown in Fig. 15-1. Here the output Y can be 1 only if both Q_1 and Q_2 are cut off; that is,

$$Y = \overline{X_1 X_2 X_3 X_4 X_5 X_6}$$

For this circuit, wiring the outputs together has the effect of extending the number of inputs of a single gate. No electrical problems occur in this case.

Care must be used in employing the wired-function method, because electrical problems arise for certain circuits. Consider, for example, the case of Fig. 15-2a, in which two DTL gates with active pull-up have their outputs wired together. Suppose $X_1 = X_2 = X_3 = 1$, but $X_4 = 0$. Q_1 will be saturated, and Q_2, Q_3, and Q_4 will be cut off. However, with Q_1 saturated, the emitter of active pull-up Q_5 is forward-biased. A large current will therefore flow in Q_5; this is not a transient current but a static current which can be large enough to cause several unwanted effects. It may cause Q_1 to unsaturate, giving an erroneous output, and it may cause destruction of some devices. Even if no devices are destroyed, the increased power supply current may cause a malfunction. Therefore, wiring of outputs should not be used if active pull-up devices are present.

In certain MOS circuits, wiring of outputs must also be avoided. Consider the CMOS circuit of Fig. 15-2b, in which two inverters have outputs wired together. Suppose $X_1 = 0$ and $X_2 = 1$; then Q_2 and Q_3 are off, but Q_1 and Q_4 are on. A large current will now flow between Q_1 and Q_4. In general, CMOS circuit outputs cannot be wired together because in each circuit there is always one output transistor on, regardless of the state of the logic. This disadvantage can be overcome by use of the transmission gate, as shown in Fig. 15-2c; here the inverter can be disconnected from its load by applying the disable signal D, which causes the transmission gate to cut off. Circuits using the transmission gate in this manner

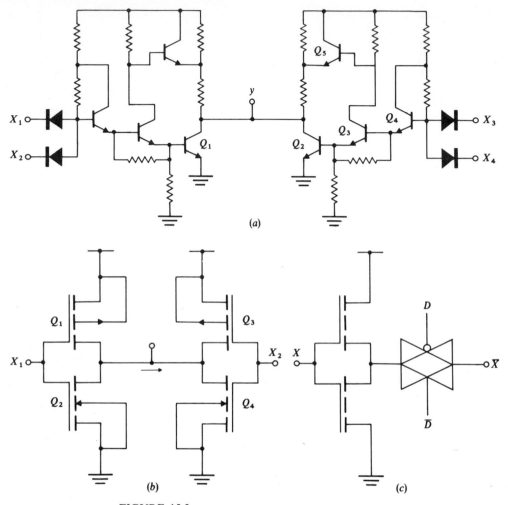

FIGURE 15-2
(a) Wiring of outputs of circuits with active pull-up is not permitted; (b) CMOS
outputs cannot be wired together because large currents flow; (c) use of a transmission gate with a CMOS circuit.

are called three-state logic circuits, since the transmission gate effectively provides a third state.[1]

Exercise 15-1 A DTL circuit with no active pull-up performs AND-OR-INVERT with each AND having three inputs. Its output is wired to the output of a standard three-input DTL NAND circuit having no active pull-up. What logic function is obtained at the output?

Exercise 15-2 Can ECL gates have their outputs wired together? If so, what function is obtained where the OR output of a three-input gate is wired to the NOR output of another three-input gate?

15-2 ADDERS

A basic building block for the arithmetic unit of a digital computer is the adder. We consider for simplicity only the serial adder, which performs addition in much the same manner as a human does, except binary rather than decimal numbers are used. To add two numbers, the least significant digits are added, producing a sum and a carry. The next digits are added, and their sum is added to the previous carry, producing another sum and carry. This sequence is continued until all digits have been added. Clearly there are three inputs: the digit A from the first number, the digit B from the second number, and the carry C from the previous sum. The outputs are the sum Q_S and the carry Q_C. Logic for the adder can be written in several forms; one form is

$$Q_C = AB + (A + B)C$$
$$Q_S = \bar{Q}_C(A + B + C) + ABC$$

In Chap. 14, we saw how various static logic functions could be formed with MOS devices. The sum and carry functions above for the adder can be performed with MOS devices; the logic diagram is shown in Fig. 15-3a. Implementation of the adder with n-channel enhancement-mode drivers and n-channel enhancement-mode saturated load transistors is shown in Fig. 15-3b; here for simplicity the substrate connections are not shown.[2] The number shown with each transistor is the Z/L ratio. This particular circuit uses 18 transistors and does not require the complement of any of the input variables.

Exercise 15-3 Make up a truth table for an adder with inputs A, B, carry C, sum Q_S, and output carry Q_C. Show that the circuit of Fig. 15-3b performs addition.

The adder can also be implemented with a combination of NAND, AND, and EXCLUSIVE-OR gates. This is done in the MC 4326F TTL circuit by using the logic diagram shown in Fig. 15-4a. The circuit diagram is shown in Fig. 15-4b. This circuit provides an illustration of the use of AND-OR-INVERT

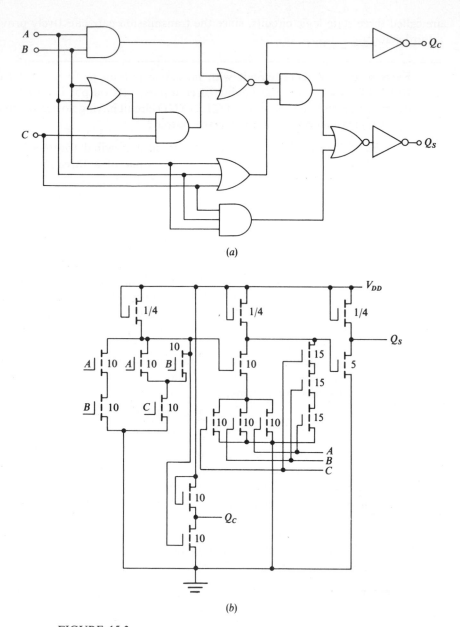

(a)

(b)

FIGURE 15-3
(a) Adder logic diagram; (b) adder circuit using n-channel drivers with n-channel saturated load transistors (after Carr and Mize[2]).

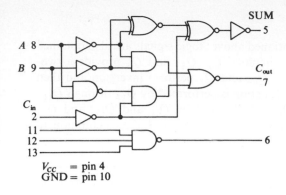

SUM
5

A 8

B 9

C_{in}
2

11
12
13

6

V_{CC} = pin 4
GND = pin 10

Truth Table

Input pints			Output pins	
8	9	2	5	7
A	B	C_{in}	SUM	C_{out}
0	0	0	0	0
0	0	1	1	0
0	1	0	1	0
0	1	1	0	1
1	0	0	1	0
1	0	1	0	1
1	1	0	0	1
1	1	0	0	1
1	1	1	1	1

Total power dissipation = 90 mW typ/pkg
Add delay = 25 ns typ
Carry delay = 13 ns typ

(a)

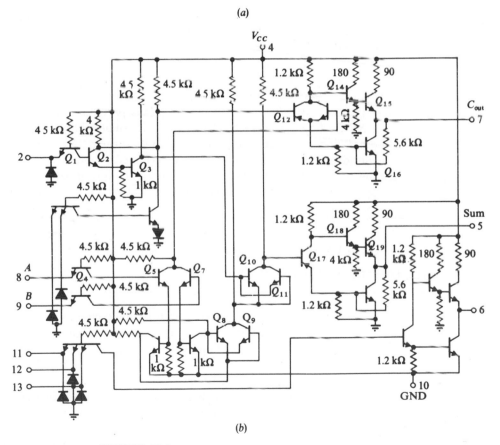

(b)

FIGURE 15-4
(a) Adder using NAND and EXCLUSIVE-OR gates; (b) circuit diagram of the MC 4326F TTL adder.

circuitry, as well as the operations mentioned above; logic signals are also obtained from intermediate points in the TTL circuits. Q_1, Q_2, and Q_3 provide a two-output inverter for C_{in}, while Q_4, Q_5, Q_6, and Q_7 perform inversion and then AND of A and B. The OR-INVERT function is performed by Q_{12} through Q_{16}. EXCLUSIVE-OR is performed by Q_8 and Q_9, and by Q_{10} and Q_{11}; and Q_{17} through Q_{20} form an inverter with active pull-up, using techniques common to MC 3100 series TTL circuits.

Exercise 15-4 Using the logic diagram of Fig. 15-4a, write the equation for Q_s and C_{out}.

15-3 FLIP-FLOPS

Flip-flops are basic 1-bit binary memory circuits; they are formed by cross-coupling appropriate logic circuits so that positive feedback exists betwen output and input. This positive feedback causes the circuit to "latch" in a particular state until a sufficiently strong forcing function is applied to cause a change of states. The use of two NOR circuits to form a flip-flop is shown in Fig. 15-5a; implementation of the circuit with n-channel enhancement-mode drivers and n-channel saturated load MOS transistors is shown in Fig. 15-5b. This particular flip-flop is called a set-reset or RS flip-flop. If R is 1, Q changes to 1; it can only be changed to 0 if S is 1. If both R and S change to 0, Q remains in its former state. The RS flip-flop has the disadvantage that the output is undetermined if both R and S are 1; for this case the state of Q will be determined by noise, circuit unbalances, etc. The RS flip-flop can also be made in CMOS form as shown in Fig. 15-5c.

Exercise 15-5 For the circuit of Fig. 15-5b, verify that if R and S are both 0, when R changes to 1 Q becomes 1 and remains 1 when R changes to 0.

Exercise 15-6 Repeat Exercise 15-5 for the circuit of Fig. 15-5c.

Other types of flip-flops can be made by incorporating logic circuits in the feedback paths. The toggle, or T, flip-flop is so arranged as to change its state each time a trigger pulse T appears. This flip-flop can be formed from AND and NOR gates, as shown in Fig. 15-6a; the circuit of Fig. 15-6b shows a T flip-flop using n-channel drivers with n-channel saturated load devices.

The problem of undetermined outputs for the case $R = 1$ and $S = 1$ in the RS flip-flop can be overcome by self-gating the flip-flop in such a way that the change of state occurs when both inputs are 1; that is, the circuit functions as an RS flip-flop except for $R = S = 1$, when it performs as a T flip-flop. Such a flip-flop is called a JK flip-flop, and its logic inputs are designated J and K instead of R and S. The MC 14027 circuit is a CMOS dual flip-flop which has provisions

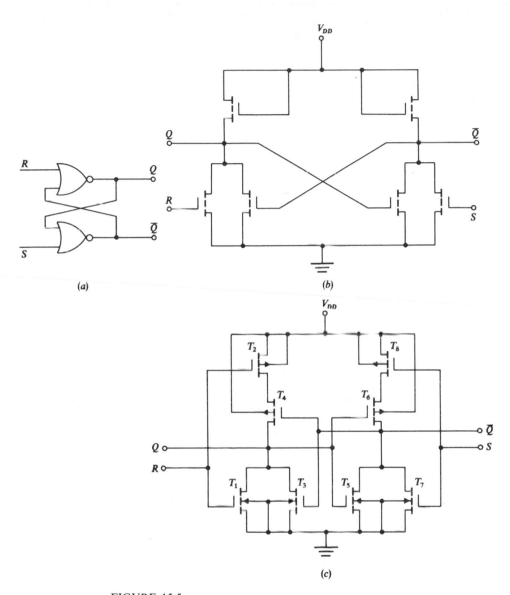

FIGURE 15-5
(a) Basic RS flip-flop made from two NOR gates; (b) RS flip-flop using n-channel drivers with saturated n-channel load transistors; (c) CMOS RS flip-flop (after Carr and Mize[2]).

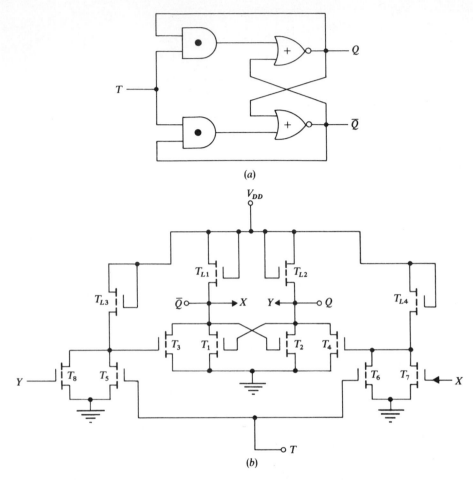

FIGURE 15-6
(*a*) Basic toggle flip-flop using AND and NOR gates; (*b*) *T* flip-flop using *n*-channel MOS driver transistors with saturated load transistors (after Carr and Mize[2]).

for *J*, *K*, *R*, *S*, and clock inputs; thus it can be operated as a clocked *JK* or clocked *RS* flip-flop. The logic diagram is shown in Fig. 15-7*a*, and the circuit is shown in Fig. 15-7*b*. Note the use of the CMOS transmission gates, labeled *TG* in Fig. 15-7*a*.

Master-slave Flip-flops[3]

In high-speed clocked logic systems, timing problems can arise in the triggering of flip-flops. This is because logic functions may change state so rapidly that the logic controlling a flip-flop actually changes state while the clock pulse is still

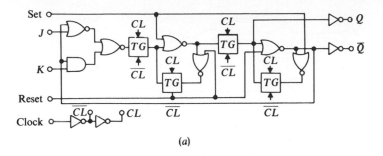

(a)

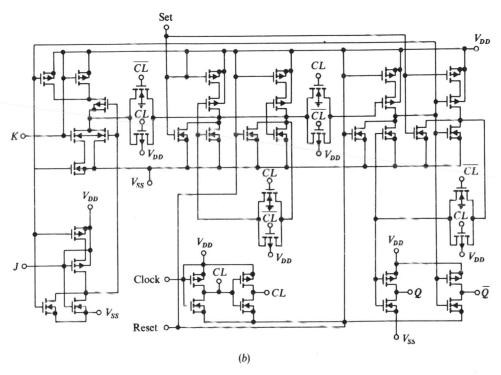

(b)

FIGURE 15-7
(a) Logic diagram of a CMOS *JK* flip-flop; (b) circuit diagram of the MC 14027.

present. This would cause the flip-flop to try to change state more than once during each clock pulse. In discrete-component circuits, capacitors can be used to provide the delays required to ensure that this does not happen. In integrated circuits, a master-slave arrangement, which is really two sequenced flip-flops, is used to prevent more than one change of state from occurring during each clock pulse. A basic master-slave *RS* flip-flop logic diagram is shown in Fig. 15-8*a*; the timing diagram is shown in Fig. 15-8*b*. Suppose that initially *Q* and *A* are 0, as are *R* and *S*. Let *S* change to 1; at the appearance of the clock pulse, the master

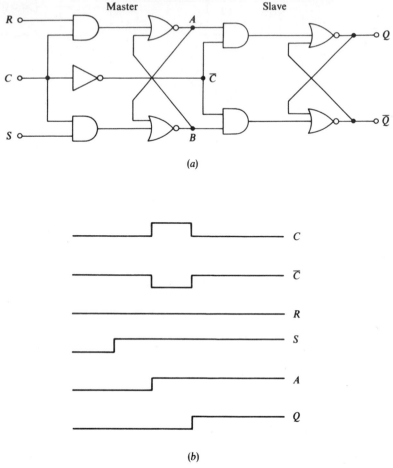

FIGURE 15-8
(*a*) Basic *RS* master-slave flip-flop; (*b*) timing diagram.

changes state and *A* becomes 1.   No change occurs in the state of the slave since $\bar{C}$ is 0.   When *C* is removed, no change of the master occurs, but now $\bar{C}$ is 1 so the slave changes state and *Q* is 1.   Note that the only outputs from the flip-flop are the slave outputs *Q* and $\bar{Q}$.   In a logic network, the *R* and *S* signals would be formed from functions involving only slave outputs.   Therefore, although the master units will change state at the appearance of the clock, the *R* and *S* inputs cannot change until the clock returns to 0.   By this time the information in the master has been transferred into the slave, so a change of *R* and *S* is now unimportant.   It is obvious that the master-slave flip-flop is really a two-phase flip-flop; by phase-sequencing the transfer of information one avoids the ambiguity occurring in a single-phase flip-flop.

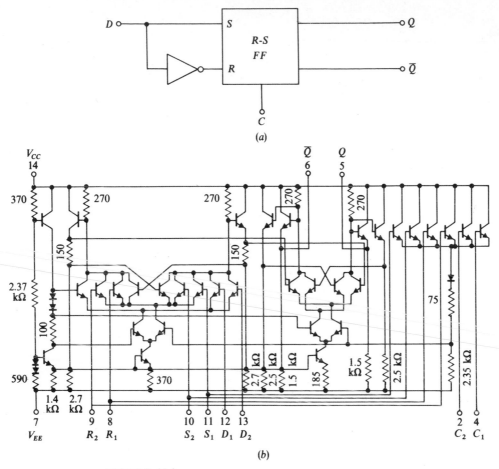

FIGURE 15-9
(a) Realization of the D flip-flop with an RS flip-flop and an inverter; (b) the MC 1022 master-slave D flip-flop.

Although we have illustrated the master-slave principle with the RS flip-flop, other types can be implemented. A delay, or D, flip-flop is one in which the Q output follows a single input by 1-bit time delay. Such a flip-flop can be realized from an RS flip-flop as shown in Fig. 15-9a. The MC 1022 is a type D flip-flop made from ECL-type gates; the circuit is shown in Fig. 15-9b. This circuit has provisions for two D inputs as well as two R and two S inputs, permitting use of the flip-flop in either the RS or D mode. The cross-coupled portions comprising master and slave units are easily identified in the circuit diagram.

Master-slave flip-flops can, of course, be realized with MOS circuits; the logic diagram for a type D flip-flop using CMOS circuits is shown in Fig. 15-10.

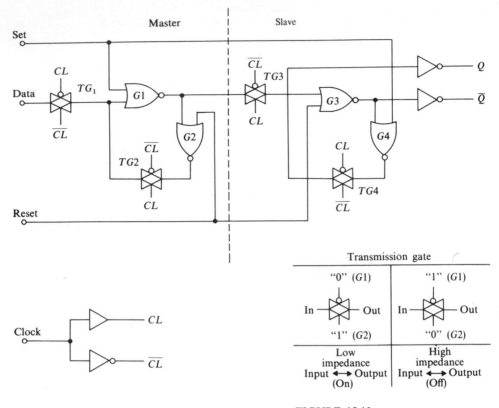

FIGURE 15-10
The CMOS type D master-slave flip-flop.

Exercise 15-7 Using the logic diagram of Fig. 15-10, draw the circuit diagram for the CMOS flip-flop.

Exercise 15-8 Explain why it is necessary to use transmission gates in the flip-flop of Fig. 15-10.

15-4 Shift Registers

A shift register can be regarded as a cascade of 1-bit-time-delay memory stages in which the input of each stage is controlled by the output of the preceding stage. Shift registers have many uses, among which are serial-to-parallel converters and circulating memories. Perhaps the largest use for shift registers is in hand calculators; here the operator is performing calculations in real time, and since human response is slow, the time limitations of serial computation in the calculator are relatively unimportant.

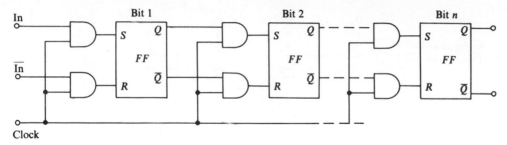

FIGURE 15-11
Basic shift-register structure.

A shift register can be constructed from gates and flip-flops, as shown in Fig. 15-11. If input information is entered into the first stage one bit at a time, after n bit times the n bits of input are stored in the register. Parallel readout can be accomplished by sampling all outputs simultaneously after the n bits have been stored.

A circulating memory can be made by feeding the output of the last stage back to the input through appropriate gates. If a control signal is provided to gate the input data, a digital word can be shifted into the register. The control signal can then disable the input gates and enable the feedback gates. The word in the register will be continually circulated in the register until the control signal allows new information to be entered.

Exercise 15-9   Show a block diagram for a 4-bit circulating register.

The MC 4012L circuit is a 4-bit shift register on a single chip. It can accommodate either serial or parallel input information, determined by a mode-control signal; the block diagram is shown in Fig. 15-12a. When the mode control is 1, the parallel input signals D_p are entered into the register at the time of the strobe signal. For serial operation with right shift, the mode control is set to 0, and input data at D_S is transferred into the register at the time of the clock signal. The MC 4012L uses TTL circuits; the circuit diagram for input gating and a typical flip-flop is shown in Fig. 15-12b.

Shift registers of the type described above are called static shift registers because the information stored in the register remains as long as power supply voltage is applied.

Shift registers are ideally suited for MOS implementation. Because of the small size of MOS devices, many shift-register stages can be realized on one chip. A static shift register using CMOS circuits can be realized by employing type D master-slave flip-flops; the MC 14015 is a 4-bit register of this type. The circuit for a single stage, together with appropriate buffer circuits, is shown in Fig. 15-13.

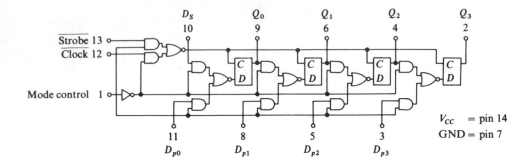

V_{CC} = pin 14
GND = pin 7

(a)

Input gating   Typical flip-flop

(b)

FIGURE 15-12
(a) Block diagram of the MC 4012L shift register; (b) MC 4012L circuit diagram.

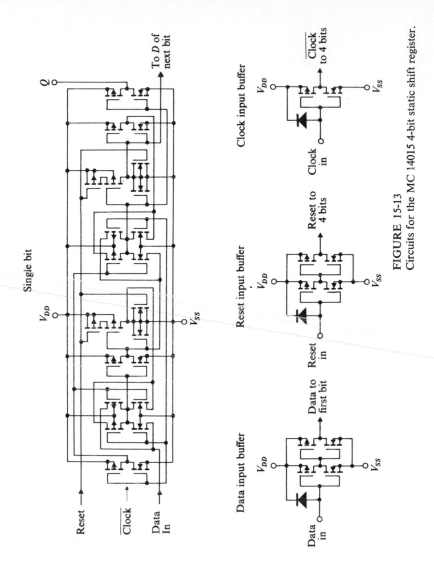

FIGURE 15-13

Circuits for the MC 14015 4-bit static shift register.

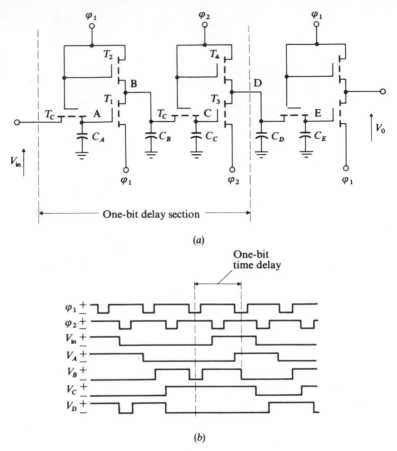

FIGURE 15-14
(a) Dynamic ratioless shift register; (b) timing diagram (after Carr and Mize[2]).

In Chap. 14, we saw that a dynamic ratioless circuit could be made of MOS devices all having the same channel type, and that 1-bit delay could be obtained by using the device capacitances for storage. A dynamic shift register can be made by cascading these dynamic ratioless circuits, as shown in Fig. 15-14a; here all devices are p-channel enhancement-mode devices with equal Z/L ratios. The timing diagram is shown in Fig. 15-14b.

15-5   MEMORIES[4,5]

READ-only memories (ROM) are memories in which writing is not part of the normal memory cycle. A fixed program is stored in the memory and only the READ operation is performed; typical examples of applications of the ROM are

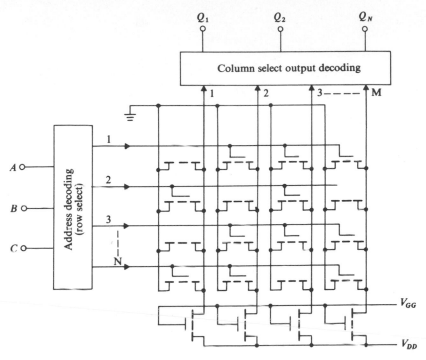

FIGURE 15-15
Wired program MOS ROM (after Carr and Mize[2]).

the storage of programs for trigonometric function calculation in hand calculators, and algorithms for polar to rectangular coordinate transformation. Special-purpose computers, such as hand calculators, make extensive use of ROMs.

There are two basic types of ROM. the wired-program ROM and the electrically alterable ROM. A wired-program ROM can be made from MOS devices as shown in Fig. 15-15;[6] here an array of devices is laid out, but thin oxide and gate metal are used only at those device locations where a 1 is to be stored. In Fig. 15-15, only the devices with a gate electrode shown have thin gate oxide and gate metal; thus only those devices can have an inversion channel. In Fig. 15-15, current will flow in column select line 2 when column 2 and row 3 are selected, but not when column 2 and row 2 are selected. Wired-program ROMs can be made with CMOS devices also; such a memory is the MC 14524AL. Since the program cannot be altered after fabrication, the user must provide appropriate program information when ordering the memory.

Electrically alterable ROMs can be fabricated so that special signals, quite different from the logic signals, are used to program the memory. In this case, the method must ensure that the conditions established by the programming

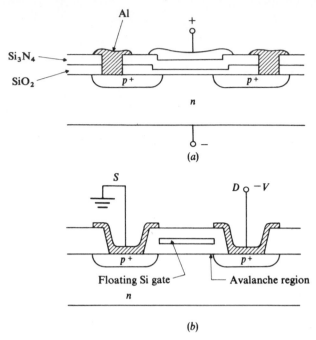

FIGURE 15-16
(a) MNOS ROM cell; (b) FAMOS ROM cell.

signals will remain when power is removed. One method of electrically program-ming an MOS ROM is to use the MNOS device structure of Fig. 15-16a, and apply a large positive voltage between gate and substrate. Tunneling occurs, causing a negative charge Q_{SS} to appear at the Si_3N_4-SiO_2 interface; this Q_{SS} produces a low-threshold MOS device. If a large negative voltage is applied, a positive Q_{SS} is induced at the interface; this produces a high-threshold MOS device. In this manner 1 or 0 can be programmed at the desired locations. The interface charge remains when power is removed.

A second method is to use a floating-gate avalanche-injection MOS device (FAMOS), as shown in Fig. 15-16b.[7] Here no contact is made to the floating silicon gate. A junction voltage of -30 V applied between drain and source causes avalanche breakdown at the drain and the injection of high-energy electrons from the surface of the avalanche region to the floating silicon gate. Since the gate is floating, this electron current through the oxide causes a negative charge to accumulate on the gate, which in turn produces an inversion channel at the silicon surface. Once the applied voltage is removed, no path exists for discharge of the accumulated charge. The memory can be programmed by applying the avalanche bias through the READ-select lines; the program can be erased by ultraviolet light.

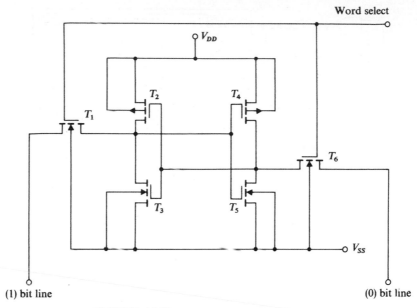

Word select

V_{DD}

(1) bit line

(0) bit line

V_{SS}

FIGURE 15-17
Bistable CMOS circuit for static RAM cell (after Carr and Mize[2]).

Random-access Memories

A random-access memory (RAM) is one in which the normal memory cycle includes both READ and WRITE operations, with electronic access to any cell in the memory. A typical RAM organization consists of a set of X and Y select lines in which the X lines, for example, are used to select the desired word, and the Y lines are used to select the bit. As was the case with shift registers, RAMs can be either bipolar or MOS, and either static or dynamic. A static RAM retains the information written as long as power is applied. Dynamic RAMs generally use some form of capacitor storage, so it is necessary to "refresh" the memory content if storage longer than a few milliseconds is required; therefore a dynamic memory cycle will include REFRESH as well as READ and WRITE operations.

For a static RAM, each cell can be made as a bistable circuit. A representative example of such a memory cell is the CMOS circuit of Fig. 15-17 which uses a cross-coupled CMOS inverter with two n-channel transistors for word and bit selection. The MCM 14505 64-bit memory uses a cell similar to that of Fig. 15-17, but p-channel devices are used for selection; a block diagram for the MCM 14505 is shown in Fig. 15-18.

Bipolar devices can also be used to form a bistable memory cell; Fig. 15-19a shows an emitter-coupled cell, which will be recognized as a conventional Eccles-Jordan flip-flop configuration using multiple-emitter transistors.[8] In standby operation, the emitter current of the ON transistor is carried by the word line,

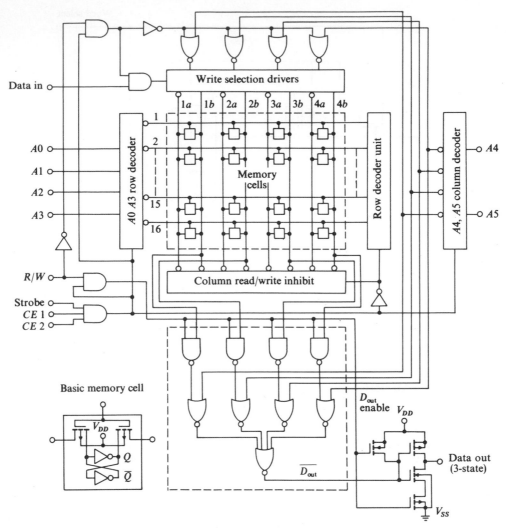

FIGURE 15-18
Block diagram of the MCM 14505 64-bit static CMOS RAM.

which has a reference level of $+0.3$ V. When the word line is raised to $+3.0$ to READ, standby current is transferred to the other emitter of the ON transistor and flows in that bit line, where it is detected by the sense amplifier. To WRITE, the bit line of the ON transistor is raised to $+3.0$, while the word line is high. When the word line is lowered, the cell regeneratively switches to its other state.

In this cell, the READ current is limited by the power supply and resistor values, and the switching speed is limited by the regeneration speed. An improvement of READ current and switching speed can be achieved with the cell

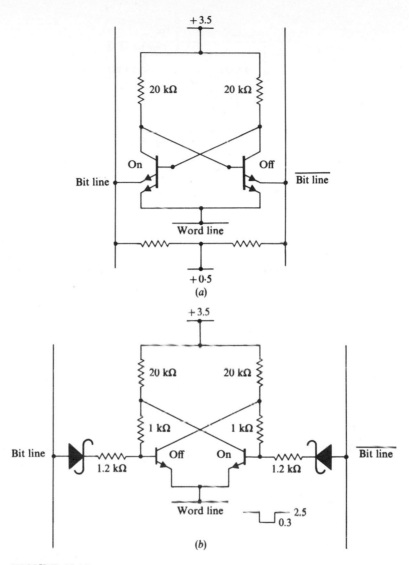

FIGURE 15-19
(a) An emitter-coupled bipolar cell; (b) a cell coupled by Schottky-barrier diodes.

of Fig. 15-19b, in which Schottky-barrier diodes are used for coupling. Here the word line is normally at +2.5 V and the bit lines are at +1.6 V. During the READ operation, the word-line voltage is lowered to +0.3, forward-biasing the diode connected to the ON transistor. The current flow in the corresponding bit line is determined by the 1.2-kΩ resistor rather than by the power supply voltage and the 20-kΩ resistor. To WRITE, the appropriate bit line is raised to 2.8 V and

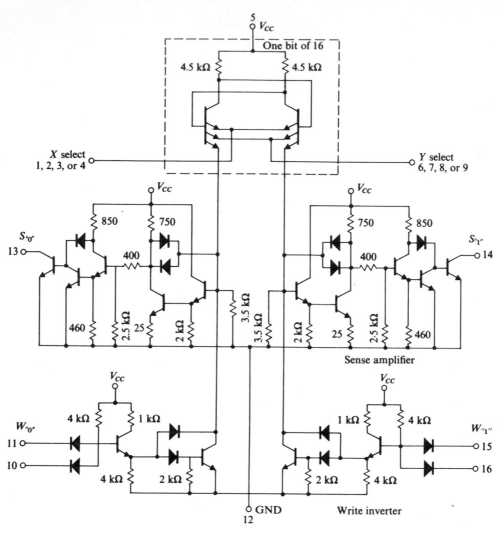

FIGURE 15-20
The MC 5484L 16-bit scratch-pad memory.

the word line is lowered to +0.3 V. The voltage produced at the base of the OFF transistor is sufficient to turn that transistor on. Regeneration is not necessary to switch the cell, since sufficient overdrive can be supplied by the drivers to cause the cell to switch to its correct state.

An emitter-coupled cell is used in the MC 5484L 16-bit scratch-pad memory, shown in Fig. 15-20. In this memory, an XY selection method is used; to READ, the X and Y select lines which are normally less than +0.25 are raised to +0.8.

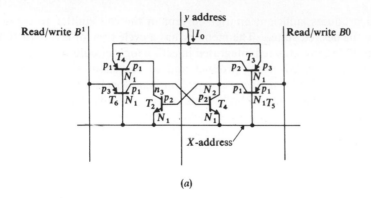

(a)

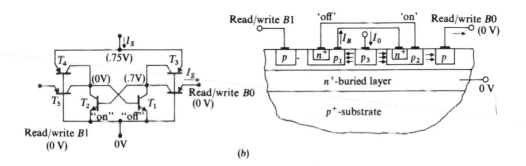

(b)

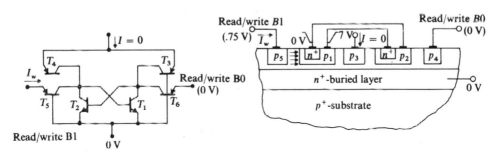

FIGURE 15-21
(a) I^2L memory cell; (b) READ operation; (c) WRITE operation (after Wiedmann[9]).

In a cell receiving both X and Y signals, the cell standby current is transferred to the emitter of the ON transistor and thence into the appropriate side of the sense amplifier. If only one of the X or Y signals is present, the current flows in the emitter of the line *not* selected, and is prevented from reaching the sense amplifier. To WRITE, the X and Y signals are applied, and a "WRITE 1" or "WRITE 0" signal is also applied. The WRITE inverter then overrides the sense amplifier and produces sufficient emitter current in the cell emitter to cause the cell to assume the desired state.   The memory has a cycle time of less than 100 ns.

In Chap. 12, we investigated integrated injection logic circuits and found them to have a very good speed-power product and to use small area on the chip. Memory circuits can also be realized in injection logic form; a basic bistable I^2L memory cell is shown in Fig. 15-21a.[9]   This cell is a complementary flip-flop in which n-p-n transistors T_1 and T_2 are switching transistors, and p-n-p transistors T_3 and T_4 act as loads.   Coupling from the READ-WRITE bit lines is provided by T_5 and T_6.

The cross section of the cell is sketched in Fig. 15-21b; here the N_1 layer serves as the emitters for T_1 and T_2, with N_2 and N_3 forming the collectors.   The p layers p_1 and p_2 form the bases of T_1 and T_2 as well as the collectors of T_4 and T_3. Layer p_3 serves as the emitters of T_3 and T_4, while p_4 and p_5 are the emitters of T_5 and T_6.   The collectors of T_5 and T_6 are also p_2 and p_1.

The READ operation is as follows.   Assume that T_2 is on and T_1 is off; the forward-biased emitter of T_2 injects carriers into the n region adjacent to layer p_2. T_5 now operates in the inverse mode, and region p_4 collects some of the injected current as I_s.   Thus the state of the cell is sensed by the presence or absence of I_s.

To understand the WRITE operation, assume that I_1 is off in Fig. 15-21c and is to be turned on.   The cell current in region p_3 is switched off and a current pulse I_s is applied to region p_5, causing T_6 to turn on in the forward mode; this injects carriers into region P_1 and turns T_1 on and T_2 off.   The circuit remains in this state when the WRITE current is turned off and I_0 is restored.   Considerable delay can be tolerated between I_w and I_0 because the junction capacitances will provide temporary storage of the flip-flop state.

Although the circuit diagram of Fig. 15-21a appears to contain many devices, it must be recalled that many of them share common regions; therefore it is possible to realize the cell in a small area.   With standard technology using 5-μm line width and spacing, the I^2L cell requires only 3 mil^2.   With a technology such as Isoplanar II, only 1.1 mil^2 are required.   Each cell has a standby power less than 100 nW.

Measurements and calculations show that a 64×64 array using standard technology would require a chip size of 160×150 mils, and would have an access time of about 50 ns.

REFERENCES

1   "McMOS Integrated Circuit Data Book," pp. 26–28, Motorola, Inc., 1973.
2   CARR, W. N., and J. P. MIZE: "MOS/LSI Design and Application," chap. 4, McGraw-Hill Book Company, New York, 1972.
3   HILL, F. J., and G. R. PETERSON: "Introduction to Switching Theory and Logical Design," 2d ed., pp. 230–234, John Wiley & Sons, Inc., New York, 1974.
4   HODGES, D. A. (ed.): "Semiconductor Memories," IEEE Press, New York, 1972.

5 LUECKE, G., J. MIZE, and W. CARR: "Semiconductor Memory Design and Application," McGraw-Hill Book Company, Inc., New York, 1973.

6 CARR and MIZE: op. cit., p. 197.

7 FROHMAN-BENTCHKOWSKY, D.: A Fully Decoded 2048-bit Electrically Programmable FAMOS Read-only Memory, *IEEE J. Solid-State Circuits*, vol. SC-6, pp. 301–306, 1971.

8 LYNES D. J., and D. A. HODGES: Memory Using Diode-coupled Bipolar Transistor Cells, *IEEE J. Solid-State Circuits*, vol. SC-5, pp. 186–191, 1970.

9 WIEDMANN, S. K.: Injection Coupled Memory: A High-density Static Bipolar Memory, *IEEE J. Solid-State Circuits*, vol. SC-8, pp. 332–338, 1973.